THE POLITICS OF AIR POWER

The Politics of Air Power

From Confrontation to Cooperation in

Army Aviation Civil-Military Relations

RONDALL R. RICE

University of Nebraska Press

Lincoln and London

Library of Congress Cataloging-in-Publication Data
Rice, Rondall Ravon.
The politics of air power : from confrontation to cooperation in
army aviation civil-military relations / Rondall R. Rice.
p. m. — (Studies in war, society, and the military)
Includes bibliographical references and index.
ISBN 0-8032-3960-2 (cloth : alk. paper)
1. United States. Army. Air Corps—Political activity.
2. Air power—United States—History—20th century.
3. Civil-military relations—United States—History—20th century.
4. United States—Politics and government—1919–1933.
5. United States—Politics and government—1933–1945.
I. Title.   II. Series.
UG633.R47 2004
322'.5'097309042—dc22
2004013676

For "My Girls"

Annette, Kathleen, and Rachel

Contents

Acknowledgments

A work of this magnitude requires the assistance of many. I must first thank the United States Air Force, the Air Force Academy, and the Academy's Department of History for allowing me the time and opportunity to complete this work. Under the short time allotted me out of the "line of the Air Force," I needed understanding and valuable help from many different organizations to complete the research and writing. I visited the Washington DC area the most, and many people extended helping hands. Undoubtedly, the most valuable assistance came from Mitchell Yockelson at National Archives II. He helped me navigate the dusty boxes of records and uncover sources I may never have otherwise found. Kate Snodgrass likewise guided me through the maze of congressional records and sources at the downtown archives. Unfortunately, I do not have the space to thank all of those friendly and knowledgeable individuals at the Library of Congress's Manuscript Division, without whose help this work would be far from complete.

Two presidential libraries provided invaluable sources, and their associated organizations offered much-needed funding. At the Hoover Library, Patrick Wildenburg and Lynn Smith shared their expertise and knowledge of the sources and gave me valuable advice on every aspect of my visit. I also very much appreciate the research and travel grant provided by the Hoover Presidential Library Association and the aid of Patricia Hand, Manager of Academic Programs. They all made my visit to West Branch, Iowa, rewarding and enjoyable. I will mirror all of those same accolades to the Roosevelt Library and staff and the generous grant given by the Franklin and Eleanor Roosevelt Institute. The only thing that rivaled working in the library with the wonderful staff was being in the Hudson Valley at the height of the fall colors.

The United States Air Force Academy Library Special Collections Branch was the final major source I tapped. The director, Duane Reed, and his assistant, John Beardsley, made me quite comfortable at my alma mater and stood ready to assist me via mail and on my summer visit. Although my visits to other places were not as long, I wish to thank the staffs of the Center for

American History at the University of Texas at Austin and the Rare Book, Manuscripts, and Special Collections branch of the Duke University Library.

I owe a special debt of gratitude to the members of my doctoral committee at the University of North Carolina at Chapel Hill and to those assisting from Duke: Professors R. Don Higginbotham, William E. Leuchtenburg, Alex Roland, and Tami Davis Biddle, and Professor Emeritus Gerhard Weinberg. Most of all, I thank my adviser, Dr. Richard H. Kohn. Beyond his vast knowledge and the instruction and mentoring he provided, I am indebted for his understanding of the intense pressures and extreme time restrictions placed upon those of us on an active-duty military advanced degree assignment. He guided me through the requirements of the Carolina curriculum but always made sure to tailor the program to further enhance my value to the Air Force and my service to the nation. Professor Kohn made sure my entire doctoral experience gave me the tools necessary to be a professional historian and a better Air Force officer. His assistance has not ended since my graduation, as he again offered his valuable time, keen eye, and mastery of the discipline in my efforts to update and improve this book for publication.

I would never have been where I am today without the advice and assistance of one special friend, Dr. Mark Clodfelter (Lieutenant Colonel, USAF, Retired). He took a discouraged sophomore engineering major at the Academy and helped me develop my until-then latent passion for history. He was a model Air Force officer and historian, and he became much more than just a mentor, role model, and friend. He helped open doors that would have otherwise remained closed, and supported my career goals of getting a master's degree, teaching history at the Academy, and getting a doctorate from Carolina. I am sure he did not envision all of the time and effort he would spend on my behalf when he agreed to be my academic adviser back in 1987. Instead of just helping me during my last years at the Academy, he has been my "adviser" for the past sixteen years and counting. My thanks to him also extend to Donna Clodfelter, and I very much appreciated the room and board during my many visits, as I conducted research in the Washington DC area. I can never repay them for all they have done. I once joked with them that I owed them my firstborn, and after seeing them interact with our girls, I know that they are already part of our family. Thanks Clods, and "Go Heels!"

For their indispensable assistance in bringing this book into being I cannot pass without offering my thanks to the staff at the University of Nebraska Press. Also, the work may never have reached their eyes without the help of Professor Peter Maslowski. I treasure him as another mentor, who likewise did not know what that his agreement to serve on a master's committee would turn into such a long-term commitment.

I have always been blessed with constant support and encouragement from my family. My mother and father constantly supported my educational desires, and my thirst for knowledge springs from their encouragement and love. From my sometimes-homesick days as a young cadet at the Air Force Academy to the pressures of a doctoral program at Carolina, I could always count on them for the little extra motivation and supportive words when my confidence would flag or I would tire of the constant grind. From my parents, to my brother and his family, and to Annette and my own girls: I relied on you more than I think you will ever know or understand.

It is to "my girls" that I give my most thanks. Annette has balanced the demands of the household and parenthood with those of critic and editor with equal aplomb for many years now. She has done so all while following me across the United States, to Europe and Japan, and enduring long days alone with the girls while I was deployed to the Middle East. I have a wonderful wife and two beautiful daughters, Kathleen and Rachel, and our time at Carolina was special indeed. All of these blessings in my life have truly enriched this tome.

From Confrontation to Cooperation

Perhaps no subject of the American interwar period has received more attention from military historians than that of aviation. Library shelves are filled with studies of Army aviation for those years. The majority of these works document the air arm's transformation from being a small section under the Signal Corps through its status as the Army Air Forces, trace the development of air doctrine or weapons, present biographies of the major air leaders, or weave together some combination of these topics. In one way or another, these histories documented the struggles of Army aviators to gain acceptance of their role in national defense and the codification of that role through increased autonomy and eventual independence as an air force.

Yet none of these works analyzed the interwar Army air leaders' push for their goals as a study in civil-military relations. In fact, one prominent historian declared: "Whatever the flaws of aviation development, they were not problems of civil-military relations at any level. . . . [C]ivilians tended to shy away from interservice and intraservice disputes on aviation roles and missions."[1] To the contrary, as this book demonstrates, early aviation reformers made alliances with politicians and worked with civilian business in order to advance aviation, gain additional roles and missions for the air arm (e.g., coastal defense), and increase funding. Both conflict and cooperation occurred between civilian and air service leaders during the interwar period, with cooperation normally following a period of conflict.

For the better part of six years—from the end of World War I until the court-martial of Brigadier General William "Billy" Mitchell in 1925—a group of the Army's Air Service officers engaged in what amounted to an insurrection. Led by Mitchell, a flamboyant, popular, and larger-than-life personality who used propaganda and worked behind the scenes with powerful allies in Congress and the press, the air officers pushed for an independent air force. They wanted it immediately and seemed willing to go to whatever lengths necessary to get it approved. Mitchell's mounting frustration led to his 1925 outbursts and court-martial. His actions led to his subsequent downfall, but his behavior challenged the tradition of American civil-military relations in unprecedented ways.

The framers of the Constitution, so steeped in the Radical Whig-inspired fear of standing armies, set up a system to maintain civilian control over the military. While they sought to provide an adequate defense, they expected to rely on the militia, believing a citizen army would not usurp its own republican system.[2] The structure they created contained four categories of safeguards against the standing army's ever-present threat to liberty and democracy: keep the need for forces as low as possible; rely on the militia rather than professional military forces in the event of war; maintain state militias as a counterbalance and safeguard against the professional army; and ensure that the military remained under control of those who were "politically responsible."[3] The ascent of the United States as an empire and then as a major influence on the world stage gave rise to a larger and more influential regular army. During the first decades of the twentieth century, with the regular army asserting more and more control over the state militias and the development of a force far larger than the nation's founders envisioned, high-ranking military officers became more important figures in peacetime America.[4] Sometimes professional military officers became involved in the political process—a development the founding fathers would have rightly frowned upon. As military historian Richard Kohn once asked, what if politics became too intertwined with the army, and elements of the latter "infiltrated the process"?[5] During the interwar period, politics and the armed forces became entangled in numerous controversies, but none were more spectacular or more undermining of civilian control than those involving the Army's air arm.[6]

Aircraft technology and its integration into society rapidly advanced during interwar years. Consider how the Wright brothers made the first powered flight at the end of 1903; after twenty years the technology remained improved but essentially the same. At the end of World War I, aircraft still used wood, fabric, and wire, and, although vastly superior to the Army's first airplane, remained a clearly visible descendant of the Wright Flyer. Twenty years later, at the beginning of World War II, the entire aircraft industry had changed dramatically. Designers and engineers had taken quantum leaps in every aspect of aviation technology: all-metal planes ruled the sky, with engines providing far more horsepower but at much lighter weights; retractable landing gear, ailerons and flaps, enclosed cockpits, and streamlined designs improved aerodynamics; and weapons integration advanced to include internal bomb bays and machine guns integrated into the wings (and tails, for some bombers).[7] Arguably, aviation advanced technologically more dramatically in the 1920s and 1930s than during any other period so far in history.

The rapid pace of improvement spurred interest in the military use of aviation and theories of its integration into the armed forces. Unfortunately for those espousing the virtues of military aviation, their theories and recommendations usually advanced faster than the technology and the government's ability to purchase aircraft and integrate the new air weapon. This developmental lag between theory and practice gave rise to the great debates over air power. The air power enthusiasts argued that the new technology allowed the nation to defend itself from others who could or would possess the technology, while also improving coastal defenses. Should another European war come, the theories they espoused would offer a way to avoid the nightmare of trench warfare. Beneath these defensive words lay the desire to develop strategic bombardment in order to cripple an enemy from the air. The air proponents needed to publicize this mission in careful words, stressing the "defensive" nature of strategic bombing, since the American mood would not allow the development of such blatantly offensive tactics and weaponry.

Americans were not the only ones debating air power's status within the military. Similar deliberations occurred within the governments of the other Great Powers, though often without the emphasis on developing "defensive" forces. Military men worldwide agreed on the importance of air forces but differed on issues of independence, doctrine, and roles. Air forces that were subordinate to other services, including those whose leaders sought independence, remained tied to the tactics of the ground and navy commanders.[8] The Japanese kept their air forces divided between the army and the navy, with each concentrating on supporting the mission of its parent service. The different air forces developed independently and neither coordinated nor cooperated with each other.[9]

The Red Air Force of the Soviet Union was structured in much the same way as the Japanese, though in reverse proportions, having a larger army air arm than the one supporting the fleets. Of the major powers, only the Soviets built a large aviation industry throughout the interwar period, and their production numbers dwarfed those of all other nations. However, in typical Stalinist fashion, they emphasized quantity over quality. Soviet production lines concentrated on close-support aircraft, and the Soviets controlled the world's largest air fleet throughout the decade leading up to World War II.[10] In developing their air forces, the Soviets first favored strategic bombing, but their observations of the Spanish Civil War led them (and others) to instead develop a force more focused on supporting ground forces.[11]

Yet obtaining an independent air force did not automatically sever ties to the armies. The German Luftwaffe officially came into existence in 1935, subordinate to the Ministry of Defense and coequal with the German army and navy. The second most powerful man in the Reich, Hermann Goering, was its commander. Luftwaffe leaders fought for developing a strategic force and Goering strove for more individual power, but official policy and Reich-swehr tactics continued to view aircraft as extensions of army firepower to support the rapid advance of ground forces. Air force procurement heavily favored small and medium bombers and fighter-bombers to support the army, though the Germans did not totally abandon independent air operations.[12]

Even being home to the greatest of the early air theorists and one of the first countries to form a separate force did not help the Italian air force forge independent air tactics. Benito Mussolini came to power in 1922 and created the independent Regia Aeronautica and made himself Air Minister. The Italian air force garnered a respectable aeronautical reputation during the interwar years, primarily due to native son Giulio Douhet. Although the Italians built a sizable force (around twenty-six hundred front-line aircraft by June 1940) and ostensibly relied on Douhet's principles, they relegated their aircraft primarily to ground support, resembling German blitzkrieg tactics.[13]

The French formed the independent Armée de l'Air in mid-1934, but they too remained tied to the auxiliary use of air power. The French tradition of unity of command and army domination of the defense organs kept aviation tied to the army. Another reorganization in 1936 placed all air units (naval excepted) under a unified command structure. In addition, the French government nationalized the aircraft industry, but neither effort significantly spurred French aviation doctrine or development. Air forces concentrated on supporting the troops at the main battle area, and the army commanders scoffed at using aircraft to attack rear forces.[14] As late as 1939, General Maxime Weygand exclaimed that "bombardments of defenseless people behind the front lines smacks of cowardice which is repugnant to the soldier."[15]

The British had no such compunctions regarding the use of its air force. Their World War I experiences and worries of continued continental entanglements spurred the creation of the world's first independent air force. More importantly, the Royal Air Force (RAF) truly embodied a separate armed service, not simply a legal severance. Britain's air leaders stressed independent air operations and limited the influence of their terrestrial sister service. The RAF did not discount ground-support operations, but its doctrine stressed massing air power for deep operations against the enemy's economy and rear areas—the RAF was more Douhetian than the Italian air force.

Whereas the French military espoused unity of command to keep their air forces closely tied to the army, the RAF countered with the military commandments of mass and economy of force to buttress their need for autonomy. Furthermore, British air leaders demanded retaining command of air units supporting ground operations and stressed the inherent flexibility of aerial operations, rather than parceling aircraft to different battlefield sectors and commanders.[16] Proponents of strategic bombing used the abhorrence of World War I trench warfare and Great Britain's island separation from potential enemies to further solidify the RAF's need for independence. Unlike their American counterparts, however, the RAF had the advantage of a population and government more attuned to the horrors of the recent war, and thus British air leaders obtained the requisite support from key civilian leaders.[17]

In the United States, the more traditional Army leaders wanted to keep the new weapon as a valuable asset for observation and reconnaissance, while also retaining control over air assets for ground support. Official War Department policy, as explained to the President's Aircraft Board by then Assistant Chief of Staff Hugh A. Drum, relied on "common sense, not sensation; on concrete conditions, not on visionary aspirations," and Drum chastised air officers for espousing a "new gospel of the conduct of war" based on theory and not proven doctrines.[18] Drum's reactions, and those of the Army leadership overall, were typical. Political scientist Barry Posen notes that when faced with changes in military doctrine, especially ones involving new technology that has not been tested in war, military organizations often first attempt to "graft new pieces of technology on to old doctrines."[19] Posen also remarks that military hierarchies will tend to suppress new ideas originating from the lower levels, and the Army's air leaders were trying to speed up drastic changes in an organization that traditionally adapts more slowly.[20]

Military historian Harold Winton agrees, contending that taking emerging technologies and projecting their actual employment on the battlefield requires "informed and imaginative military judgment."[21] Additionally, Winton underscores how a "service culture" affects a military institution's ability to reform, with the most critical element being the "ability to tolerate dissent and balance such dissent with the ever-present requirement for discipline and obedience."[22] The air proponents attacked Army leaders for lacking imagination, yet they did not approach sweeping changes with balance and discipline. Conversely, the Army's service culture not only resisted change but became even more recalcitrant due to the disobedient and undisciplined approach of the more ardent air advocates.

Yet the War Department's resistance involved an additional and critical underlying motive: money. Changes in military doctrine usually occur due to technological improvements, but money is a key ingredient needed to implement the modifications.[23] If an independent air force developed, budgets would not increase to fund the new arm; instead, existing miserly appropriations would become further subdivided. To avoid showing the "selfishness" of dividing the budget pie further, the War Department used such arguments as air expenditures' not being within the president's economic program and tried to show that the air service was not a "stepchild" but rather a "favored son."[24]

The air proponents' need for money to fund expensive weaponry and its necessary supporting infrastructure could not have come at a worse time. Although the early 1920s witnessed a return to prosperity, neither the public nor its legislators wanted to lavish funds on the military—especially since no country threatened the United States. By the time Congress and the president appeared ready to increase spending after the post–World War I lows, the Great Depression shifted spending priorities. The 1926 Air Corps Act mandated a Five-Year Plan to build up the Air Corps, but due to the lateness of the legislation (effective on 2 July 1926), implementation of the program could not begin with the fiscal year 1927 budget. Therefore, the buildup was planned for fiscal years 1928 to 1932. Thus the heart of the improvement program came during the Great Depression, and neither the Herbert Hoover administration nor the first term of Franklin D. Roosevelt's New Deal had room for increased Army spending. Politics, public support for disarmament, and the depression combined to create an even more austere environment for military men needing money for weapons of war.[25] Although Mitchell pointed out how the cost of one battleship could fund a vast air force, the isolationist mood caused many to echo Calvin Coolidge's rhetorical question, "Who's gonna fight us?"[26]

The American public remained ambivalent. The traditional reliance on a strategy of mobilization and the lack of a threat caused a rapid reduction in postwar budgets and forces. Still, Americans loved the airplane and those who piloted these new wonders.[27] Air shows and barnstormers fascinated the public, and flamboyant heroes like Mitchell, Eddie Rickenbacker, and Charles Lindbergh were celebrated. Still, public support for military aviation came with exceptions; Americans seemed enthralled with the *idea* of having an air force, but with no visible threat on the horizon they did not see the immediate need for a large, new armed service or for spending the money required to obtain and maintain it.[28] In order to sway public opinion—and thus, it was hoped, the two elected branches of government that held the purse strings—airmen often lived on the edge of proper conduct with their civilian masters. Aviators used propaganda in an attempt to arouse public sentiment, skirting military and civilian superiors to appeal directly to the public and to responsive and like-minded congressmen. The Army airmen tried to turn the battle into one pitting the innovative and progressive against the unimaginative and conservative.

The air arm's unique and recent arrival within the military establishment actually assisted its development of a cadre of politically astute officers. When Army aviation emerged during the tenure of President Woodrow Wilson, its officers mirrored the youthfulness of the hazardous service. The lieutenants who took to the air developed with the service and understood its intricacies and myriad needs. Due to their experiences and expertise, they also learned to testify before many congressional committees and War Department inquiries of the immediate pre- and postwar years. They combined their knowledge of aviation and the Army with their theories and vision of what the future of air power should be, while also making friends on Capitol Hill and learning the intricacies of political action.

When the first legislation for a separate air service appeared in the House (before World War I), those who would later fight zealously for such legislation testified against it. Lieutenants Benjamin D. Foulois and Henry H. Arnold, both future heads of the Army's air arm, believed that independence at such an early point would actually be detrimental to aeronautical development. Mitchell, then a captain, agreed, pointing out how the primary roles of the new technology—reconnaissance and communication—still belonged to the Signal Corps. As the air arm grew slowly and matured, so did the political astuteness of its youthful but influential officer corps. Their comfort with political contacts encouraged an affability with Congress they would come to see as natural, but depending on how it was used, they could—and often did—cross the boundary between appropriate and improper subordination to their military and civilian superiors.

Military officers' proper relations with elected representatives included providing information and advice when called. When requested by the civilian lawmakers, officers could recommend specific legislation and even draft proposals and submit them to Congress, after routing the recommendations through their chain of command (which included the Secretary of War). During the early years of the aviators' struggle for independence, they often circumvented these established procedures. The servicemen should have provided counsel and expertise to Congress without overtly lobbying the lawmakers—often a difficult and blurry distinction still today.[29] The president, with the advice of his administration and in partnership with Congress (when necessary), defined military policy and carried it out. The War and Navy Departments implemented the president's orders through established chains of command.[30] Congress provided a check on the president's power and influenced military policy through lawmaking, through its advice and consent role for military promotions, through oversight and investigatory work, and through its appropriations functions—with the last providing the most leverage.

In pursuing their proposed policy desires, air leaders sometimes crossed what were at the time considered the normal boundaries of civilian control, and the norms described by political scientist Louis Smith, writing in the late 1940s. Three of Smith's five elements of civilian control state that professional military leaders remain under constitutional and effective control, that departmental control of military officers rests with civilian managers, and that elected representatives control the formulation of military policy.[31] Yet studies of American civil-military relations demonstrated that this delicate balance often rested as much on personalities and power relationships as on institutions.

In his landmark work, Samuel P. Huntington decries the reliance on "subjective civilian control," defined as a civilian group or groups maximizing power at the expense of other civilian groups by involving the military in political struggles, rather than on institutional "objective civilian control" based on a professional officer corps functioning essentially autonomously within its professional sphere.[32] Richard Kohn identifies three causes of inherent tensions that give rise to strained civil-military relationships, one being grave crises or disagreements between strong personalities.[33] The air power debates definitely fall under this category. The civilians, from Coolidge and Weeks to President Franklin D. Roosevelt and his Secretaries of War and the Treasury Harry H. Woodring and Henry Morgenthau Jr., clashed at different times and on different topics regarding air power with the equally well known and vibrant personalities of the airmen, from Billy Mitchell to "Hap" Arnold. At the different points of tension, the Army Chiefs of Staff, from Pershing to Generals Douglas MacArthur and George C. Marshall, weighed in on the debate on different sides. Kohn proposes that civil-military relations are more "situational" and dependent upon the people, issues, and political and military forces involved.[34]

Political scientist Michael Desch expands upon the situational aspects to include how perceived internal and external threats affected civil-military relations. He theorizes that civilian control of the military weakens when a nation is not threatened externally. When combined with a low internal menace to security, conflicts can emerge that pit civil and military groups against each other.[35] Such a state existed early in the interwar period, as airmen and their congressional supporters clashed with the different administrations and ground Army leaders over aviation.

Army officers understood the restrictions on their words and deeds because these limitations were spelled out both in writing and in unwritten traditions of accepted conduct passed down in the service. As General Omar Bradley later related, "Thirty-two years in the peacetime army had taught me to do my job, hold my tongue, and keep my name out of the papers."[36] Servicemen knew they had the responsibility to question in private a policy they considered wrong and to discuss the proposal in the proper forums, but this responsibility did not include the right to challenge the wishes of the president or his administration publicly. When partisan wrangling entered into a military issue, military officers were supposed to defer to the political appointees over the services, who would work through the issues.[37] The norms of proper behavior did not permit officers to use congressmen as leverage to oppose presidential policies.[38] In fact, George C. Marshall, usually considered the paragon of proper military subordination, believed that such resistance to the commander in chief, in the words of his biographer, "weakened the fabric of a democratic society."[39] From the conclusion of World War I until near the end of President Franklin Roosevelt's first administration, the Army's air leadership came perilously close to doing just that.

Thus this study will examine how the leadership of the Army's air arm worked with—and against—the country's civilian and military leaders over the two decades of the interwar period. As such, this work also represents a case study of American civil-military relations during a turbulent period in American history as the airplane evolved into an important military instrument. Billy Mitchell's trial represented the high point of the air insurgency and a good starting point to begin analyzing the ebb and flow of interwar civil-military relations between the Army's air arm and the elected branches of government. His court-martial exposed the differences of opinion and arguments surrounding the air power debates and also represented the most blatant and public rift between civilian policy makers and the military leadership—it was the zenith of the period of confrontation. Yet the conflict over Army aviation had raged within and around the War Department since the end of World War I, and the foundations of the air insurgency were laid during the six years before Mitchell's trial.

After the trial and Mitchell's resignation in early 1926, Coolidge and the War Department quelled the uprising and ended the period of air officers' use of overt confrontation to obtain their goals. Everyone from the higher-ranking officers, who would take command positions immediately, to the then field-grade officers, who would lead the air arm into World War II, witnessed or experienced the demise of the defiant tactics and used this knowledge and experience to push for their goals in a more subdued fashion. Mitchell's antics raised public awareness about the state of the air arm, and perhaps even enabled his more moderate successors to compromise and gain the concessions from the civilian leaders necessary to advance military aviation. Moderates such as Major Generals Mason M. Patrick, James E. Fechet, and Oscar Westover stepped in and set a different tone, one of cooperation with the Army and the administration to gradually improve the service and its ability to perform a mission that so far existed only in pages of doctrine and theory. They and the even more ardent air power proponents like Foulois and Arnold believed that before the step to independence could be made, the air arm must first have the equipment to carry out strategic bombing and demonstrate that mission as being singularly important enough to require a separate service. Yet as this moderate attitude developed and took hold within the service's leadership in the years between Mitchell's trial and World War II (and aided for seven years by the appointment of a civilian air secretary), the air leaders from time to time would still maneuver and use whatever political levers they could in what at the time and since would be considered inappropriate behavior from the standpoint of civilian control of the military.

Foulois stood out as the most notable leader who reverted to the politics of confrontation. He revived the Mitchell-like tactics during Franklin Roosevelt's first administration, and for doing so he was forced out of the service early. Like Mitchell, though, Foulois' problems and public downfall allowed another moderate leader, Oscar Westover, arguably the air service's most conservative flying officer and the epitome of cooperation with the Army and the government, to direct and advance the Air Corps. Westover returned the Air Corps to a more subordinate and less confrontational relationship with civilian leaders.

Thus the history of Army aviation civil-military relations during the interwar period can be seen as one developing unevenly from confrontation to cooperation, from the extremism of Billy Mitchell to the political astuteness and clever maneuvering of the former radical "Hap" Arnold. Although they sometimes disrupted civil-military relations and acted inappropriately, the confrontational leaders focused public and political attention upon the air arm. Following each period of conflict and overt confrontation, a more conciliatory personality took charge and reestablished proper military subordination to civilian authority, while at the same time gaining needed concessions and advancing aviation's status in the Army. By the late 1930s, with President Roosevelt confirming his belief in the need for a strong Air Corps and supporting the spending of large amounts of money for military aviation, the airmen no longer felt the need to stir controversy or attract attention, and a strong civilian leader asserted his control over a corps of essentially cooperative military officers.

THE POLITICS OF AIR POWER

The Trial of Billy Mitchell

It was the "trial of the century" before that term became overused to describe the many that came later, from Bruno Hauptmann's 1932 conviction for the kidnapping of Charles Lindbergh's infant son to the 1995 media circus of the nine-month trial of Orenthal James "O.J." Simpson. Like those famous civilian trials, the court-martial of Colonel William "Billy" Mitchell attracted large galleries, and the media attention helped captivate the American public. The airplane, still a relatively new technology that entranced many Americans, added spice to a national drama replete with stars and heroes. The *New York Times*, commenting on the sensationalism that swept Washington and interested the nation, captured the mood: "Rarely has the national capital been so stirred as it is over this prosecution of the officer who has charged both the administration of the air services of both War and Navy Departments with extreme inefficiency. Telegraph instruments in the old warehouse are clicking out what goes on in the court every minute of the day. A national convention could hardly be more completely 'covered' than the trial of the Wisconsin officer who rose from the ranks to the second highest command in the aerial defense of the nation."[1]

Billy Mitchell understood this dramaturgy. In fact, he embraced it and played to it. The trial was his stage. He had fought for seven years to increase the role of aviation in the United States' national defense, from his many appearances before congressional committees to writing books and articles and giving speeches before civic groups. These efforts had failed to produce his desired outcome of a separate air service. Now, Mitchell hoped to use the highly publicized trial to force the president and the War Department to take action; he wanted *them* put on trial before the American public, and he thought he would win.

The American people knew Billy Mitchell. He had obtained hero status as a brigadier general in World War I, working on the staff of General of the Armies John J. Pershing, and as the field commander of the Army's Air Services in France. He organized and led the air attacks during the American

Expeditionary Force's (AEF) only two major campaigns, St.-Mihiel and the Meuse-Argonne. During the first battle (12–16 September 1918) he directed the greatest concentration of air power assembled during the war, some 1,481 aircraft.[2] Mitchell's men loved him, and he encouraged the dashing portrayal of the aviator, in part to encourage more young men to take up the dangerous business of military aviation. He wore nonregulation uniforms, established his headquarters at a château on the Marne (a former hunting lodge of King Louis XV), and drove around at breakneck speed in a Mercedes racing car.[3] These touches of flair and the American public's fascination with aviation made him a household name.

Mitchell used his newfound prestige to further the cause of aviation. He wanted an independent air service because he believed it would best serve the needs of the country's defense, and he remained captivated by the new technology and its possibilities. However, he was not above promoting himself to be the leader of any new service and to improve his own standing and place in history.[4] He used every opportunity to advance his views of an independent air service, separate from the Army and Navy, and led by a flyer as opposed to nonflying generals and admirals. Those men, he believed, who reached command by a more traditional route, remained trapped in an archaic paradigm of warfare on the earth's surface, which assigned only small and complementary roles of reconnaissance and observation to the air forces. The older Army and Navy officers did not disparage the use of aircraft, but they wanted to ensure that their use remained firmly under their control and as an auxiliary to the main forces. As Assistant Chief of the Air Service beginning in 1921, and one of the youngest generals in the Army, Mitchell relished every opportunity and invitation to appear before congressional committees, be they the regular appropriation hearings or the multitude of hearings connected with bills to separate the air arms from the Army and the Navy, which President Calvin Coolidge disparagingly called "Mitchell Resolutions."[5]

The more Mitchell crusaded for an independent air service, the more radical he became after each effort failed in Congress or stalled within the web of the General Staff. He became impatient with those who disagreed with him, and he believed the Army and Navy brass who opposed him conspired to protect their privileges and authority. Mitchell accused them of being hidebound and lacking the vision, understanding, or imagination to implement fully the latest revolution in military technology. Further frustrations caused him to call those who opposed him either stupid or immoral.[6] The administration finally "banished" Mitchell to an obscure assignment in San Antonio and reduced him to his permanent rank of colonel; with the kindling thus laid, only a final spark was needed.[7]

Instead of a spark, the nation received a double dose of air tragedies with the crash of the Navy dirigible *Shenandoah* on 3 September 1925, only two days after the disappearance of a Navy PN-9 aircraft en route from the West Coast to Hawaii. The day after the airship accident, Mitchell promised a statement and gave a preview of the defiant words to come. He assured the press that he would reveal the "truth" behind the Air Service situation and would "rip the cover off these deplorable conditions."[8] Clearly wanting to fight the administration, he issued a challenge: "If the War Department wants to 'start something,' so much the better. Then I can get the case before Congress and the people."[9] Three days later, Mitchell gathered reporters together and handed them a nine-page statement that would set off the hoped-for firestorm.[10] The press release loosed all the venom Mitchell could muster, and he clearly intended to incite a showdown with the administration. Mitchell accused the Navy and War Departments of "incompetency, criminal negligence and almost treasonable administration of the National Defense," and he laid the deaths of airmen at their feet, calling them "pawns in their hands."[11] He also accused the administration and the military leaders of giving false, incomplete, or misleading information to Congress and of coercing airmen to provide false or distorted information.[12]

Although Acting Secretary of War Dwight F. Davis responded that Mitchell would be dealt with by "action and not words," the actions did not come immediately. Everyone understood, however, that a court-martial was inevitable.[13] Meanwhile, Mitchell continued his crusade in the press, and it took two weeks for Major General Ernest Hines, the commanding general of the Eighth Corps Area and Mitchell's immediate superior, to relieve him of duty.[14] Davis, not wanting to give Mitchell his podium to preach air power, tried to keep everyone focused on the matter as one of insubordination. He extolled the case as "one of discipline" and asserted that the court would not delve into the "controversial questions respecting air policy raised by Colonel Mitchell."[15] President Coolidge, aware that a court-martial would provide such a forum for Mitchell and his followers, took a surprise step by announcing the formation of the President's Aircraft Board to investigate the state of the nation's air defense.[16] The President's Board began meeting on 21 September, and Mitchell was ordered to Washington to testify before it and to meet his court-martial, which would be held in Washington instead of Texas. He arrived to a tumultuous celebration at Union Station four days after the board began meeting, and he testified on 29 September 1925.[17] Mitchell stayed in Washington to prepare for the trial.

On 22 October, six weeks after Mitchell's San Antonio tirade, the War Department finally served him with the court-martial papers. Placed under technical arrest (but free to move locally around Washington), he received the formal indictments the next day. The eight specific charges included insubordination, conduct prejudicial to good order and military discipline, and conduct of a nature to bring discredit upon the military service.[18] War Department officials, trying to limit the size of the gallery and dampen Mitchell's "show," chose an interesting venue for the trial. The Emory Building, an empty warehouse across from the Capitol, formerly home to the Census Bureau and vacant for two years, housed the proceedings, instead of rooms within the War Department itself. As trial members inspected the building, they stepped over stagnant pools of water and ducked under fallen timbers. Although workers shored up the room, the Army limited standing room for spectators, supposedly to avoid collapsing the floor.[19]

The various participants in the trial added flair and drama. President Coolidge, through Acting Secretary of War Davis, proffered the charges against Mitchell himself (though this fact was not revealed until the second day of the trial), and Mitchell chose Representative Frank R. Reid, a freshman Illinois Republican, as his civilian defense counsel. Reid, who was an experienced trial lawyer but knew little about military law, agreed with Mitchell's view of air power and had previously displayed his agreement at various committee hearings. All twelve of the Army officers originally seated as Mitchell's judges had fought in World War I, and all except Brigadier General Ewing E. Booth had graduated from West Point; none of the officers represented the "flying Army." The panel included officers well known to the American public: Major General Charles P. Summerall, who had previously controlled all Army forces in Hawaii and currently commanded at Governors Island; Major General Fred W. Sladen, Superintendent of the Military Academy; and Major General Douglas MacArthur.[20] As high-ranking officers in a small peacetime Army, the accused and the judges knew each other well, which created an odd atmosphere. Mitchell, upon arrival at the courtroom, smiled and greeted them all. Commenting on the trial from the Philippines, where he served as Governor-General since his recent retirement, General Leonard Wood summed up the unique situation when he assured jury member Brigadier General Frank R. McCoy, "I do not envy you your detail on the court."[21]

Three of the officers obtained a reprieve from the duty as the defense quickly challenged and had them dismissed. Reid and Mitchell objected to both Summerall and Brigadier General Albert J. Bowley for "prejudice and bias." During the previous week, the latter had made statements demonstrating his dislike of Mitchell and the Air Service. As for Summerall, Reid read an unflattering inspection report by Mitchell of Hawaiian defenses during Summerall's tenure, as well as the future Army Chief of Staff's reply to his superiors. Summerall admitted to the statements and then characterized the airman's report as "untrue, unfair, and ignorant." Reporters followed Summerall out, where the general spat, "Now it's all over. We're enemies, Mitchell and I." Reid then ousted General Sladen with his allotted summary challenge.

As this dismissal did not require an explanation, one can only speculate as to the exact circumstances.[22]

The intricacies of the seven-week trial need no lengthy recapitulation here. After the defense asked for a dismissal due to jurisdiction, believing the case properly belonged in Texas, the government reluctantly revealed that the president had ordered the court-martial as commander in chief, thus retaining jurisdiction. Mitchell and Reid announced a defense based on the First Amendment, and they cited President Coolidge's June address to Annapolis's graduating class, where he assured students that "officers of the navy are given the fullest latitude in expressing their views before their fellow citizens."[23] However, the major defense testimony served only to reiterate Mitchell's views on air power. Witnesses included many Air Service officers (a number of whom would later rise to prominence), including Carl A. Spaatz, Horace Hickam, Harold George, Ira Eaker, and "Hap" Arnold.[24] The Mitchell defense used another famous airman, America's "Ace of Aces," Edward "Eddie" Rickenbacker. The defense even called to the stand Rear Admiral William S. Sims (Ret.), one of the Navy's greatest reformers and critics. Over the years, if Mitchell had upset any organization more than the War Department, it would have been the sister service. Navy secretaries from Josephus Daniels to Curtis D. Wilbur, and especially Mitchell's archrival Rear Admiral William A. Moffett, Chief of the Navy's Bureau of Aeronautics, cursed the airman and his efforts and to demonstrate the superiority of the airplane over battleships. However, Sims, who wanted the Navy to move away from battleships and embrace aircraft carriers, stood apart from his more traditional naval officers. He agreed with Mitchell that "ignorant and unfit" officers controlled aviation.[25]

On the last day of the trial, the prosecution entered damning evidence of previous instances of Mitchell's insubordination and defiance of civilian authority. Earlier that year, then–Secretary of War John W. Weeks, in a letter to President Coolidge, had condemned the flamboyant airman for being disobedient, lawless, and a publicity hound and recommended that he not be reappointed as Assistant Chief of the Air Corps.[26] Although the public long knew of the animosity between Weeks and Mitchell, such damnation came as a heavy blow, and the prosecution presented it late to maximize its impact on both the court and the press. It brought back to the surface the central reason for the trial and helped chronicle Mitchell's repeated episodes of inappropriate conduct. Major A. W. Gullion, one of the prosecutors, summed up Mitchell's insubordination and how he had tainted the Air Service, and he asked the court to dismiss the airman "for the sake of the young officers of the Army Air Service whose ideals he has shattered and whose loyalty he has corrupted."[27] Mitchell may not have shattered their ideals, but he had indeed helped fragment the loyalty of a group of air officers, many of whom would remain in the air arm throughout the interwar period and lead the service into World War II.

As with any such event, the press sensationalized the matter to bolster interest and sell newspapers. Reid clearly played to this feature, combining humor with some of his objections and often causing the courtroom to erupt in laughter—even the military guards posted about the room.[28] Running accounts appeared all over the country, though the trial lasted longer than anyone had predicted and interest waned somewhat as it wore on. Although the American people seemed to like Mitchell and agreed with his ideas, they, like the economy-minded Coolidge, seemed unwilling to spend money in a time of peace for a "defensive" weapon against unseen aggressors. As Arnold adroitly understood, "public enthusiasm . . . was not for air power—it was for Billy."[29] Overall, the press handled matters evenhandedly. Editors agreed with Mitchell that the nation needed to alter its current deplorable condition as far as commercial and military aviation development, but they clearly admonished the renegade general for his approach. The *New York Times* remarked: "If Colonel Mitchell has rendered a great public service by his agitation, he will get credit for it, even if he is declared insubordinate for language used toward his superiors. On the other hand, if in charging them with criminal negligence and almost treasonable conduct he has maligned them, he will surely suffer in public esteem."[30]

As much as Mitchell tried to avoid the facts, neither the administration, the Army, nor Congress faced prosecution for the state of the national defense. He alone faced the court on the eight charges of insubordination and detrimental conduct, and it seemed that he alone believed he was not guilty. Even the Air Service officers, called "Mitchell's Boys" by one historian and, conversely, "Mitchell and his worshippers" by a less supportive peer, knew he was guilty as charged.[31] Arnold understood that Mitchell was "licked" from the beginning and that everybody knew it except the air power prophet himself.[32] The court handed down its verdict on 17 December, the twenty-second anniversary of the Wright brothers' first flight, and found Mitchell guilty on all charges. Much speculation surrounded how the court members voted, but MacArthur, years later, virtually conceded that he was the lone dissenting voice for acquittal.[33] The military court sentenced Mitchell to suspension from rank and command and forfeiture of all pay and allowances for five years—a sentence they considered lenient due to his service in World War I.

Mitchell's congressional friends immediately tried to intervene. Representative Fiorello H. LaGuardia, a New York Republican who had served during World War I as a major and commander of the air forces on the Italian-Austrian front, introduced a measure to protect military members who had earned the Congressional Medal of Honor or the Distinguished Service Cross

(Mitchell's highest award) from courts-martial under the ninety-sixth Article of War. Democrats joined in, though most understood that it was more for "the purpose of annoying the administration" than from any heartfelt agreement with the cause.[34] Representative Thomas Blanton of Texas submitted a resolution that not only would have promoted Mitchell and made him Chief of the Air Service but also would have demoted two Army generals down to captain and suspended two others, including Drum, for five years.[35]

As with most of the aviation legislation, though, the calls of the more radical members did not coalesce into action. The *Army and Navy Journal* perfectly understood these ramblings and characterized the congressional actions of "Colonel Mitchell's friends" as purely political: "Resolutions have been introduced in the House, restoring him to rank and pay, abolishing courts martial, and the like. Southern Democrats are particularly busy in this direction. None of these politicians has any other motive for their activity than a desire for publicity."[36] Along a more reasoned line, most congressmen probably agreed with New York Republican Jonathan M. Wainwright, an Army veteran and former Assistant Secretary of War. Wainwright agreed with the decision and echoed the reasoning of the court, saying that "respect for supreme authority is the very keystone of discipline."[37]

The majority of the press also agreed with Congressman Wainwright's sentiments, except for William Randolph Hearst's publications, which normally backed Mitchell and the aviator's cause.[38] The *New York Times* echoed Wainwright, noting, "The verdict is a vindication of army discipline, and the deliberation with which it was arrived at deprives Colonel Mitchell of the pose of martyrdom."[39] Another New York paper understood how Mitchell meant for the trial to further his cause but disagreed with his methods: "Whatever his motives, whatever the notions of his fellow aviators, whatever the merits of his plan and that plan, the inescapable conclusion remains that Colonel Mitchell violated his soldierly obligations, and violated them in gross and outrageous fashion."[40] Even Mitchell's home-state newspaper agreed that the officer had overstepped his proper bounds, and it criticized his "extreme charges" against the administration.[41]

Coolidge took more than a month to announce his verdict, and he did so by taking the unusual step, for "Silent Cal," of issuing a long statement. The president's legally mandated review upheld the verdict but reduced the sentence. Mitchell received half pay, a total of $397.67 per month, instead of none. However, Coolidge once again reminded all of the true nature of Mitchell's transgressions, and why they could not be erased, since the country expected its officers—especially those of high rank—to follow the strict rules of law and proper subordination to their military superiors. The president also explained why such a breach of discipline could not go unpunished. If military officers became prone to such undisciplined relationships with their superiors, they "would not only be without value as a means of defense but would become actually a menace to society."[42] The administration designed the punishment to keep Mitchell in the service and not allow him to claim martyr status through a dismissal. Instead, Coolidge wanted to force him to resign, and then move to restore discipline. Mitchell wanted the dismissal, but six days after Coolidge's final adjudication came down, he resigned.[43]

Despite the efforts of the Army and the civilian administration, Billy Mitchell *would* become a martyr. Those aviators who supported him, whom one historian called "insurgents," then decided not to fall upon their own swords but to work from inside the service for the ends they sought and Mitchell championed.[44] After all, to a man they realized, as would any impartial observer, that Mitchell was indeed guilty of the charges. However, the crux of Coolidge's statement went unnoticed by many, especially those rebellious Air Service officers. Mitchell had threatened the balance of proper civil-military relations in a nation founded upon civilian control of the military. For years he had publicly crusaded against the announced programs of the administration and had derided those in power in the War and Navy Departments. However, by focusing attention upon himself and overstating and sensationalizing to promote himself and his cause, Mitchell's crusade became more about his insubordination than a true debate on national defense and the place of air power.[45] He had circumvented established procedures for relations with Congress and then used this access for far more than giving advice. A generation of young airmen witnessed these acts and saw how Mitchell had reset the bar for the limits of appropriate behavior in pursuit of his visions.[46] Mitchell's self-righteous attitude left an indelible impression on the cadre of air officers who adored him, despite (or in some cases because of) his faults. He advised his young fliers to continue the fight for an independent service, thereby intensifying the rivalry between the "traditional Army" and the fliers, and he urged them not to be satisfied with anything less than independence from the Army.[47] This dangerous state was Mitchell's immediate legacy and the one that has often been overlooked as airmen and historians alike have focused on his sacrifices for the advancement of air power.

Mitchell failed, and he would not live to see an independent United States Air Force, yet his interpretation as a martyr lives on within the service.[48] Billy Mitchell's court-martial stands as the most sensational military trial in American history, but until recently it has received attention merely for his martyrdom and as part of the story of the rise of American air power, rather than for what it really was: a successful prosecution of an insubordinate officer who had long since violated the traditional standards of officer behavior under the norms of American civil-military relations.

CHAPTER ONE

Billy Mitchell and the Politics of Insurgency, 1919–1923

"In a sense, for Billy, the Armistice was an untimely interruption—as if the whistle had ended the game just as he was about to go over the goal line."[1] With these words Colonel Henry H. "Hap" Arnold summed up the frustration of the man he admired as an air power pioneer, Brigadier General William "Billy" Mitchell.[2] Mitchell emerged from World War I sure of the airplane's dominant role in future conflicts, and although he never commanded the Army's air arm, he became the nation's air power prophet and led the charge for a separate service until his court-martial drastically reduced his influence both within the Army and in the public eye.

To the flying officers, Mitchell became a martyr, but even his most ardent supporters often questioned his methods. He used a charming personality, his athletic abilities on the polo fields and hunting grounds, his family name and influence, and a highly cultivated public persona to attract attention for the air power crusade. Those he enticed included politicians and some friendly media, and he used them to further his cause. He targeted the American public and the legislature because he knew he would not get results from within a traditional military structure controlled by generals from the combatant arms, the "Old Army" officers, and a Navy that worshiped battleships and the writings of Alfred Thayer Mahan.[3] Instead, his numerous published works and congressional testimony sought to circumvent the traditional chains of command. Therein lies the unexamined paradox of Mitchell's actions as a military officer: he overtly denounced the old power structure and inside politics of the influential Army officers with the administration and Congress, yet he himself openly violated rules appropriate for military officers and acted outside the traditional boundaries of proper civil-military relations. Although his actions never advocated or approached the level of a coup, he did subvert the stated policies of his civilian masters. For the first seven years of the interwar period, Mitchell stood at the center of the air power controversy, and he pushed the limits of appropriate conduct until a court-martial ended his government service and his immediate influence within the Air Service.

A FOUNDATION FOR INFLUENCE
AND AN INTEREST IN AIR POWER

Mitchell's life was based on a "fair foundation," as biographer Alfred Hurley called it. Three elements composed Mitchell's character and set the tone for his career and his activities: his personality, a distinguished family and privileged upbringing, and his experiences in the Army during a time of great transition.[4] These components allowed him to do what few other Army officers could—walk the halls of Congress with an air of comfort, familiarity, and self-assurance. He undoubtedly learned the political skills he would use with ease, and sometimes abandon, from a paternal line of congressmen and his upbringing between the pillars of power in Washington and the outdoor life in Wisconsin.

Although Mitchell's rise within the Army occurred through his own hard work and intelligence, his career began with an early push of political influence. His grandfather epitomized the American dream and the robber baron. Alexander Mitchell was born in Scotland, immigrated to America twenty-two years later, and made his fortune in banking and the railroads. He represented Wisconsin as a Democrat in the Forty-second and Forty-third Congresses (1871–75) and seemed popular in his adopted home, refusing candidacy for the next Congress and declining a nomination for governor three years later.[5]

John Lendrum Mitchell carried on his father's tradition and followed a recognizable path for a son born into wealth and privilege who would rise to political influence. He attended a Connecticut military academy but later studied in Europe, returning to fight as an officer in the Civil War. After later serving in the Wisconsin legislature, he turned to the national scene and was elected as a Democrat to two terms in the House of Representatives, resigning during his second term upon his election to the Senate in 1893. John served only one term, declining candidacy in 1898, wanting to return to Europe to study and engage in agricultural pursuits in Wisconsin.[6] Born in 1879 in Nice, France, Billy Mitchell did not see Wisconsin until he was three, but he then spent his early boyhood there at the family estate, learning the skills of horseback riding, including polo, and marksmanship.[7] His father became a congressman as Billy entered his adolescent years. Tiring of Wisconsin, Billy persuaded his parents to bring him to Washington DC during his father's tenure in the Senate. Thus, Billy Mitchell's father entered Congress when his son was twelve, and they remained there during Billy's teenage years. Although none of Mitchell's papers or biographies dwell on this fact, one cannot fail to appreciate how his understanding of the world and its workings would have been molded by spending these years in a political environment.

Political influence and Army life merged early, and probably seamlessly, for Billy Mitchell. He enlisted in the Army as an eighteen-year-old during the early enthusiasm for the Spanish-American War, but he served only three weeks as a private before his father helped him gain a second lieutenant's commission. His abilities and adaptability gained the notice of important Army and congressional figures, including Senator William Jennings Bryan and Major General Adolphus Greely, the Chief Signal Officer of the Army. When Mitchell noticed other lieutenants gaining promotion, many of whom he believed were of lesser abilities, he again called on his father. Billy Mitchell once wrote, "influence cuts a larger figure in this war than merit."[8] These brief examples highlight the development of a man who would clearly demonstrate exceptional talent and abilities and who performed best as a commander in the field, yet who clearly understood politics and how Washington insiders could affect the military.

Mitchell became acquainted with Washington at a relatively young age and early in his career. Only a thirty-two-year-old captain in March 1912, he became the youngest officer and sole Signal Corps representative on the War Department's General Staff. As the Signal Corps then included the Army's aviation (all four aircraft), Mitchell set out to learn more about this unexplored element of his branch. To do so he befriended Lieutenant "Hap" Arnold, an instructor pilot at College Park, Maryland. Mitchell also sought to create a solid foundation to assist his career and political influence. He used his natural abilities and athletic talents to work himself into the inner circle of social life—which also meant access to the halls of political power. He participated in high society through horse shows and playing polo as well as involvement in the proper clubs. As an Army captain's pay could not possibly support these endeavors, he called upon his mother for financial assistance. He also got to know several influential congressmen, including Virginia Democrat James Hay, chairman of the House Military Affairs Committee.[9]

At this early stage, Mitchell viewed the air arm as necessary for the Signal Corps' role of reconnaissance and communication. He opposed a 1913 bill introduced by Representative Hay proposing the creation of an "air corps" coequal in the Army with the other combat branches, remarking that aircraft had not yet proved their offensive capabilities.[10] Hay's first attempt to legislate an air service never left the committee. One year later he reintroduced the bill with minor changes (which Mitchell claimed he drafted) that kept aviation in the Signal Corps but created a new Aviation Section. If Mitchell had a hand in the legislation, he acted without War Department approval; more likely he merely advised Hay on essential elements. A reading of the bill's proposals suggests that Mitchell did not draft it alone, especially since the proposal excluded Mitchell from flying, as the law limited flying duty to unmarried lieutenants under thirty years of age.[11] With an eye toward war in Europe, the National Defense Act of 1916 eliminated these limitations, and

Mitchell, then deputy head of the small Aviation Section, paid for his own flight training. The next year he received orders to France as an aeronautical observer, and his fascination with flight would emerge, influenced by the foremost British proponent of air power, Major General Hugh Trenchard, commander of the Royal Flying Corps.[12]

Mitchell's contacts and experiences in Europe sparked a desire to bring aviation in America, both commercial and military, to a point corresponding to the nation's growing international world stature. He worked with, and learned from, the British (who would soon create an independent service), the French, and the Italians. After taking an intensive course in aeronautics taught by the best Allied aviators only four days after his arrival in Paris, he began to formulate his views from what he learned and what he would soon experience. He believed that a combination of British methods of employment (aggressive fighting and a desire to carry out long-range bombing) and superior French aircraft constituted the ideal force.[13] He also began to learn about Italian air power ideas from their main bomber manufacturer, Gianni Caproni, a friend and proponent of the great theorist Giulio Douhet.[14] Using these ideas, Mitchell formulated his own concepts of air power and subsequently organized the two combined aerial offensives involving the American Expeditionary Force (AEF) during September and October 1918.

While in Europe, Mitchell also made enemies and tangled with men who would compete for influence and power in the postwar service. Prior to the arrival of an American military staff in Europe, Mitchell had established the contacts that made him valuable to the overall American commander, Major General John J. Pershing. Pershing originally made Mitchell the AEF's Aviation Officer, followed by appointment as the commander of the Air Service in the Zone of the Advance, and finally the top air combat commander, Chief of the Air Service, First Army. Then, in an appointment seemingly made in Washington and not by Pershing in France, Brigadier General Benjamin D. Foulois, one of the Army's original aviators (he had learned to fly from the Wright brothers), became Chief of the Air Service, much to Mitchell's chagrin.[15] A conflict ensued between Foulois, who resented Mitchell for not "having one minute's flying time" as an official Army pilot (due to his civilian instruction course), and Mitchell, who resented Foulois and his "incompetent lot" for arriving late and taking over. Pershing soon became dissatisfied with the conduct of both and brought in a Corps of Engineers officer and trusted West Point classmate, Major General Mason M. Patrick, to take command of the air forces in the AEF. Foulois dropped to command First Army's aviation, and Mitchell moved down to command the air services of the First Brigade.[16] This animosity probably drove Mitchell to accept a job in Washington after the war that he had originally declined. If the future of the air arm was ripe for change, he wanted to be the one in power.[17]

PEACE IN EUROPE, WAR IN THE
WAR DEPARTMENT, CONGRESS IN THE
MIDDLE: THE WILSON-HARDING YEARS

Billy Mitchell did not get the top air job he so desperately wanted and which his young cadre of airmen wanted him to have. Instead, he took the subordinate post of Director of Military Aeronautics, while Major General Charles T. Menoher, another Military Academy classmate of Pershing's, headed the air arm. A renowned disciplinarian and efficient administrator, Menoher was chosen by the War Department largely out of the hope that he could keep Mitchell in line. An air chief who didn't fly much, Menoher likewise did not share enthusiasm for independent air operations. As a commander of the Forty-second (Rainbow) Division in France, he resented not having air support during the Aisne-Marne campaign, and he believed that support of ground operations represented the proper use of aircraft. If the General Staff and the War Department wanted to limit agitation from the air officers, they seem to have picked the right man—one of their own—to command a section they believed needed discipline and not autonomy.[18]

The two years after the Armistice included a flurry of investigations. With the war over, the country wanted a return to a small peacetime army, Congress sought economy, and the Army needed to shrink. One year after the war's conclusion, the Army's air arm retained only 220 regular officers, and by 1920 the service totaled 10,000 officers and men, a 95 percent reduction of its wartime manning.[19] Wanting to keep its air arm and protect its shrinking budget, the Army tried to counter with its own inquiries and in-house solutions. Pershing appointed a board led by Major General Joseph T. Dickman to identify lessons about aviation from the war. Composed of ground officers, the board predictably concluded that aviation must remain an auxiliary force since future wars depended upon massive ground armies. They also raised the major argument the Army would use during the interwar period against a separate air service: unity of command.[20] Secretary of War Newton D. Baker, who agreed with the Army generals' conclusions, decided to order an even broader examination of air power's role in national defense, to include its effects on civilian aviation. In May 1919 he appointed Assistant Secretary of War Benedict C. Crowell to form a commission to study the aviation of the allied countries in the war. Known both as the Crowell Mission and the American Aviation Mission, it submitted its report two months later, with findings running counter to Baker's program and beliefs.

In a letter to Chief of Staff General John J. Pershing, Colonel Mason M. Patrick (then in New Orleans, having rejoined the Corps of Engineers) announced his dislike for members of the Crowell Mission, believing they "were all practically committed in advance, in their own minds, to a separate Air Service . . . and they sought in every possible way to find arguments in favor of this organization." Patrick also stated his opposition to a separate department. He believed that those officers pushing for independence probably did so out of personal interest, as a separate promotion list would speed their rise within the military and give them important positions.[21]

Supporting the calls of Mitchell and his coterie, Crowell recommended a unified air service consisting of all military and commercial aviation and coequal to the Navy and War Departments, as well as the opening of an air academy to mirror the missions of West Point and Annapolis. Dissatisfied with the Crowell report, Baker nonetheless did not suppress it, nor did he keep Crowell from testifying before Congress on the various bills regarding aviation.[22] However, Secretary Baker clearly disagreed with its findings, adding a dissent when he sent the findings to Capitol Hill: "The mission has, in my judgment, gone too far in suggesting a single centralized Air Service."[23]

Congress entered the fray over the size and status of aviation in a variety of ways: investigation, appropriation, and legislation. First, the lawmakers began their own investigations into air power. Congressmen especially wanted to know why the billion-dollar aircraft construction program had not succeeded in putting American aircraft over the European battlefields. Congress also pondered the future of aviation generally. In control of Congress beginning in 1919, Republicans wanted to investigate various aspects of the Wilson administration's handling of the war effort, primarily for partisan purposes. Second, Congress legislated the size of the Air Service (and the Army overall) through appropriations. Finally, those congressmen interested in the air arm, and in connection with the flying service's influential members, introduced a flurry of legislation relating to aviation. From 1919 to 1920, members introduced eight different bills providing for a Department of Aeronautics.[24] Noting the 1918 creation of the British Royal Air Force, American airmen likewise wanted independence, and they began to work with congressional investigations and on legislative actions. Airmen participated, in some degree, in all three types of congressional action. They helped draft bills, testified before committees, and kept the pressure on the War Department. To keep aviation in the public eye, they performed aerial stunts and demonstrations, and a favorable press printed their exuberant statements. Mitchell coordinated the efforts, believing that "changes in military systems come about only through the pressure of public opinion or disaster in war."[25]

Military officers of all branches and services participated regularly in the legislative process on defense matters. In coordination with the War Department and with the administration's support, officers assisted in the drafting of military-related legislation and regularly testified before committees so as to provide legislators with the information they needed to make informed decisions. For example, if the War Department noticed a problem with the promotion system, officers could draft legislation to incorporate changes they viewed as necessary and present them—through proper War Department channels—to a congressman for introduction. Officers could then go before Congress to give their appraisal of the system and to say how legislation was needed to improve the situation. Although this may seem inappropriate from today's perspective, drafting legislation and offering testimony regarding current or proposed operations within the military did not represent a challenge to civilian authority when such actions occurred with the War Department's approval. Air officers, however, often acted outside normal procedures and in concert with anti-administration congressmen. By promoting their own agenda, the flyers became involved in partisan battles in Congress and supported programs and policies opposing those of President Woodrow Wilson.

The Wilson administration's priorities for the Army were to return the Army to prewar strength as quickly as possible while spending as little as possible. The War Department staff worked within these boundaries. The Secretary of War, after all, represented the president's wishes for Army policy and served as the executive official directing Army affairs. His General Staff provided supervisory and coordination functions and served as advisers to the secretary.[26] Under Baker's stewardship, the War Department supported the president, while still trying to preserve what it could of the Army.[27] Baker and his Chiefs of Staff, General Peyton C. March and later General of the Armies John J. Pershing, also wanted to develop aviation as far as possible, but they wanted to keep it subordinate inside the Army and as a complementary force to ground operations. With Menoher at the head of the air arm, they had a willing accomplice. Although he would fight for his share of the budgets, Menoher would not in any way crusade for a separate service or increased autonomy.[28] In fact, he had earlier chaired a board of officers who rejected the arguments of the air separatists, and his report became the backbone of the War Department's response to the "air insurgency."

The Menoher Board met from August to October 1919 to "study" aviation proposals then before Congress and make recommendations. Including Menoher, the board consisted of four major generals—all artillery officers by branch training and all opposed to a separate air arm. The board's findings, quickly endorsed by Secretary Baker, believed that a separate air service would cost too much and would violate "unity of command" in war, as its primary purpose remained support for ground forces. Air officers, especially Foulois, believed the board to be a whitewash and its conclusions written by the ground generals in order to keep aviation subordinate to the Army.[29] With an intransigent General Staff and an administration unwilling to push the cause, the air officers took their efforts directly to Congress.

During the last Congress of Wilson's administration, the Republicans gained controlled of both houses. The GOP held a forty-six-seat margin in the

House but only a two-vote majority in the Senate. Those congressmen who regularly supported pro-aviation legislation came primarily from the Republican side of the aisle. President Wilson and Secretary Baker both opposed air autonomy.[30] During the last year of the Wilson presidency, Baker's power and an overall congressional desire to cut spending killed most air independence bills. Still, a pattern emerged. A small number of air-minded Republican congressmen, supported by Army air officers, would pursue legislation to help the insurgent air officers. During the initial postwar years, however, the air-friendly legislators could not garner enough Republican support, and Democrats retained party discipline to keep any bills from reaching Wilson's desk.

The post–World War I Congresses were friendlier toward aviation than their predecessors, and airmen found allies on the Hill. In their respective chambers, two Republicans, Representative Charles F. Curry of California and Senator Harry S. New of Indiana, led the way, with notable assistance from Republican representatives Kahn, Fiorello LaGuardia, and Harry E. Hull of Iowa. LaGuardia intimately shared the desires of the Air Service officers due to his prior Army aviation service. Within three weeks of the Crowell report, two aviation bills surfaced, one in each house, from Curry and New. Both bills proposed a variation of the recommendations of the Crowell Mission.

Curry's bill, introduced on 28 July 1919, was the more detailed of the two. The Californian called for a Regular Air Force with administrative and support units, and a Reserve and National Guard force. Senator New proposed his own legislation three days later, and although it was less specific in its provisions, he called for a United States Air Force, but assigned to a Department of Aeronautics, which would also control all aviation matters for the Army, Navy, Coast Guard, Post Office, and other government departments. The Director of Aeronautics, appointed by the president, would control and assign the units as needed. These two bills never emerged from the respective Military Affairs Committees, but both congressmen reintroduced them in the fall session, with minor revisions.

Predictably, air officers supported the bills, while the War Department and the ground generals opposed them. Senator Gilbert M. Hitchcock, a Democrat from Nebraska, summed up the general feelings of the minority and opposed air service independence because of the proposed costs. On the other side of the aisle, James W. Wadsworth Jr. of New York, chairman of the Senate Committee on Military Affairs, joined other Republicans in supporting the bill on the grounds of improving national defense. However, not all Republicans supported the measure. Noted isolationist and arms control supporter Senator William E. Borah of Idaho stood as the most notable Republican against the bills.[31] Due to the narrow margin in the Senate, any

Republican defections doomed legislation not supported by Wilson and the Democrats. Still, the GOP majority and support for the legislation within the Military Affairs Committees allowed hearings to proceed.

LaGuardia's subcommittee on aviation held hearings on the United Air Service bill, not as formal testimony on the revised Curry bill but for "general information."[32] The former aviator coordinated the hearings as a forum for the air officers, and the other members of the subcommittee reflected a strong pro-aviation bias.[33] In contrast to the formal hearings on the New and Curry bills, where only Foulois and Mitchell carried the weight of the pro-aviation testimony for the Army, LaGuardia brought in eighteen Army air officers and also included Navy fliers. Most of the Army aviators supported Mitchell and his beliefs, and the flamboyant airman led them into the hearings preceded only by the top pro-aviation civilian in the administration, Benedict Crowell. The air officers testifying included Colonels Arnold, Bane, and Thomas Milling, Lieutenant Colonel Leslie MacDill, Major Foulois, and the publicly popular "American Ace of Aces," Captain Eddie Rickenbacker (Ret.). Menoher and Baker countered the airmen and testified against any unification plan. LaGuardia's subcommittee did not call any of the prominent or high-ranking ground officers to testify. In an obvious move to allow Mitchell and his followers to take the spotlight, the pro-air legislators had provided the opportunity for the aviators to make their case and limited the ability of the administration's representatives and the ground Army officers to counterattack.

From the immediate rash of air independence bills, only Senator New's revised bill emerged from committee. Yet New realized from the short floor debate that his bill would not pass, so he received approval for its resubmission to committee for further study. The bill never reemerged.[34] The fight for these bills was not in vain, however, as an omnibus military reorganization bill eventually came out of the Senate, and, after the usual conference committees and amendments, the president signed the Army Reorganization Act of 1920 on 4 June. The aviation section of the law created an Air Service, separate from the Signal Corps and headed by a major general chief and a brigadier general assistant, with 1,514 officers and 16,000 men, and prescribed that flying officers must command flying units.[35] The airmen and their Republican allies did not get all they wanted, but they did advance their air agenda one step forward.

The air officers had the authority to speak their mind freely before Congress and, in fact, were encouraged to do so. Secretary Baker continued a previously established policy of allowing all officers "to testify with the utmost freedom as to their own opinions and beliefs on the policy in question" when summoned by Congress.[36] He did believe, however, that the Army would be better served if differences of opinion could be worked out in-house prior to presenting them to Congress and if an agreed-upon "official judgment" could be provided on matters of military legislation.[37] Behind each of the congressional actions, air officers sometimes assisted in drafting the legislation and used the hearings to provide testimony and circumvent an unfriendly War Department and administration.[38] Coming so soon after the war, and with friendly members of Congress available, the air officers pressed their case with more fervor than was seen during the prewar period, setting the precedent for relations between their branch, the administration, and Congress.

During this early period, Mitchell suppressed his wrath for the War Department and his civilian masters when appearing before Congress, although he continued to press for a program inconsistent with the administration's stated goals. He presented his air platform and defended its validity, reserving the majority of his venom for the Department of the Navy. Instead, he worked behind the scenes with the help of his devoted young officers. Mitchell used newspapers to publicize his views on national security and to sway public opinion. He would contact air-friendly publications and identify the legislation he wanted them to highlight. He and his staff followed the press, and Mitchell sent letters to editors if he favored their stand.[39]

Mitchell's officers in the field also kept abreast of local situations, asked his advice for their actions to further aviation in the press and politically, and informed him if they believed he needed to use his influence to sway influential men. In one instance, Colonel Thurman Bane, stationed at Mc-Cook Field in Dayton, Ohio, wanted Mitchell to persuade Orville Wright to testify in favor of the Air Service bill then in Congress. "I tried to make it clear to him," Bane reported, "that we on the inside, who were working under the General Staff, were thoroughly convinced that we would get no proper development until we were separated from them."[40] According to Bane, Wright declined because he felt the current situation was satisfactory and that the General Staff could not hold back the inevitable development of aviation. Major Arnold went even further. Not only did he ask for Mitchell's influence in passing civilian aviation legislation in California, but he helped write the bill, admitting that "we were very careful in drawing up the Bill to do, as we thought, everything that you would desire of us."[41] In return, Mitchell would keep the flyers informed as to which legislation he would "get behind." In the initial rash of postwar air bills, for example, he sent letters to Air Service officers all over the country to show that of all the different air proposals, he was supporting the Curry bill.[42]

With young officers stationed all over the country and in command of flying units at relatively junior ranks, Mitchell kept them abreast of his efforts in Washington. In turn, they acted as the grassroots level of the air insurgency. Spread about the country, these officers acted as a veritable intelligence-gathering and -dissemination organization for the aviation agenda. Meno-her's attempt to control Mitchell and limit his influence by keeping him out of Washington actually contributed to Mitchell's closeness with the younger officers. His frequent inspection tours took him to the airfields, which were commanded by majors due to a shortage of higher-ranking officers.

Mitchell became very close to one of these officers, Major Carl A. Spaatz. Spaatz moved the pursuit unit he commanded from its base in Houston to Selfridge Field outside Detroit in 1922, and Mitchell visited Spaatz at least seven times from August 1922 to February 1923.[43] Sometimes the trip represented an official inspection; at other times Mitchell took advantage of Spaatz's local contacts to go hunting and fishing. Mitchell's assistant, Lieutenant Clayton Bissell, kept all of the geographically separated officers in touch, which could have been accepted as normal except that these direct communications often skipped the intermediate corps-area commanders.[44] The information in the letters did not always cover routine operational topics. In one handwritten note, Mitchell urged Spaatz not to become discouraged and said that he, Mitchell, would push during the coming winter to solve the personnel and equipment problems by "absolute and specific focus" upon an "Air Service proper," meaning independence.[45]

Although he coordinated with his officers and press around the country, Mitchell focused the majority of his efforts on Capitol Hill. His friendship with some legislators, such as Curry, shone unequivocally, but others Mitchell courted in a variety of ways. He sent copies of his book *Our Air Force* (1921) to several members and assisted others on their inspection trips.[46] Arnold also sent Mitchell advance notice that a new senator from California had expressed an interest in the Air Service and was a proponent of independence. The politically astute major urged Mitchell to meet the senator to "secure his unqualified support" and obtain valuable influence in the future.[47]

Also interesting were Mitchell's many replies to requests for assistance. Quite often congressmen would ask for air assets to conduct some type of goodwill trip, present an air show, or provide logistical help for something in their districts. They would also intervene for transfers of Army personnel to the flying service or to different bases. Mitchell did not handle all requests equally. The usual response from the Air Service office would decline the requests as "not being in accordance with the policies of the Secretary of War." The form-letter reply became predictable, and that type of letter was sent to Senator Kenneth D. McKellar, a Tennessee Democrat whose congressional assignment included the Committee for Post Offices and Post Roads.[48]

On the other hand, if the requestor held an office of influence and might be able to help the Air Service, the response reflected a different attitude. Although he had to deny a similar request from Senator William J. Harris, a Georgia Democrat, Mitchell did not send the regular form letter. Instead he wrote: "With the shortage of funds, equipment, and personnel which now exists in the Air Service, the policy has been adopted that this season it will be impossible for the Air Service to participate in these different exhibitions."[49] Both McKellar and Harris were southern Democrats requesting similar activities within a month of each other. Undoubtedly, the different language in the reply stemmed from Senator Harris's position on the powerful Appropriations Committee. These moves reveal a military officer comfortable working the political inside who understood how to turn an unfavorable reply into a political statement. During these years Mitchell did not have to rant and rave to receive attention—that he left to Benjamin Foulois.[50]

Less politically astute, and undoubtedly stung over having reverted to the permanent rank of major while others retained stars on their collars (including his peers on the General Staff), Foulois provided the congressional fireworks. During the hearings on the Army reorganization bill, he forthrightly stated: "In my opinion the War Department through its policy-making body, the General Staff of the Army, is primarily responsible for the present unsatisfactory, disorganized, and most critical situation which now exists in all aviation matters throughout the United States."[51] He went on to give what he believed to be historical evidence that the Secretary of War and the General Staff had intentionally limited the Air Service's development and had overstated the air arm's limitations.[52] During the hearings on the United Air Service bill, Foulois provided the clearest statement of why Air Service officers took their controversial political actions: "In my own experience I think I fully appreciate the fact that when we lay our cards on the table we can get better results from Congress, and always have. . . . In order to get legislation such as this Air Service legislation before Congress the practical fliers have been called disloyal, but in order to get our case before you we have dared to openly express our opinions in a manner which is not entirely customary in Army routine."[53] To Foulois, in other words, his desired ends of aviation policy justified the means. Like the other pro-independence air officers, Foulois did not appreciate, or care, that their vision was that of a specialized minority and countered the president's stated policy. Instead, the flyers continued to push for what they believed was best and used friends in Congress to help them by providing the soapbox.

Like Mitchell, Foulois often reserved even more defiant words for the Navy. At one point, both men would oppose Assistant Secretary of the Navy Franklin D. Roosevelt. Foulois characterized the future president as "rich, influential, and pro-Navy all the way," and he countered Roosevelt's testimony as "wrong in his facts and so biased in his opinions."[54] On the naval testimony, Foulois did, prior to his appearance, properly inform his chief about the prepared remarks—something Mitchell would not do. The air officers sincerely believed that their efforts would improve the nation's defense and that Congress wanted honest testimony.

The administration, while disagreeing with the air activists, did not intentionally stifle their appearances. The War Department did, however, expect to be informed of any testimony in advance, in accordance with accepted procedures. In this instance at least, Foulois provided his chain of command with an advance copy of his testimony. He recalled that although his remarks received a "very chilly" reception within the War Department, he could not "be accused of going behind anyone's back."[55] Although this represented the proper way for airmen to give their opinion to Congress when disagreeing with the administration's stated policies, advance notice to their superiors seemed the exception rather than the rule.

Due to the air officers' testimony on the immediate postwar aviation bills (and other problems within the department), Baker took steps to limit future insurrectionary actions. Problems with the Quartermaster Corps, not the air arm, actually motivated him to act.[56] The Quartermaster General of the Army, Major General Henry L. Rogers, embarrassed Baker in the early spring of 1920 by attempting to influence legislation and personally arrange the appropriations for his department. Baker issued Rogers a personal reprimand. In a separate letter that the secretary decided not to send—but in which he unequivocally expressed his opinions—Baker asserted that Rogers's action "tends to create, and has in fact created, in some minds the impression that officers in charge of staff corps are making combinations and arrangements among themselves rather than relying upon the presentation of their views to the Congress or the Secretary of War." Baker noted how the incident not only embarrassed him but caused outsiders to believe that the department was ruled by inside "army politics," not the civilian leadership.[57]

Responding to this latest incident, and probably foreseeing the tempest to come with the pending air legislation, Baker explicitly outlined the behavior expected of officers when they interacted with Congress. In a memorandum to the War Department, he reiterated that when properly summoned before Congress, every officer could testify without reserve as to his own opinions, though he hoped that any disagreements could be properly raised and handled internally (outside public and congressional view). He furthermore ruled that he could impose "safeguards" (in effect, limits) as he saw fit on testimony of a confidential nature. He further ordered that "all other efforts, direct or indirect, on the part of officers to influence legislation affecting military policy will be at once discontinued and not resumed."[58] The War Department later codified these instructions into Army General Orders. Thus the instructions became binding on all Army officers, and failure to follow these measures for obtaining approval prior to appearing before Congress could result in punishment under military justice procedures.[59] Since elements within the War Department, including the air arm, had ignored the unwritten rules of conduct, Baker had moved to outline explicitly the proper civil-military relationship between officers and Congress.

Only four weeks after Baker issued the memorandum on conduct with Congress, Mitchell made inflammatory statements before the Senate Military Affairs Committee. He accused the War and Navy Departments of duplication of effort costing an extra $11 million. Acting in his position as chairman of the Aeronautical Board, General Menoher sent a memorandum to the navy and war secretaries alerting them to Mitchell's breach of Secretary Baker's order.[60] Instead of receiving an official reprimand, Mitchell was subjected to a slew of memos dashing back and forth across the desks of War Department officers. The Adjutant General told Menoher, as Director of the Air Service, to take "such steps as you deem necessary" to ensure compliance by air officers.[61] Menoher subsequently required each officer in the Air Service (including Mitchell) to read Baker's memorandum on conduct and to sign a statement stating that he understood the policy. The signed statement was then kept on file.[62] Mitchell did not receive an official reprimand or any other punishment, but Menoher was clearly open to using any incident to entrap him. Mitchell did receive a memo reminding him to pay attention to the "attitude of the War Department and of this office [Air Service] on such matters," and the Air Service staff notified the Adjutant General of the actions taken.[63]

Most Air Service officers initially believed they had reason to celebrate the election of President Warren G. Harding in 1920. Along with a Republican-controlled Congress, there might be a breakthrough in altering national defense organization. Harding fed these hopes with a pre-inauguration promise to abolish the War and Navy Departments and create a single department of defense, providing aviation equal status with the older services. The president never followed through on these promises, which Mitchell blamed on congressmen's fearing loss of committee stature and seniority (due to the combining of the military and naval affairs committees) and on the powerful lobbyists.[64] Admitting to being surprised by Harding's reversal on aviation, Mitchell remained hopeful because of congressional interest.[65]

The Republicans increased their power in Congress on the strength of Harding's victory. The GOP enlarged a 46-seat majority in the House to a margin of 168 seats.[66] In the upper chamber, Republicans added ten more senators and expanded what was formerly only a two-seat edge to twenty-two.[67] The Democrats gained many of those seats back two years later, as Republicans suffered a serious setback in the midterm elections. In 1922 the GOP lost seventy-five seats in the House and clung to an eighteen-seat margin.[68] In the Senate, the Democrats added five senators and halved former the Republican majority from the previous Congress.[69] Despite these fluctuations in GOP successes at the polls, the core group of air-friendly Republican lawmakers remained in Congress, and some Democrats, in what would become a trend during the three interwar Republican presidencies, would vote for air legislation seemingly to irritate the administrations. During the first two years of Harding's term, however, the Republicans held such a large majority in Congress that Democrats could not swing enough GOP rebels to make a difference and pass any bills opposing the administration's air policies.

The Harding administration wanted tax reduction and economy in government and recognized the mood of the country for disarmament. Republican congressmen hoped Harding would allow them more power, reversing what they believed was a twentieth-century trend toward presidential meddling and a loss of "constitutional government." The GOP contingent also believed that Harding would provide them more leeway due to his recent service in Congress and his friendliness and compromising personality.[70] Harding opened his doors to congressional leaders, but he did not allow them free rein. During his two years as president he doggedly pursued his goals of "normalcy" and took no action to increase government spending, especially on defense. In a telling demonstration of his policy, Harding took the politically risky move of vetoing the Soldier Bonus Bill in 1922. Adamant about reducing the deficit, he risked Republican support and powerful veterans' lobbyists to cancel a program that would have added more than 16 percent to the national debt.[71] Harding came from a different political philosophy than Wilson, but the Republican's actions gave the airmen little reason to believe the change would speed their dream of independence. The flyers' early hopes that their situation would improve under the new president evaporated.

Even before tackling Mitchell's views on air equality, the Harding administration, and especially Secretary of War John W. Weeks, faced controversy with the popular officer. Weeks was not a newcomer to military policy. An 1881 graduate of the Naval Academy, he served two years on active duty with the Navy. During his one term in the Senate (1913–19) he served on the Military Affairs Committee.[72] Only three months into Weeks's tenure, General Menoher asked the secretary to dismiss Mitchell as the deputy of the Air Service. Mitchell did not hide his dislike of Menoher's leading the service, due to the latter's non-flying background and opposition to a separate air arm. According to the *New York Times*, one cause of the rift lay in the interpretation of the National Defense Act of 1920 as to the number of non-flying officers permitted in the service.[73] The law decreed that the number of officers in each grade below brigadier general not exceed 10 percent.[74] Menoher wanted to interpret the rule as applying to 10 percent of the entire officer corps (and not each grade individually) in order to keep more non-flying officers, who were loyal to him, in the service (Mitchell, of course, took the opposite view since he wanted to limit the number of non-flying officers assigned to the air arm).[75] The *Times* also touched upon what later emerged as the crux of the spat: Menoher perceived Mitchell's advocating a separate service as insubordination and strongly objected to his actions.

Known as a disciplinarian and demanding the utmost obedience to superiors, Menoher believed that Mitchell's opposition to the administration's wishes undermined proper conduct. Mitchell should be removed.[76] In fact, Menoher accused him of deliberately antagonizing the Navy in order to influence legislation and of using "undesirable publicity" to further Mitchell's acclaim while undermining Menoher's own prestige. Most of all, Menoher decried Mitchell's push for an aviation program at odds with the stated goals of the administration they both served.[77] Weeks smoothed over the conflict and both men remained in their offices, but by retaining Mitchell, Weeks had weakened Menoher's authority and probably encouraged Mitchell's disrespectful behavior toward his superior. Weeks did admit, according to press reports, that Mitchell's "enthusiasm might at times have led him to indiscreet utterances."[78] Weeks could have been influenced not to dismiss Mitchell due to congressional rumblings to intervene and hold hearings, but most likely, being new to the position, Weeks wanted to avoid moving either of the two decorated World War I veterans and just looked for a way out.[79]

The *New York Times* continued to lean toward supporting Mitchell, but not to the point of extremism. The influential paper noted that if Weeks could not negotiate a settlement he must comply with Menoher's request and remove Mitchell. The *Times* backed Menoher in this situation due to his being a senior officer trying to enforce the required discipline upon his subordinate. However, the paper expressed the hope that Weeks could keep the peace so as not to lose such an aviation expert and talented officer.[80] Weeks indeed retained Mitchell, but he probably came later to regret the opportunity to get the controversial airman out of Washington.

The summer of 1921 gave Mitchell his greatest triumph and the platform from which to preach further the gospel of air power. The Navy became his primary target, literally and figuratively, as he secured the use of derelict American battleships and the "unsinkable" German dreadnought *Ostfriesland*, taken as part of the postwar settlement, for bombing tests. The Air Service sank the unsinkable and used the press to further attack the Navy. The success also emboldened Mitchell's supporters in Congress, most notably air-friendly Representative Charles Curry, who praised Mitchell for standing tall though "ridiculed and damned" for his theories and who promised that together they would "fight harder than ever for a great American Air Force."[81] As Mason Patrick noted, "General Mitchell's horn was greatly exalted and quite loudly blown."[82]

Although congressional aviation supporters celebrated the bombings, the

administration commented very little. Weeks watched the attacks and commented on the good outcome for the Army fliers and their abilities, but he
admitted that the controlled nature of the test did not prove the obsolescence
of capital ships. Likewise, Menoher did not laud the important demonstration but kept a seemingly orchestrated low tone by the administration and
the War Department, most likely to avoid embarrassing the Navy.[83] The
War Department undoubtedly also wanted to avoid applauding Mitchell's
efforts.

Mitchell followed the bombing tests with staged mock raids on several
large coastal cities to make his point of America's vulnerability and especially
that of the Atlantic Fleet's naval bases. Mitchell gave his report to Menoher, and when the apathetic chief refused to publicize it, Mitchell leaked it
to the press. At the same time, Mitchell blasted the official report, which
was very pro-Navy and talked more about the conditions being ideal and
the test setup favorable for the bombers. The triumphant airman not only
reiterated his calls for a Department of National Defense but further antag-
onized the Navy by asserting that aircraft should take over responsibility for
America's coastal defense out to a two-hundred-mile perimeter.[84] Once again
the administration did not comment on the airman's outbursts, but the War
Department immediately became involved in the associated imbroglio with
Menoher. Mitchell proudly called his now-public report a "bombshell," but
for Menoher it was the last straw.[85]

Menoher demanded to Weeks that either he or Mitchell must go. Due
to the recent success of the bombing tests and a press lauding a triumphant
Mitchell, Weeks had to support his controversial airman.[86] Mitchell remained
in the Air Service offices, and Menoher was reassigned. Menoher publicly
stated that after three years in the Air Service he desired to return to the line
of the Army.[87] Even though Weeks supported Mitchell in this instance, the
secretary also realized the need to control the brash airman. Weeks especially
did not want to elevate Mitchell to become the Chief of the Air Service
and even further strengthen the airman's political clout, nor did he wish
to appear to reward Mitchell's inappropriate behavior. As Pershing had in
Europe, Weeks turned to Major General Mason Patrick. Once again, both
men looked to Patrick to control the air insurgency led by Mitchell.[88]

General Patrick acted immediately to break Mitchell's power. Aware of
air officers' criticism of his predecessor and despite being sixty years old, the
new chief immediately learned to fly. He also broke up Mitchell's "cabal" and
brought a discipline to how the Air Service conducted business that Menoher had only promised.[89] Patrick denied a reorganization plan that would
have given wider power to the Assistant Chief, and he forbade Mitchell's
issuing any orders without Patrick's prior approval, telling his admonished
subordinate that he would remain Chief of the Air Service "in fact as well

as in name."[90] Patrick knew his assistant chief well, describing him as having a "highly developed ego" desiring the public limelight, while also being "forceful, aggressive, spectacular."[91]

For Mitchell, Patrick's promotion was the beginning of the end of his influence, only weeks after achieving his most impressive triumph. Even before the public announcement of Patrick's promotion, which would be well received by the press, Mitchell tendered a letter of resignation. The Adjutant General put that request on hold, but when Patrick assured Mitchell that Patrick would accept a later offer, Mitchell withdrew it. The new chief had made his point of who would be in charge.[92] He also omitted an entire section from a memorandum submitted by Mitchell outlining the new organization and duties. Attempting to secure his own position, Mitchell wanted to have access to the Secretary of War "whenever I deem it necessary," after first reporting to the chief. Mitchell also asked to be the approving authority over everything concerning equipment. If he could not become the chief in name, he tried to reorganize himself into the position of power. Patrick immediately rejected this paper coup.[93] The new chief received the solid support of the conservative and ground-oriented Army generals, and many Air Service officers also approved the selection. Arnold and others understood the appointment: "We all recognized that the new Chief's experience with air power was a secondary consideration in his appointment. In the eyes of the General Staff, it was experience with Mitchell that counted. . . . [T]o control him, Patrick was the man."[94]

CHANGES ALL AROUND

With Patrick in charge, a fact of which Mitchell was often reminded, the next three years passed rather quietly. The chief kept his energetic assistant out of Washington as much as possible, inspecting air bases and performing other duties in the States and abroad. During these years, Mitchell influenced air tactics and doctrine more than he did national legislation. To further curb Mitchell's influence, Secretary Weeks forbade him from publishing any articles without official clearance.[95] Weeks's health would decline, eventually forcing him to resign, but his assistant secretary, Dwight F. Davis, would continue the administration's policies. Health problems also caused a change in presidential leadership.

On 2 August 1923, President Harding collapsed and died of heart failure, and Calvin Coolidge became president the next day. The personality difference was drastic: a quiet, cool Yankee replaced an affable midwesterner. Coolidge immediately vowed to retain Harding's programs and cabinet, and he pursued economy in government with even more passion than his predecessor.[96] Coolidge shared no love for air power or defense spending overall. Being of a reserved nature, he also resented Mitchell's unorthodox methods of promoting a program against presidential wishes. Arnold called Coolidge and Mitchell exact opposites and believed that "there can have been few of his citizens who aroused a testier feeling in him [Coolidge] than Billy seemed to do personally."[97]

Coolidge's relationship with Congress continually deteriorated. Without even the "presidential honeymoon" offered Harding, a Republican insurgency, led by Wisconsin senator and presidential hopeful Robert M. La Follette, constantly blocked Coolidge's legislative programs. Four months after Coolidge delivered his annual message to Congress, the legislature had not acted upon a single proposal.[98] The president's December 1923 message barely mentioned defense, although he insisted that Congress hold the line against further reductions. For the Air Service, he noted almost in passing the need for additional aircraft.[99] He followed the defense budget and spending meticulously, as evidenced by his complaint of the extravagant cost of hiring a band for a recruiting promotion in Portland, Maine.[100] The president would routinely reply, or his Budget Bureau often did for him, to monetary requests as "not being within the financial program" of the president.

The 1921 Budget and Accounting Act had created the Bureau of the Budget under the president. The bureau coordinated the president's fiscal policies with the different executive departments and allowed the chief executive a means to present a coordinated budget for congressional action.[101] In order to protect the budget process from "disruptions," the statute specifically forbade federal agencies from exerting direct influence upon Congress, unless Congress asked for clarification.[102] Thus, military officers appearing before Congress could not press for more money than the approved program submitted to the bureau and approved by the president, but they could answer questions from congressmen in relation to that budget and in the process make clear their personal opinion that more resources were needed. The War Department Budget Officer worked closely with the War Department's Legislative Branch to ensure that any legislation, regardless of whether it originated within or without the department, agreed "with the financial program of the President."[103]

The Army took steps to clarify the legislative process and to remain within the president's financial program. The Legislative Branch reviewed all proposals concerning the War Department, even if the measures did not propose higher expenditures. If any part of the department, including the Air Service, wanted new legislation, that department coordinated its efforts through the Legislative Branch and the Secretary of War. If the legislation originated outside the War Department, this branch coordinated a study of the bill to ensure that it agreed with the organizational and financial program of the president.[104] For example, if a congressman proposed new air legislation, it would be sent to the Legislative Branch, which would then send it to the

Air Service for study and recommendations, and the bill would return to Congress via the same route. Therefore, the department retained control over all Army matters and prohibited subordinate branches from going to Congress without approval. Although the Legislative Branch may not have originated due to Mitchell's activities, the airman certainly induced a refinement of the controls. Meanwhile, the restrictions on the renegade officer's Washington activities had further tightened.

Freed from the Washington grind, Mitchell found himself again among his fellow flyers. He flew regularly and conducted bombing tests, though these became less frequent because of the rapid deterioration of the Air Service caused by miserly budgets. It was during this respite that he also became aware of Douhet's published theories, which helped to amplify his own. Arnold characterized Mitchell as "down in the dumps" in 1923, frustrated because all of his actions so far had failed to move the country toward supporting aviation the way he believed it needed to be supported. The airman believed that the people supported a strong air presence, but leaders in Washington displayed only apathy. Despite all of Mitchell's work, "air power doesn't seem to be getting anywhere at all."[105]

Looking at the lack of legislative proposals or inquiries from 1922 to 1924 as compared to the previous three years, Mitchell's assertion was indeed valid. With Patrick in charge and his pretentious assistant effectively muzzled, the "back door" to air-friendly congressmen seemed closed, and the new chief's personal contacts with congressmen lacked the intimacy and consistency of Mitchell's.[106] The major legislative fights centered on budgets and personnel, and the Air Service fought these through regular channels. However, with an economy-minded administration and Congress, little support existed for expansion of the Air Service. Still, and to Patrick's credit, the air chief kept the service as viable as possible, and its reductions remained lower than those in the Army at large.[107]

Patrick's style moderated between the Mitchell's separatists and the unity proponents of the General Staff. One congressman applauded Patrick for taking an evolutionary rather than revolutionary approach.[108] Patrick disagreed with Mitchell's drastic contention that the airplane made other arms of defense obsolete, but he believed that aviation should assume a more important role than merely observation and reconnaissance. With a talent for organization and a clear understanding of obtaining the probable before stretching for the distantly possible, Patrick laid out a plan for his first two years. He wanted to make the Air Service a workable organization that could complete its current missions with competence. Patrick wanted not only to establish credibility but also to show his superiors the woeful state of the service.[109] He noted in 1922 how the Air Service, in cooperation with the president's mandate to reduce costs, returned $800,000 to the Treasury—

just over 4 percent of the $19.2 million appropriated. However, Patrick emphasized that while economy in government was good, it should not go so far as to curtail the effectiveness of national defense.[110] His first two annual reports did not mention a separate air service or increased autonomy. Instead, he concentrated on asking for increased appropriations for aircraft, bringing the service up to its authorized enlisted strength, and correcting the officer promotion system to place Air Service officers on competitive status with their colleagues in the other branches.[111]

Patrick believed that a lack of discipline caused the most problems within the service. The air arm was predominantly a young man's organization. As such, youthful and inexperienced officers took command at a rate unprecedented in other branches.[112] Many of these young officers constituted Mitchell's "cabal" and looked to him for leadership and for setting the example. "It is the youth and inexperience of its officers," Patrick wrote to a fellow general, "whom it is necessary to place in responsible positions that are largely the causes of the trouble which is found."[113] With Mitchell under control, Patrick now had to rein in the enthusiastic younger officers. He firmly believed that before he could do anything in and with the service, he must reestablish discipline. None of those intimately involved in the Air Service's problems (Patrick, Weeks, and Pershing) viewed the actions of Mitchell and his supporters as an improperly functioning civil-military relationship. Instead, they all believed it was an internal failing of discipline. Patrick handled it as such. He wanted the officers and men to concentrate on their jobs, not on publicity.[114] He followed these efforts by a masterful restructuring of the flying corps' assets and organization. Then Patrick concentrated on changing air doctrine and establishing the support and funding for this change from Congress and the War Department. Most importantly, though, in contrast to Mitchell, Patrick used official channels to implement his changes and advance his ideas.

The air chief urged change primarily through his annual reports, lectures at the Army War College, and approved appearances before Congress.[115] Unlike Mitchell, Patrick was always very careful to avoid any appearance of using Congress improperly, and he even cautioned Weeks once for the appearance of political connections with the Air Service. Patrick had once confiscated pictures of Weeks presenting a signed contract to the congressman from the district gaining the government's money when the congressman wanted to use the pictures for his reelection bid. Patrick did not like the implications of politics in Air Service dealings. Weeks agreed and added, "We are not going to play politics in the Air Service." Weeks did not forget Patrick's astuteness, and it solidified an already solid rapport between the men.[116] In another example of the effect of Patrick's style, approach, and moderate views, Air Service officers began writing speeches concerning aviation for delivery by Assistant

Secretary of War Davis. The speeches, usually originating in the Information Division of the Office of the Chief of the Air Service and touting the positives of aviation without separation polemics, would be routed through Patrick before being sent to Davis.[117] By keeping Mitchell under control, Patrick won the approval of Secretary Weeks and the General Staff; Patrick built political clout where he thought it best served the interest of the service.

Patrick's conduct allowed him to gain some concessions from the only major air inquiry between 1921 and 1923. Patrick had presented a blistering assessment of the state of the service at the end of 1922. Despite new advances in design and technology, the service still retained World War I–vintage equipment. Additionally, even though the service did have some of the newer aircraft, notably the Martin MB-2 bomber, the mobilization status would never have allowed the production of a force necessary for combat operations. Patrick also advocated a plan for dividing the force into an "air service," which would include support aircraft for reconnaissance and observation, and an "air force" of tactical units. He believed the air force should account for 80 percent of the force, whereas it currently accounted for only 20 percent.[118] Weeks asked for an enlarged study to recommend remedial actions. Patrick agreed, but he wanted Weeks to accept the premise of his organizational structure, where "air service" units (i.e., observation and reconnaissance) would make up only 20 percent of the flying arm, the rest being assigned the offensive role. Patrick then proposed his plan for air force organization and doctrine. He opposed the assignment of offensive units directly to field army commanders, as he wanted air units commanded by air officers. Patrick also noted the urgent need for legislation to increase the Air Service's officer and enlisted strength. Trying not to alienate the Navy, he argued that Army aviation should defend the continent from land bases and allow the Navy to use its assets at sea.

Instead of stonewalling the report, Weeks recommended a wider investigation and on 17 March 1923 convened the Lassiter Board, led by Major General William Lassiter. It initially appeared as if this board, consisting of predominantly non-flying General Staff officers, would go the way of previous inquiries. After five days it issued a report supporting Patrick's ideas and advocating a ten-year expansion program for the Air Service. Five weeks later, Secretary Weeks approved the report. Although the findings did not instigate any immediate legislation, the results served as the conceptual basis for Air Service organization and operation.[119] More importantly, the board's quick work and favorable ruling resulted from the new attitude brought in by Patrick and his maintaining control of Mitchell. Still, neither Weeks's compromising attitude nor the muzzle on Mitchell would last.[120]

During the first five years after the end of World War I, the actions of Mitchell and his supporters represented a coordinated effort to overturn the administration's stated military policy. The different presidents did not support an independent air service, and neither did the majority of Congress. Yet as a flurry of legislation on air service issues came before Congress, air proponents happily appeared before committees to support the various pro-air bills and push for appropriations. The airmen took even more controversial and questionable actions by prodding congressmen and senators to support pro-air legislation, using the press to stimulate public pressure, and even using Air Service operations and maneuvers to arouse public support. The aviators' actions overwhelmingly conflicted with the official policies of the presidents and their War Departments. Of the "cabal" of officers Mitchell led—most of whom wanted him to succeed in his crusade for an independent air service— most did not recognize the impropriety of his words and deeds.

Air officers often acted out of self-interest and in a manner they believed best suited national defense. In doing so, they crossed the line of proper conduct with respect to their civilian masters. Although Mitchell, Foulois, Arnold, and the others may have been working to improve the military, they did so outside the appropriate bounds. They especially operated outside proper and accepted spheres when they coordinated their efforts with opposing political elements in Congress. From the end of World War I until Wilson left office, the airmen supported and worked with Republican congressmen. After Republicans took control of the White House and Congress, the air officers still worked with air-friendly congressmen who opposed Harding's and Coolidge's aviation policies. Many airmen remained determined to support their own agenda instead of supporting the president's policies and goals. One could argue that these air officers needed to take the measures they did in order to bypass the opposition within the War Department, since they believed their efforts would improve the nation's defense. But their circumvention of civilian authority and efforts to force presidential administrations to alter their military policies, which were in line with the mood of the country and the majority in Congress, clearly violated the tenets of civilian control as understood at the time. Only the politics of the situation—the popularity of aviation and the party struggle in Congress—prevented civilian authorities from coming down hard on the airmen, Mitchell especially.

By using propaganda and access to influential congressman to propose and gain support for air-friendly legislation, Mitchell and the early air proponents violated the norms of civilian control. Instead of providing advice and working within the chains of command, the airmen circumvented established procedures and violated the code of proper behavior. Perhaps it was because Mitchell was so comfortable with both the civilian power structure and the military life that he would not, or could not, later keep them separate. More likely, he did not see (because of his background) or else chose to ignore (because of his egotism) how at the highest levels of interaction military officers were supposed to recommend military policy but not cross the line into actively pursing legislation and forming policy—especially when the policy conflicted with the stated programs of the president and his administration. Either way, in order to advance his cause and to publicize his crusade for air power, Mitchell succeeded in moving outside the bounds of political behavior considered acceptable for a military officer.[121]

Patrick had temporarily halted the Air Service's movement, led by Mitchell, toward even more confrontation with civilian policy makers. But by 1924, Mitchell would reemerge with his congressional allies, and Patrick, frustrated by inaction on the Lassiter proposals, would call for an independent air force under a single Ministry of Defense. Confrontation, publicity, and alliances with politicians would once again rule in the Air Service.

CHAPTER TWO

The Politics of Investigations, 1924–1925

For the officers and men of the Air Service, the legislative front during the first few months of 1924 looked as bleak as it had during the previous two years. Due to disagreements between the Secretaries of War and the Navy over percentages of appropriations for their services, the Lassiter Board's recommendations never coalesced into legislation.[1] From Patrick on down, Army air officers became more and more frustrated. Those who worked for expansion and improvement of the current system, like the previously moderate Chief of the Air Service, became frustrated by inaction and moved closer to those, like Mitchell and his followers, who still wanted a separate air service.

Events would build until they exploded in a political firestorm at the end of 1925. Patrick, ever prudent, would not openly oppose the administration and the General Staff, although he became more aggravated by legislative inaction and miserly appropriations and seemed more willing to contact certain congressmen. Mitchell, again with help from air-friendly congressmen and renewed public interest, would reemerge during a long, public, and politicized investigation. Buoyed by these events, he would resume his old ways and outrage Secretary of War Weeks and President Coolidge. Due to the popularity of Mitchell, and aviation generally, and a less-than-comfortable majority in the House, the Coolidge administration could not simply suppress, discipline, or banish the airman and his supporters for expressing their views, even though the flyers did so in inappropriate ways. At least, the administration apparently felt that it could not.

While the congressional committee prepared its report, aviation matters became even further entwined in politics and partisan maneuvering. As the Lampert Committee finished its work (most understood it would side with Mitchell), two air tragedies occurred within days of each other. Mitchell threw all caution to the wind and set out on a public rampage and openly condemned the administration and the Army generals. Knowing that he must court-martial the insubordinate but popular airman, Coolidge moved to quell what he knew would be an attempt to use the trial and the coming congressional report for a renewed push for an independent air service. Coolidge opposed air independence because he believed that the British experiment with a separate aviation branch had not proved entirely successful. He also feared duplication of effort on the part of the different services, which weakened national defense while increasing the budget, the latter being something he would absolutely oppose.[2] Coolidge quickly organized his own commission to study air matters and make recommendations, while also having the War Department focus upon the trial as one of an insubordinate officer, not of the administration's policies.

By the end of 1925, Coolidge had regained control of his War Department and cowed the Air Service. With Mitchell cashiered and the findings of the President's Aircraft Board guaranteed to support Coolidge's program, the public and Congress soon followed Coolidge's lead. Patrick and the Air Service made the arguments for an independent force during this pivotal year, but they also demonstrated a willingness to accept incremental steps. Mitchell's crusade would go down in flames during the last month of 1925, and with it the tactics of confrontation—at least temporarily. However, Mitchell left behind a tumultuous political landscape, with his congressional allies battling the Coolidge administration.

CONGRESS TAKES CENTER STAGE

The almost year-long congressional inquiry known as the Lampert Committee, which looked into all aspects of the country's air operations, stemmed from a disgruntled inventor and former lieutenant in the Naval Reserve, James V. Martin. He accused the Air Service of using dishonest procurement practices (the service favored the Barling bomber, which failed performance tests, and inexplicably destroyed Martin's aircraft).[3] Wisconsin Republican John M. Nelson took the inventor's case before the House and charged the Air Service and the Manufacturer's Aircraft Association with monopolistic collusion and using appropriations to subsidize the industry. Specifically, Nelson condemned the different government agencies' air services with inefficiency, for he believed that negotiated contracts wasted money and possibly infringed on patents.[4] Representative Jonathan M. Wainwright, a former Assistant Secretary of War and generally a supporter of aviation, rejected Nelson's claims, but to little avail.[5] Nelson's accusations led to the formation of the House Select Committee of Inquiry into Operations of the United States Air Services, commonly known as the Lampert Committee for its chairman, Republican Florian Lampert. The committee, consisting of nine members from both the House Military and Naval Affairs Committees, was directed to investigate the operations of all the government's aviation services, including the Army Air Service and the Navy Bureau of Aeronautics.[6]

The investigation's scope officially related to procurement and contracts, but the committee went beyond these initial constraints and recommended vast changes in the country's air forces.[7] Reflecting the House Republican majority, Republicans filled five of the nine committee seats.[8] Reflecting the youth of the Air Service, both in its relative age as a force in military affairs and in the age of its fliers, the committee's composition must have pleased those in the Mitchell camp. Older congressmen may have been more inclined to support the ground Army and not want to take a chance on a new and relatively unproven implement of war. Only two committee members, Indiana Republican Albert H. Vestal and California Democrat Clarence F. Lea, had served in Congress during World War I, and both came to Washington with the Sixty-fifth Congress in March 1917. Lampert was the only other committee member who had been elected prior to 1920.[9] Four of the members were in their first Congress, and two of the Republicans were halfway through their sophomore term.[10] Only three men on the committee had reached the age of fifty, and the youngest, Rodgers, was not even a teenager when Orville Wright took flight at Kill Devil Hill, North Carolina. From a political standpoint, the committee also seemed destined to support Mitchell. The Democrats could use the podium to counter Coolidge's policies, and the majority of the Republicans were noted aviation enthusiasts.[11]

Created in March 1924, the committee did not begin hearings until October. Billy Mitchell became the center of the next five months of investigations and testimony. In most of his previous writings and speeches, Mitchell backed a single cabinet-level Department of Aeronautics that would include all of the national aviation assets and planning, including Army and Navy aircraft and the air assets of all other government agencies. The department would have two major divisions under it, one each for civil and military aviation.[12] Mitchell had always pushed for this departmental concept, and Representative Curry's legislative proposals mirrored the "Mitchell plan." Now, however, Mitchell also supported the alternative, which had been proposed in legislation by others, including Senator New. The other arrangement called for a single department of national defense, with a cabinet-level secretary and three subsecretaries for land, air, and naval forces.[13] The main provisions of either proposal suited Mitchell: an independent service unhindered by the Army and with cabinet representation.

Lower-ranking air officers continued to support Mitchell from the field. The younger airmen also persisted in acting outside the proprieties of both written and unwritten rules for proper behavior in political matters. In 1925, Arnold and Spaatz asked influential businessmen to lobby Congress to further the Air Service's goal of independence. In one instance, Spaatz wrote William Stout, the head of Stout Aircraft Company in Detroit, urging him to support the Curry bill and to rally other prominent citizens to the cause.

Enclosing a copy of the proposed legislation, Spaatz told Stout how passage of such a bill depended upon the support of the country and people informing Congress of their pro-aviation wishes. "Knowing your relationship with the prominent citizens of Detroit on matters aeronautical and the weight which your thoughts on such matters carry," the airman pleaded, "I am writing to see whether you can assist the cause by inducing some of the more prominent citizens like Mr. Henry Ford, to publicly espouse [support for legislation] . . . along the lines of the Curry Bill."[14]

Whether Stout acted on Spaatz's request is not known, but aviation industry leaders generally supported a stronger air force if only for the possibility of increased business and profits. Air leaders understood these dynamics, and the ardent Mitchell supporters contacted business leaders to influence Congress and pending legislation without the knowledge or approval of the General Staff or the War Department. However, the tide had clearly changed in the Air Service, and Mitchell's influence, even among most air officers, continued to wane as Patrick further asserted himself.

By his discipline, straight talk, and moderate approach to the needs of the service, Patrick spoke with ever increasing support. During his testimony before the Lampert Committee, he recommended the creation of an Air Corps under the Secretary of War, which would place the air forces on equal footing as a combatant arm within the Army and would provide what Patrick called a "rather long" step in the direction toward independence.[15] He also offered to eliminate duplication by defining the roles of the Air Corps and the Navy (and eliminate the feuding caused by Mitchell), limiting the air arm to performing coastal defense out to two hundred miles. While Mitchell wanted independence immediately, Patrick demonstrated a grasp of the possible and accepted the moderate position of taking intermediate steps toward independence.[16] "The ultimate solution of what I call the air defense problem is the concentration of responsibility therefor [*sic*] on one head," he asserted before the committee, "but I do not believe the time for that is quite ripe yet."[17] Patrick's conciliatory line even caused an ardent pro-Mitchell congressman to treat the air chief roughly.

Representative Frank Reid, although a noted aviation supporter, tried to manipulate Patrick to state more extreme opinions during their exchanges in the committee room.[18] Reid first tried to coax Patrick into disagreeing with and disparaging the Navy and Secretary of the Navy Wilbur.[19] At one point, Reid wanted Patrick to discuss Wilbur's aviation experience, to which the air chief replied, "Please, Mr. Reid, I do not know anything about that." With Patrick refusing to play Reid's game, Reid tried to entice the general into a discussion of which service, the Army or the Navy, should take precedence when discussing air matters. Patrick again politely deflected Reid's invitation to attack the seagoing service.[20] Later, and on other matters, Reid became confrontational with Patrick and tried to make him contradict his own state-ments.[21] During the relentless grilling, Patrick at one point became frustrated and confused. "I am doing poorly, I know," he conceded, "but I am doing the best I can."[22] After a recess for lunch, Reid took a softer line with Patrick, conceding, "I cut you off this morning in some of your answers, General." When asked if he wanted to add anything to his morning answers, Patrick declined. Reid's afternoon questioning took a more civil tone.[23] Most likely Reid had been approached by the more moderate members of the committee during the recess, men who probably disagreed with his treatment of a dis-tinguished officer and who wanted to move air power forward, just not on Mitchell's timeline. After all, the committee did want to push the air power agenda, and Patrick supported Mitchell's ultimate goal, just not the more controversial airman's methods.

The testimony of other Air Service officers (not Mitchell) generally sup-ported this evolutionary approach and backed the Patrick plan. All of the airmen testified that the Air Service suffered under the conditions then exist-ing and that Congress should mandate some sort of change. Spaatz admitted that he knew between 60 and 70 percent of the Army's fliers, and the "gen-eral feeling is that under present conditions we are not getting anywhere."[24] Spaatz clearly supported a separate air force, calling it "absolutely essential for the air defense of the United States," and he called the air corps concept "a start toward the ultimate end which is necessary."[25] What is surprising is that none of the influential young cadre or the more outspoken veterans, save Spaatz, testified before the air-friendly committee. Arnold, the most outspo-ken of the group and an ardent Mitchell supporter, did not testify despite being stationed in the District during the hearings. He may have been too busy trying to complete his coursework at the Army Industrial College, and he may have felt that his testimony would have added little, since it only mirrored Mitchell's beliefs.[26]

Foulois, an outspoken officer who hated Mitchell and his tactics, was another notable absence. The committee, being inclined to support Mitchell and including several of his intimate supporters, may have passed over Foulois in order to avoid a clash of views and personalities. Foulois himself took a lower profile at this time. Like Arnold, the elder airman remained very busy in pursuing the requisite military education for promotion, and academics did not come easy to Foulois. Also, he may have preferred to keep quiet and thus make himself more palatable to the administration for future advancement and possibly the top air position; everyone knew that Patrick had to retire in 1927 and that another senior air officer would be picked as a replacement.[27] Undoubtedly, anyone who supported Mitchell or employed inappropriate tactics or espoused policies conflicting with those of the Coolidge administration would not be considered for the post.

Mitchell took center stage at the Lampert hearings, testifying on more occasions, at more length, and on a wider variety of topics than any other witness.[28] The congressmen also took quite a few unusual steps that demonstrated their favoritism toward Mitchell. On the afternoon of his first day of testimony, the committee invited Representative Curry, an unabashed Mitchell supporter and air power advocate, to appear at the inquiry table and ask questions. During the current congressional session, Curry's most recent bill advocating a united air service sat before the Military Affairs Committee. The committee welcomed Curry and allowed him to question Mitchell and lead them down the path most wanted to go. Mitchell announced his support for his concept of independence, which was the united air service with all aviation under one cabinet head—not coincidentally the formulation of the current Curry bill (H.R. 10147).[29]

During Mitchell's testimony, an interesting and probably not coincidental line of questioning arose regarding the quashing of "free speech" by the War Department, the so-called muzzling of officers who would otherwise testify to Congress about the inadequacies of national defense and proposed remedies. Mitchell and his congressional supporters on the committee, again the most overt and outspoken being Reid, allowed Mitchell to express his views. Mitchell spoke against his military superiors, the War Department, and the president. The committeemen feigned amazement that the airman could expect retribution, including not being reappointed to his current position as Assistant Chief of the Air Service, for "exposing" the conspiracy to suppress new ideas and the officers promoting them.

The reappointment topic first appeared during Mitchell's second visit to the committee.[30] In an obviously leading statement, Representative Randolph Perkins of New Jersey casually suggested that Mitchell would be associated with Army aviation "for some time to come." Instead of honestly admitting that he would remain in the Air Service in some capacity, the airman stoked the flames by replying that his tenure as Assistant Chief would end on 26 March and that he had not yet received notification of reappointment. Perkins, playing the tune for Mitchell to dance, then coaxed him into saying that the reappointment rested with Coolidge, who at that time had already renewed other officers in similar positions.[31] After Perkins lamented how the loss of Mitchell's expertise would hamper the Air Service and the country's defenses, the congressman moved into an area long touted by the airman as a major problem: how the administration threatened and coerced officers not to provide information to Congress that would promote the air arm. The first thread of "evidence" concerned how Mitchell would probably not be reappointed due to his testimony. Onlookers and the press witnessed the following exchange:

MITCHELL: I imagine that it [my reappointment] may not be made
on account of the evidence that I have given before the
committee.

PERKINS: Well, you have been before this committee at this committee's
request, and not at your request, and you have come again
this second time particularly at my request. I trust that there
is no indication that the failure to nominate you is due to
evidence before this congressional committee?

MITCHELL: I would not be surprised.

PERKINS: How can a congressional committee get evidence from the
various branches of the service unless some one comes and
tells what he thinks?

MITCHELL: It can not. It is impossible.

PERKINS: Well, now, for instance, I think you have been rather an
outspoken witness, and the committee always likes to have
some one who is outspoken tell us what he really thinks. . . .
Have you any reason to think that your being outspoken has
something to do with the failure to send your name in?

MITCHELL: Oh, I think so. I have been told in times past that it would
operate that way. However, as far as I am concerned I will
give exactly the evidence that I think is proper for the country
to know in every case before committees of Congress.[32]

Mitchell had begun to sharpen his martyr's sword. He also probably hoped
to use the committee as his armor against removal from the position he
occupied, and perhaps even as a boost to the office he so desperately wanted.

In the questioning and testimony that immediately followed, Perkins enticed Mitchell to relate how the War Department took actions against certain
officers who testified before Congress in a manner not consistent with department policies. According to Mitchell, members of the War Department
and the General Staff would read transcripts of congressional testimonies and
call the officers in to explain any disputed statements and to substantiate any
unorthodox views. After prodding, Mitchell exposed Secretary Weeks as having forced Mitchell, via "confidential communications," to confirm his past
testimony. Perkins made it clear that the committee would seek out Weeks
to explain these communications. Perhaps in an effort to cover his tracks in
directing this "revelation," Perkins added, "This is a rather interesting branch
of the inquiry, that we have chanced to get accidentally." He closed the present
topic of questioning by restating, "I had not meant to get so far off the line
of inquiry. I happened to get there by not understanding some things."[33]
From there Perkins went on to other topics, but he was soon interrupted by
another committee member who wanted to take the issue further and discuss

free speech within the department and before Congress. Mitchell was more than happy to comply.

Representative Charles L. Faust of Missouri wanted to pursue a line of inquiry about discipline versus open testimony. Was it acceptable to "tell the truth" and give opinions about what the officer believed was in the nation's best interest even when those views countered War Department policy and General Orders?[34] Mitchell asserted that officers had indeed faced disciplinary measures when they testified against stated War Department policies. Although he could not name any officer besides himself who received a reprimand, he affirmed that officers in both services feared coming before the inquiry. Mitchell pointed out that British officers could state their opinions and even participate in politics, which he believed allowed them to express more honest opinions and get better results. He declared that since American air officers did not have that same freedom of action, Congress would not get the best and most honest opinion on military matters, and especially those regarding air power. Mitchell conceded that the military required discipline but stated that discipline should not be used to undermine honest testimony.[35] Both Perkins and Representative Lea jumped in on the questioning. Mitchell admitted knowledge of "general rules" not to give out views without War Department permission.[36] He probably used "general rules" instead of the terms "standing orders" or "General Orders" to demonstrate his disagreement with those rules that forced officers to agree with War Department policies when speaking outside the department. Mitchell knew of the Baker memo and of the two General Orders, which codified the positions of Baker and Weeks.

Weeks had not retracted the Baker order, but he added further provisions and clarifications to it by issuing General Order No. 20 in 1922. The latter authorized "public discussion on appropriate occasions by officers" on department policies. However, Weeks urged Army officers to support the War Department, adding that the organization "expected that this support will be freely given when the undoubted merits of the policy are understood and when attention is called to the burdensome and dangerous alternative that must be faced [if department policies were not followed]."[37] The orders also required officers to obtain permission from the Adjutant General's office if a "different presentation" will be given.[38] Upon further questioning by the committee, Mitchell acquiesced that these "rules" had been officially adopted, but "without permission."[39]

During the follow-up questions by Lea, Mitchell accused the War Department of muzzling officers and threatening to discipline those who publicly opposed official policies. The airman emphasized that War Department retribution continued "in spite of published orders allowing freedom of testimony."[40] Mitchell agreed with Lea that because of these threats Congress could not get a true answer on the air questions.[41] The "unexpected" branch of the inquiry intrigued the congressmen, and they quickly questioned the Assistant Chief of Staff, Brigadier General Hugh A. Drum, and called Secretary Weeks to reappear before the committee to answer the charges.

The muzzling charges allowed the board to broaden its inquiry but also clearly demonstrated an affinity for, and defense of, Billy Mitchell. While the members overall seemed inclined to support Mitchell in his battle against the War Department, Congressman Reid stood out as the attack dog. When Drum took the stand, Perkins began the inquiry and tried to get Drum to concede Mitchell's status as an outstanding officer. Perkins even tried to use deductive reasoning to corner Drum. After obtaining Drum's agreement that Mitchell had the most knowledge about air power of all Army officers, Perkins maneuvered to have Drum admit that Mitchell must have known what was best for the Air Service, but Drum deftly avoided the issue.[42] Interrupting Perkins's questioning after a heated exchange, Reid commented on Drum's evasive answers and wisecracked, "He can beat an airplane on maneuverability."[43]

Reid hammered away at Drum during the duration of his testimony, even interrupting other lines of inquiry. At one point, the Illinois Republican even asked that Drum be excused from the witness stand for bringing up "other matters."[44] On the topic of officers' testimony before Congress, Drum held to the department line and asserted that officers would not receive reprimand or punishment for stating their own opinion. Drum also countered the "conservative" moniker given the General Staff. He asserted that the generals actually encouraged forward thinking but that officers must support the administration and use discipline and discretion.

However, at one point Drum seemed to disregard the General Orders concerning congressional testimony. When asked what actions the War Department should take with an officer who states his opinions about air power "out of harmony" with those of the General Staff, Drum answered that the officer could testify without restraint under any circumstances.[45] Drum's answer demonstrated the fine line that he and the War Department walked. They had to allow officers to testify frankly before Congress, which allowed the legislature to make informed decisions, but they also desired to control the occasional "renegade" element from proposing changes not in step with the administration's policy or intentions. Drum agreed that officers could testify freely, even though General Order No. 25 seemed to limit such freedom, and he went so far as to call the Coolidge administration "especially liberal in encouraging such things."[46]

Weeks took the stand for the second and final time one month after the "interesting branch" of inquiry on muzzling began with Mitchell's accusations.[47] The committee sent Weeks a memorandum with eight questions, and his reply, along with his subsequent testimony, directly confronted the controversy over free speech and muzzling.[48] Weeks's written response did little beyond reiterating the appropriate General Orders. He did, however, try to counter the negative "conservative" label given to his department and staff: "If the statement that Army leaders are conservative is meant to indicate an attitude the opposite of radicalism, the charge of conservatism might be sustained; but that does not mean that they are not progressive. . . . In determining policies affecting the Air Service, the Chief of Air Service is always given an opportunity to express his views and he is regarded as the air expert of the Department, his views almost invariably being approved."[49] Weeks took a thinly veiled swipe at the committee's previously expressed view of Mitchell as the "air expert."[50] The secretary astutely sidestepped, with the "almost invariably approved" language, the issues of the Lassiter Report's not being implemented and Patrick's repeated pleas for action on a variety of issues in his annual reports.[51] The muzzling controversy took center stage, though, and the committee opened their questioning on this issue, with Weeks stating that he eagerly looked forward to clarifying the matter.

Secretary Weeks vehemently denied any muzzling, calling the charge "unfounded" and "untrue," declaring to the inquiry, "If there is any officer in the Air Service or in the Army who has not had an opportunity to speak his views under proper conditions I do not know it."[52] As for these "proper conditions," he added that if his rules denied any officer the right to speak, that officer must have been a "timid soul."[53] He then brought up Mitchell's past remarks and the reason the airman had to submit articles for publication. That rule derived from "propaganda" publications Mitchell made following the initial bombing tests, which greatly upset the Navy. During the fall of 1924, Mitchell went directly to President Coolidge for permission to publish articles in the *Saturday Evening Post*. The president granted this permission, subject to approval from Mitchell's supervisor, but Mitchell had skipped the chain of command and sent articles directly to Coolidge, no doubt because the president would have had the least amount of time to read the articles and investigate the contents. Weeks called this action outside "the usual course of procedure."[54] Mitchell had not sought the permission of those he skirted (Patrick and Weeks) but rather published the articles forthwith.

Representative O'Sullivan then asked Weeks: If there was no muzzling in the War Department, why should Mitchell submit his articles? "That has been done a good many times in cases where there were controversial matters, and there was no intention to muzzle General Mitchell," Weeks replied. The secretary then stated that he was only trying to keep the peace with the sister service (Mitchell's articles slandered the Navy for its faith in battleships and its stand against air power).[55] Under continued questioning by Representative Prall, Weeks denied ordering Mitchell away or telling him to resign his office.[56] Later, O'Sullivan also asked about the holdup of Mitchell's pending reappointment, which Weeks deflected, noting that the matter rested with Coolidge.[57]

Later during Weeks's testimony, Congressman Lea returned to the question of the General Orders and the limits on officers who tried to influence legislation. Weeks agreed that he interpreted the orders not as restraining officers from giving their opinions but as intended to keep confidential information (i.e., War Department secrets) from becoming public and to restrict officers from lobbying for legislation not approved by the War Department. Lea then asked Weeks whether he knew of any instances of officers trying to lobby Congress. For example, Lea inquired, "Would you construe a man in the Air Service, for instance, who would go to New York, or any other city, and make a speech to the chamber of commerce advocating legislation to reorganize the War Department and establish a separate air service contrary to the plans advocated by the General Staff, as violating this section [section V of General Order No. 25]?" Weeks once again avoided answering the question, and talked about support for some changes in the orders. However, Lea persisted in trying to use actual events as examples to trap Weeks into calling these instances lobbying, but Weeks would either provide a vague answer or say he did not know of any cases. Lea pressed further, and through pointed and leading questions he asked about lobbying for changes in the Air Service.[58] Weeks continued to deny knowing of any officers lobbying for legislation, asserting that the order sounded much more strict in its wording than how the War Department actually applied its prohibition.[59]

None of the congressmen took Weeks to task about the past few years of officers pushing for the Curry bill, or lectures and articles by Mitchell, or any of the known speeches in front of civic organizations—all of which advocated a position counter to presidential policies. While Weeks may not have known of these efforts (like Arnold's in support of California legislation and the Curry bill), his assertion of ignorance was indeed a stretch. Congressman Reid was not present this day. As hard as he had taken Drum and Patrick to task, Reid would have certainly applied maximum pressure to the ailing secretary. As it ended then, the committee took Weeks's testimony without judgment and added it to the voluminous pile of evidence under consideration that would await the final report more than nine months later. The committee ended its hearings within a few days of Weeks's testimony and began its deliberations.

All impressions pointed to a report that would support air power in some form and probably push Mitchell's program. In addition to the favorable treatment of Mitchell and other air proponents, and the sometimes unfavorable treatment of those opposed (like Drum and Wilbur), it seemed the committee only called others outside the affected departments to testify on the air independence question if they were known to support the proposal. The inquiry also allowed an air-friendly congressman to question Mitchell, and the committee members intentionally interviewed others on Capitol Hill known to support Mitchell.[60] At one point in the testimony, Congressman Reid offered his opinion that the group should go on record as supporting Mitchell. "If you let this fellow [Mitchell] be punished for this [coming before the committee]," Reid predicted, "you will never get a man to come up here and say green is green." When the committee denied the idea of a vote of confidence, pending the outcome of the inquiry, Reid indicated that Mitchell would "be back in the sticks [by] then."[61] The Illinois Republican foresaw the possibility that Mitchell would be punished by the War Department, removed from his position as the second-ranking airman, and demoted to a less important post. Even though the committee did not take any proactive steps to protect Mitchell, the close of the inquiry in March 1925 gave the controversial airman and his supporters hope. The Mitchellites seemed headed for a victory, but much would pass within the next nine months.

For the previous three years Mitchell had carried out the agreement he had made with Patrick in the fall of 1921 to take a lower profile and submit items to the chief for approval before releasing them publicly.[62] Even up to December 1924, Patrick approved of Mitchell's conduct and recommended his reappointment as assistant chief, as his four-year term would expire in the coming spring. The first two months of 1925 radically altered Patrick's thinking, as the outspoken Mitchell reemerged, with the recently ended congressional inquiry's overt—and frequent—support. Mitchell broke all of the rules laid before him by Patrick and once again crusaded against the administration and the War and Navy Departments. The controversial airman resumed writing articles in the press pushing for an independent air service and relished his appearance before the Military Affairs Committee considering the Curry legislation, which was interspersed between his appearances before the Lampert inquiry.[63]

Because Mitchell reverted to his insubordinate ways, Weeks ordered Patrick to recommend a new Assistant Chief and, in a seven-page letter to Coolidge, detailed the reasons for not reappointing Mitchell. The letter concentrated on the airman's false and misleading testimony on the status of the Air Service and the "muzzling" of air officers before the Lampert Committee. Weeks countered Mitchell's testimony with facts, pointing out that "all this was well known to General Mitchell when he apparently endeavored to startle the country" by offering inflammatory testimony.[64] Weeks closed out his formal denunciation of Mitchell by saying: "General Mitchell's whole course has been so lawless, so contrary to the building up of an efficient organization, so lacking in reasonable team work, so indicative of a personal desire for publicity at the expense of everyone with whom he is associated that his actions render him unfit for a high administrative position as he now occupies. . . . His record since the war has been such that he has forfeited the good opinion of those who are familiar with the facts and who desire to promote the best interests of national defense."[65] With those words, Weeks began the end of Mitchell's career. At the time, the administration hoped it had heard the last of Mitchell and hoped to still his voice by assigning him far from Washington. Mitchell, reduced to the permanent rank of colonel, went to San Antonio, Texas, to oversee aviation matters of the VIII Corps area. The administration finally had removed Mitchell from the limelight and from the District. In far-off San Antonio he could not provide as much fodder for the congressional and press anti-Coolidge forces.

Although a slew of editorial cartoons supported Mitchell and ridiculed Weeks, the newspapers did not come out in Mitchell's defense. As they had during the Menoher-Mitchell squabble, the newspapers backed Mitchell's views but would not look kindly upon insubordination or any actions that sullied senior officers (virtually all of whom ranked as decorated veterans of the recent war). The papers seemed to reflect the general mood of the American public: it liked the idea of a strong air arm and wanted to see it operated in the best manner, but no one supported spending huge amounts of money. The press also seemed generally to like Billy Mitchell, but it did not support his more extravagant antics and acts bordering on disobedience. Several papers, even those that usually supported Mitchell's positions, deplored his denigrations of the Navy and of officers and civilians of both services. The *New York World* summed up the general mood by condemning Mitchell for "virtually taking the position that every officer, either in the Army or Navy[,] who opposes him is dishonest, or stupid, or both."[66]

Once again, with Mitchell in hot water, an influential congressman came to champion his name and cause. New York Republican Fiorello LaGuardia, long considered a party insurgent, introduced not one but three measures in the House to protect Mitchell. The first proposed limiting military officers from administration reprisals because of congressional testimony—including transferring the officer to another assignment, as had just been done with Mitchell. The second resolution would have forbidden reprisals against officers who were called before congressional committees or inquiries and who responded to questions or requests. This measure provided immunity when officers were called, in effect subpoenaed, before Congress. The third measure blatantly called for Mitchell's reinstatement, failing only to say his name. It created a post, Chief Flying Officer of the Army Air Service, who would also serve as Assistant Chief of the Air Service. The proposal mandated that the Chief Flying Officer must have been a brigadier general for at least the preceding five years, had served in the American Expeditionary Forces for at least eighteen months, and had flown for at least ten years.[67] In other words, the bill created a position that only Mitchell could fill and, not so coincidentally, also returned him to Washington as Patrick's assistant.

These measures did not attract support. Patrick tapped Lieutenant Colonel James E. Fechet as Mitchell's replacement. Commissioned into the cavalry from the enlisted ranks at the turn of the century, Fechet had since become a veteran flier and most recently commanded the Advanced Flying School.[68] More importantly, he brought to Washington neither political clout nor an agenda, and he could be expected to serve as Patrick's loyal assistant and not lead an insurgency from the War Department offices. Patrick needed to stock the Air Service's offices with airmen who agreed with his position of wanting an independent service but were willing to accept incremental gains toward that end and would eschew confrontational methods. Even though Fechet was a friend of Mitchell's, he did not count among the "cabal."[69] The press commented: "Few officers now on duty in Washington know Fechet intimately . . . he has seen little staff duty" in the capital.[70] Equally important to those who had failed to muzzle his predecessor, Fechet was described as "modest and unassuming" and "not the type who seeks publicity."[71] Although he lacked Mitchell's war experience, Fechet seemed qualified to fill the office while also not threatening the administration's programs or upsetting either the War or Navy Department. In other words, he was not Mitchell.

Mitchell's seeming last hope on influencing the future of the air arm hinged on the Lampert Committee's report. The nine congressmen ended their hearings on 2 March 1925 but would not issue a report for nine months, according to Congressman Perkins, because the members were all tired and wanted to forget about the whole matter for a few weeks.[72] This interim allowed Coolidge to take the political initiative and divert any initiatives by Mitchell and his supporters.

THE PRESIDENT ACTS: POLITICS AND THE "INSIDE BOARD"

By the early fall, with the Mitchell court-martial and the Lampert Committee report looming, Coolidge understood the need to act. Mitchell and his congressional allies had succeeded in raising the public's interest in air power once again with sensational charges and claims of an inept, inefficient, and neglected service. If the president let the air-friendly Lampert Committee issue its report on the heels of or during a highly public trial of a very popular officer, the political initiative would be lost and any counterproposals weakened politically. Due to the relatively narrow Republican margins in both houses of Congress, Coolidge still needed to take the offensive and assert his own position, but he had to do so in a seemingly nonpartisan, fair manner. If he sat back and did nothing, enough Republicans might join the aviation backers and the opposition Democrats to cause him more political controversy. The 1924 elections had also improved the president's political position from his abbreviated first term, and Coolidge could act a little more aggressively on divisive issues. During the Sixty-eighth Congress (1923–25) the GOP held margins of only eighteen seats in the House and eleven in the Senate. Riding the short coattails of the Coolidge's reelection victory, Republicans gained twenty-two additional seats in the House and one in the Senate.[73] Thus "Silent Cal" went on the offensive and formed a board of his own to investigate the nation's air "troubles" and make quick recommendations along lines he would support. An independent air service would be out of the question.

Only nine days after the crash of the dirigible *Shenandoah*, and one week after Mitchell loosed his famous tirade against the services and the administration, Coolidge appointed his own investigative board.[74] At the time of the surprise announcement, Coolidge also knew he would take disciplinary action against Mitchell and that the Lampert Committee would be releasing its findings within a few months. To ensure that the board's findings dovetailed with his own views, he stacked the board with his political friends and men not so friendly to the Mitchell cabal. Formally called the President's Aircraft Board, the group consisted of nine well-known members and included one senator, two representatives, a retired general, and a retired admiral.[75] Coolidge asked them to meet in four days to select their own chairman, to "proceed immediately to a consideration of the problem involved," and to report by the end of November.[76] The men chose Dwight W. Morrow as their chairman, and the group became popularly known as the Morrow Board.

Morrow's election as chairman became the first outward signal of this group's intention to stay within the bounds of what Coolidge would support. A staunch political ally of the president's, Morrow had always been a political conservative whose career to date suggested that he was more interested in the accumulation of wealth than national service.[77] Coolidge made sure that, from the outside, the board seemed balanced between political parties and views on air power. The *New York Times* lauded the committee as being "of such a character as to win public approval and place the inquiry above partisan grounds."[78] Upon first glance even Mitchell approved, saying, "The personnel of the board is a surety that the study will be painstaking and fair."[79] In truth, only Senator Bingham, a former Air Service lieutenant colonel and World War I veteran, represented aviation enthusiasts.

Bingham, a former professor of history and politics at both Harvard and Princeton, had served as a captain in the Connecticut National Guard and became an aviator in the spring of 1917. Soon after that, he organized the United States School of Military Aeronautics. From August to December 1918 he commanded the flying school at Issoudun, France.[80] His thirteen publications included the 1920 book *An Explorer in the Air Service*.[81] The day after Coolidge announced the board, the *New York Times* ran a article written by Bingham in which the senator fully explained his ideas on aviation.[82] While Bingham's article left no doubt about his support for a stronger national air service, he did not champion Mitchell's desire for a separate department controlling all air assets. Instead, he touted the middle ground of establishing a "Bureau of Air Navigation" under the Commerce Department and creating an air corps in each service on the same principle as the Marine Corps in the Navy. Bingham admitted that his own contacts with many pilots in both services revealed the flyers' desires for an independent air service, but he confessed that many changes must come first. In effect, he hinted that a gradual approach to independence might better serve the national defense.[83] These positions demonstrated support for the positions of Patrick, whom Bingham lauded in his article.

Patrick and Bingham corresponded more and more during this period, and Patrick began to blur the lines of proper relations with Congress. He had written Bingham soon after the latter took office, and Bingham had responded that it had been his pleasure to serve under Patrick in France.[84] Only two days after the announcement of the President's Aircraft Board, Patrick wrote a personal letter to Bingham congratulating him on being selected and offering preliminary information. At that time, the Patrick had already submitted his Air Corps idea (mentioned earlier as growing out of his previous annual report) to the War Department. He now slipped it, still unofficial and unapproved, to the freshman senator, prefacing his proposal by saying, "I hope you will find it interesting. . . . [I]t does probe rather deep and outlines certain policies which I think should be adopted."[85] Indicating an understanding that he was circumventing the chain of command, Patrick closed by cautioning, "In order that we may observe the proprieties, you will understand, I am sure, that this paper should not appear in your proceedings unless obtained from the War Department or from my office by a formal call for it."[86]

While inappropriate, Patrick's conduct never went as far as Mitchell's. Yet it seems that the elderly air chief came to believe that being politically inactive would not benefit the air arm either. Having a Secretary of War who generally supported the ground Army officers, Patrick still believed that the Air Service needed outside assistance or it had no chance to develop into an independent air force. The Army wanted to keep control over aviation and use it as a support element for the combatant branches, while the Air Service needed to demonstrate the capability to provide unique capabilities that could best be nurtured in a separate organization. Like Mitchell, Patrick now faced the frustrating prospect of being continually stonewalled and not advancing the more autonomous capabilities and independent operational doctrine the flyers desired. He had been Chief of the Air Service for four years, and his cooperative style had pleased the old guard only to the extent of quieting Mitchell. But Patrick could not point to any real gains in improving the Air Service's organizational status or weapons. He could only look upon the Lassiter Board as a partial success, as its recommendations sat on a War Department shelf and gathered dust. Patrick hoped the Morrow Board would offer an improvement, especially considering Bingham's presence.

The Morrow Board focused on Mitchell's ideas and the controversy he had created. Morrow's biographer believed Coolidge called the board into being only to "meet these hysterical surmises" of Mitchell's propaganda against the administration.[87] Thus, to prove the case for the president and establish a long-range aviation policy along Coolidge's frugal ideas, the board needed to counter Mitchell's facts, which were often embellished, and disprove the claim of strident conservatism among the high-ranking officers of the Army and Navy. Assistant Secretary of War Dwight Davis (acting as Secretary, with the ailing Weeks out of the office) summarized this feeling with his opening statement: "I believe that the board wants sense, not sensation; facts, not fancies; arguments, not mere assertions."[88] The board seemingly attempted to take the middle line between Mitchell's plans and the results of previous anti–air power "stacked" boards (i.e., the Dickman and Menoher Boards) supported by officers who did not want any change.[89] In other words, the board strove for moderation, which coincided with Patrick's plan and his Air Corps concept.

Before Patrick took the stand, the administration and the old guard took their predictable stances before the committee. The first morning of testimony, 21 September 1925, included appearances by Davis, Drum, and Chief of Staff John L. Hines. Davis presented no plan except for stating that the War Department supported the Lassiter Board recommendations. He did read a letter from Secretary Weeks, who once again countered the arguments for a separate air force on the basis of unity of command.[90] Hines and Drum, both staunch ground Army officers, trotted out the old boards, letters from Pershing, and internal Army studies all denying the need for an independent air force and a national aviation department. The two highest-ranking Army generals also endorsed the Lassiter findings as being able to cure the ills of the nation's air inadequacies, which they believed were not as bad as presented by air officers (Mitchell) and the press.[91] Hines, however, also struck a blow against the moderate position of an Air Corps and endorsed the status quo. "A separate Air Corps within either the Army or Navy," he explained, "is really an independent Air Services within either of those forces. Such an organization violates the needs for cooperative action and coordination."[92] Such was the impression the board left at its midday recess.

When Patrick took the stand after the break, his tone and approach differed noticeably both from his predecessors' and from that which Mitchell would bring. He began not with past boards or arguments or independence but with an honest statement of the current operations of the Air Service and its men and equipment and how he had managed the service to deal with aging World War I supplies and small budgets. He reserved his displeasure for the meager appropriations given him by Congress, which used the figures presented to the Bureau of Budget and the War Department to make appropriations. For the fiscal years 1923 through 1926, the service received only 58 percent of its requested appropriations.[93]

Patrick quickly outlined his aircraft and procurement plans before moving to the heart of his testimony, the organization and future of the service. Taking his usual moderate approach, he forcefully espoused the place of the air arm within the military without calling the other branches obsolete. He did, however, state that "the coming into being of this new arm has somewhat lessened the importance of every other combat arm of the Army."[94] Patrick stated this belief not to slight the other branches but to show how the airplane could more effectively support the Army's mission or assist its other components.[95] Without implicating the traditional ground Army officers and their civilian supporters as being treasonous, Patrick more gently and eloquently stated:

> It is human nature that under such circumstances the coming into being of this arm has not been so greatly welcomed by those whom it in a measure at least displaces, and the result is that every recommendation that I have had to put forward must be passed upon by men with that trend of thought, men trained in the old schools, and they have not realized, I think, the full importance of the air arm. I think they are being educated, and that the importance of the arm is being impressed upon them more and more every day, but it was a long time before I could get any one to recognize the fact that there is what we now call an air force.[96]

Accordingly, Patrick spoke of how the new technology must someday revise the organization and employment of the country's national defense. Here he forwarded his idea of a defense department with three coequal services of air, land, and sea, what he called "the ultimate and ideal solution."[97] Then, while touting the definitive but unobtainable, he presented the steps he believed could, and should, be immediately implemented—a semiautonomous status as an Air Corps with direct reporting to the Secretary of War, similar to the Marine Corps within the Navy.[98]

Patrick's presentation was very effective. He approached the topic in a moderate tone and an appropriate manner, in accordance with General Orders and accepted traditions of military officers testifying before Congress (although the board was not a congressional body). He demonstrated his position with facts and reason, and he disagreed with his civilian masters and military supervisors without discrediting or deriding them. Patrick continued this style even in subsequent questioning when Senator Bingham asked him about Patrick's three major disagreements with the War Department. Under questioning from the senator, which obviously had been choreographed, Patrick outlined problems with getting the War Department to grant some of his recommended alterations, including uniform changes and personnel and promotion issues.[99] In his testimony, Patrick clearly differentiated between stating facts and stating his own opinion—distinctions that Mitchell often blurred. In sum, Patrick's was an articulate and persuasive presentation that provided the information requested and required by the board, while also elucidating Patrick's own ideas free of polemics. It was the antithesis of "classic" Billy Mitchell.[100]

Mitchell himself did not present a very convincing case before the Morrow Board. By the time he took the stand on 29 September, he probably understood that this board would not receive his comments in as friendly a way as had the Lampert Committee. After answering a few preliminary questions, he launched into his "statement," which was actually a reading of his recently released book, *Winged Defense*. For the better part of two days, Mitchell read his text dryly, with only a few interruptions. Chairman Morrow, understanding that this testimony would only hurt the airman's crusade, remained overly polite and allowed him to drone on.[101] Many Mitchell supporters attended the hearings and disagreed with Mitchell's approach. Arnold lamented, "Billy's expert testimony turned out not to be the brilliant defiance we had looked for."[102] Arnold also recalled how Senator Bingham tried gently to stop Mitchell by assuring him that the committee had a copy of the book, to which Mitchell snapped back at perhaps the only air-friendly member of the board. Morrow knew that Mitchell wanted controversy, and he refused to give it to him. He remained solicitously polite and let Mitchell talk endlessly without interruption and to the airman's loss of impact and prestige.[103]

The only tense moments occurred when Representative Carl Vinson, a longtime naval activist and supporter, challenged Mitchell on his assertion that the War and Navy Departments and the General Staff had coerced air officers not to speak their minds, or not speak at all. In a reintroduction of the major controversy during the Lampert hearings, Mitchell accused the departments of limiting officers' testimony and calling on them immediately after congressional appearances to explain their positions to the secretaries. Vinson asserted that it only happened once, while Mitchell said it had happened at least four times to him personally and that he could document other instances. Perhaps feeling the pressure to assert the facts, Mitchell asked to move on and promised to provide documentation later.[104] He later provided an appendix to his testimony with six documents supporting his claim of coercion. However, five of the six documented his own problems, and three of them overlapped as they dealt with his early 1925 testimony before the Lampert Committee. The only other source of Mitchell's claims came from Admiral Sims before that same body.[105] Mitchell failed to prove his case.

As in the congressional inquiry, the assertions of coercion and of junior officers' not being allowed to testify freely became a major issue for the Morrow Board. In the former investigation, the congressmen used it to support Mitchell and show the conservative doggedness and resistance to innovation by the War Department and ground Army officers. Now the Morrow Board wanted to firmly reject the Mitchell charges and offer evidence to counter the outspoken airman and his supportive legislators and put the controversy to rest. Assistant Secretary Davis reasserted that fact in his opening statement: "Officers, enlisted men, and employees will appear before you if and when desired and will testify fully, freely, and frankly. There has not been, is not now, and will not be any 'muzzling' of Army officers in the expression of their individual testimonies."[106] In fact, the board asked a standard question of each officer who appeared, inquiring if he knew of any efforts to suppress testimony or coerce officers.[107] In this case, the junior officers either could not, or would not, support the notion of an administration conspiracy to keep testimony suppressed. The committee asked most of the testifying air officers whether they had felt coerced or had heard of other junior officers being unwilling to testify for fear of retribution. None admitted any evidence supporting Mitchell's assertions, and some even countered those claims. "On the contrary," Major Horace M. Hickam explained when asked by Senator Bingham about officers not wanting to testify, "every one would welcome an opportunity to state his opinion."[108]

The younger officers may have felt that defying the administration could jeopardize their careers, but there is no evidence that they believed it, nor did any submit anonymous statements. In fact, junior officers like Spaatz, Doolittle, and Arnold had testified candidly before Congress before, and without lasting detriment to their careers. Only Mitchell received a reprimand, and only because he presented his views in such an inflammatory and insubordinate manner, as he had done before the Lampert Committee.[109] Additionally, none of the junior officers of Mitchell's "cabal" received reprimands or warnings for supporting the controversial figure—at least not until his court-martial.[110]

It seemed that the flamboyant airman's core group of young supporters came to realize, as Patrick had previously, that the best hope of someday obtaining an independent air service was by supporting the interim measure of an air arm within the Army. Even Mitchell's most ardent supporter, "Hap" Arnold, now supported Patrick's plan. Almost echoing Patrick's words,

Arnold called the Air Corps concept "not . . . the ultimate solution by any means, but . . . a step in the right direction."[111] Major Leslie MacDill agreed, offering his opinion against an air department and of having commercial aviation under a single department (Mitchell's desires).[112] Only Major Hickam supported Mitchell's proposal for a separate service immediately, asserting that "nothing short of a department for defense with all of the elements" would work in war, and "we should not delay starting this another minute."[113] Still, Hickam believed that Patrick "has made a very reasonable start on it."[114]

Although the major part of the air officers' testimony ended, the board continued to work long hours and hard days on the other aspects of aviation. Remarkably, it concluded its work in less than a month, ending in the late evening of 15 October 1925. Political necessity forced the Morrow Board to accomplish in less than ten weeks what had taken the Lampert Committee more than eighteen months. In the six weeks before the board issued its final report, many matters would come to a head. Mitchell would soon be called before his court-martial, which began exactly one week after the Morrow testimony concluded. But before the end of the Morrow hearings, Coolidge made a major speech in Omaha that outlined his military policies and his view of the proper conduct of military officers toward their civilian masters.

The president's speech refuted rumors that he was seeking reductions in the officer strength of the Army and Navy. Coolidge soothed over fears of a reduction, but he returned to his theme of economy. With no enemies able to attack the United States or any plans for the nation to take the offensive, every dollar spent must produce the maximum effect for defense. On civil-military relations he laid down the gauntlet to the officers of both services. In an obvious allusion to Mitchell, Coolidge insisted that military power must remain subordinate to civil authority:

> It is for this reason [civilian control] that any organization of men in the military service bent on inflaming the public mind for the purpose of forcing Government action through the pressure of public opinion is an exceedingly dangerous undertaking and precedent. This is so whatever form it might take, whether it be for the purpose of influencing the Executive, the legislature, or the heads of departments. It is for the civil authority to determine what appropriations shall be granted, what appointments shall be made, and what rules shall be adopted for the conduct of its armed forces. Whenever military power starts dictating to the civil authority, by whatever means adopted, the liberties of the country are beginning to end. National defense should at all times be supported, but any form of militarism should be resisted.[115]

Although the *Army and Navy Journal* originally believed that Coolidge's target was an effort by young naval officers to force the resignation of Secretary of the Navy Wilbur, the *Journal* also mentioned the possibility that the president was alluding to Mitchell. Coolidge's audience suggested the primary target: the American Legion had long supported Mitchell and his wishes for an independent air force and had lobbied on Capitol Hill for the "Mitchell Resolutions."[116] The *Galveston News* clearly understood this concern, writing that "whatever may have been in Mr. Coolidge's mind when he penned those words, they will be construed as a disparagement of Colonel Mitchell since he is for the moment the best known type of publicity wielding man."[117]

The press overwhelmingly supported Coolidge's stand for firm civilian control. "There are members of the body the President was addressing who should take this warning to heart," the *Brooklyn Eagle* wrote, and the *Washington Daily News* admired the president for "hitting at those responsible for the tempest now ravishing the entire military establishment."[118] Papers nationwide, regardless of their political leanings, praised the president's position. Coolidge had touted the military profession as a patriotic and high calling, and he had warned against officers using propaganda and lobbying against presidential policies. Even the Democratic *New York World* noted, "coming from the average man at such a time the speech was extraordinary; coming from Mr. Coolidge, it is well nigh incredible."[119] The president succeeded in shifting the issue from aviation to civilian control. "Silent Cal" had struck a blow against the voluble Mitchell, and in the court of public opinion he had won.

Amidst all these events, Patrick published his annual report for the fiscal year ending July 1925. The tone of his report had changed dramatically from the previous four years' publications, and he seemed to press his case more forcefully.[120] He primarily noted the inaction of legislation on the Lassiter report and the fact that none of the recommendations he made in the previous year "has been followed by tangible affirmative action."[121] The Air Service shortage of aircraft stood at over two thousand under the Lassiter recommendations and one thousand short of the planned requirements. Patrick also lamented the imbalance in the service of "air force" (combat) to "air service" (support) aircraft.[122] The report also revealed an organizational change within the Office of the Chief of the Air Service.

The Information Division added two subdivisions: a Legislative Section and a Press Relations Section.[123] The former existed to coordinate legislative action and make recommendations. However, the other Army branches (i.e., Cavalry, Infantry, and Artillery) did not have separate legislative offices, since the section in the Adjutant General's office served that purpose for the entire War Department. By having a separate legislative section, the Air Service demonstrated its concern, and perhaps distrust, for the War Department's handling of air-related legislative matters. A separate office perhaps also allowed the Air Service separate access to Capitol Hill legitimately and openly. The section lasted less than two years, as it did not appear in the annual reports beyond 1926.[124] Major Arnold headed the Information Division for thirteen months, and this experience gave him insight into how to deal with Congress and how to use propaganda and favorable press to advance air power. During this short time he made political connections, lobbied for air legislation, and wrote aviation-related articles.[125] He would put this experience to use ten years later when he returned to Washington as Assistant Chief of the Air Corps.

Soon after the release of the annual report, the Morrow Board completed its work. In order to strengthen its findings and support of Coolidge, Morrow coaxed his fellow members to issue a unanimous report, which they released on 2 December 1925. Reporting during the midst of Mitchell's trial and only two weeks before the guilty verdict, the board soundly (and expectedly) countered the airman's claims. Nowhere in the final report did the members even mention Mitchell's name, but they directly attacked his allegations and plans. Even the *New York Times* noted how the board rejected the Mitchell program: "Radical proposals of the character sponsored by Colonel Mitchell are rejected, and allegations concerning the state of the military aircraft situation in this country . . . are declared to be exaggerated or baseless."[126] The board recognized that both sides contained men who firmly believed in their position, and the report forecast that no report could firmly settle those differences or change any minds, but the members asked for patience and understanding of each side's position. In a remark undoubtedly meant for Mitchell, Morrow's board hoped "men will approach it [the aviation problem] with less feeling and more intelligence."[127] As for their recommendations, the board denied independence and semi-autonomy, stating the usual Army reason of destroying unity of command.

As for the Morrow Board's conclusions, they did include some of Patrick's recommendations. The board supported a name change to Air Corps, not as a step toward semi-autonomy but to differentiate its name from one of its roles (Patrick's use of "air service" to denote auxiliary/support missions). The biggest gain for the service came when the report concluded that the Air Service needed civilian representation and suggested adding another Assistant Secretary of War for air matters. The Secretary of War then had only one assistant who primarily handled procurement matters for the Army. The additional secretary would provide the aviators with a civilian voice that bypassed the General Staff, and one who could properly engage in political maneuvering. Also, as a member of the "Little Cabinet" and a political appointee, the right person would have more direct access to the president and thus be able to influence civilian leaders more properly on aviation matters.

The Morrow Board also saw fit to recommend the creation of two additional brigadier generals to assist the air chief. However, the inquiry did not support changes in the personnel or promotion systems, both of which Patrick had mentioned as problems. Although the President's Board did not fully back all of his recommendations, Patrick had gained ground. What in Mitchell's eyes was an all-or-nothing proposal, Patrick saw as smaller steps toward an eventual goal. The chief had built a reputation as a moderate but staunch supporter of air power. Aside from his friendly contacts with Bingham, he generally followed War Department protocol and General Orders Nos. 20 and 25. He now looked forward to using his influence to achieve meaningful legislation, and he hoped the Morrow Board recommendations would not follow the Lassiter Board and languish on a dusty War Department shelf.

The nation's newspapers approved of the committee's work but wanted to see action following the words. The *Knoxville News*, the *New York Herald-Tribune*, and the *Providence Journal* all applauded the committee and its proposals but noted that Congress must quickly convert the recommendations into legislation.[128] In yet another indication of its pro-Mitchell stance, the Hearst syndicate did not support the Morrow Board's findings. On the contrary, its publications belittled the board's conclusions and predicted that the modifications would in no way alleviate the existing situation in the Air Service or improve the nation's overall aviation program.[129] Still, the Hearst papers' minor and predictable dissent did not mitigate the positive reception of the board's findings, which had served their purpose by outmaneuvering the pro-Mitchell Lampert Committee and had beaten the end of the Mitchell trial to the presses.

Following quickly on the heels of the Morrow Board, the Lampert Committee completed its drawn-out work. While its voluminous research perhaps represented a more in-depth investigation of the air services than the Morrow Board's (although biased in the opposite direction), the Lampert Committee had lost its momentum and been outmaneuvered by the latter. As in the case of the Morrow Board's recommendations, the public knew before Lampert released his report whom it would support and what it would generally propose. During the final deliberations, Representative Reid's comments assured all of the House committee's support for Mitchell and his vision. In between acting as Mitchell's defense counsel for the court-martial, Reid completed his House and committee duties. The final report bore his (and Mitchell's) stamp. Of the report's final twenty-three recommendations, Reid sponsored nineteen—and all of them aligned with the now familiar Mitchell arguments.[130] Primarily, the committee recommended a unified department of national defense with three undersecretaries for the land, air, and naval forces. It also wanted a separate air budget, a five-year aviation program, and spending of not less than $10 million annually on aviation in each department.[131]

In yet another overt act of support for Mitchell and his program, Reid appended a "Special Concurring Report" to the committee's assessment. He claimed that his more detailed recommendations and addendum were necessary because the committee did not "present to Congress an outline of how this department [of air] should be organized, and I believe that this committee, having gone into the question very thoroughly and being in possession of all the facts necessary upon which to present an outline, should have done so."[132] Reid went on to outline how Congress should organize a unified air service, "an air college" similar to West Point and Annapolis, and the other specifics that he (and Mitchell) supported.[133] The report bore the tone of not being simply an innocent addendum. Instead, Reid probably meant his addition as the opening volley for the coming legislative session where the Lampert Committee would battle with the Morrow Board for public and congressional support for aviation.

The coming months sealed the victory of Patrick and the moderates over Mitchell and his followers. With two major concurrent investigations and interest and support from both Congress and the president (even though in different directions), the Air Service knew the coming year would bring some type of legislation, though Fechet worried about the impact of Mitchell's trial on Congress. "The Air Service generally believes that General Mitchell is right but we do not approve of the method which he adopted to put his stuff before the country," the new Assistant Chief wrote a friend. "[Mitchell's] action, undoubtedly, was insubordinate and has embroiled us with the War Department, as well as lined the Navy solid against us. These facts may work to our disadvantage when Congress meets and we ask for new legislation."[134] Patrick's control of Mitchell and his supporters changed the future of the air arm and how it would interact with the Army and the civilian leadership. Mitchell's court-martial made him a martyr in the long term but destroyed his credibility and influence for the immediate future. His ardent supporters, seeing their leader fail, reassessed their approach. They saw the changing tide of the service and gauged the mood of the public, which ardently supported the idea of a strong air arm but was unwilling to spend large amounts of tax money on war implements.

Mitchell and his supporters believed that propaganda and public support for aviation could force the Coolidge administration to overrule the General Staff and give larger budgets and roles to the Air Service. To gain that end more quickly, Mitchell courted Republican aviation enthusiasts to try and build political support and leverage for an independent service. Because aviation held the public's fancy as well as the support of an important group of legislators, the Coolidge administration could not simply, and perhaps appropriately, rid itself of Mitchell. The airman was a war hero and a dashing and popular figure who held the ear of more than a few congressmen in a closely divided House. Only after Mitchell had gone too far, which the public properly recognized, could Coolidge demote the airman and temper the calls of the Air Service. Patrick himself was not a paragon of proper conduct with civilian authorities, but he was a vast improvement over the more outspoken and controversial Mitchell. With Mitchell gone, Patrick could turn to other senior officers who agreed with the more moderate tone and incremental philosophies of achieving air independence. The coming year would dramatically alter the history of American aviation, but the changes would occur without Mitchell's interference and within an atmosphere of more appropriate and acceptable civil-military relations.

CHAPTER THREE

Last Acts of the Rebels, 1925–1926

The period of overt confrontation ended with Mitchell's trial and subsequent resignation, but a pivotal year lay ahead in 1926. The administration, Congress, and the War Department needed to work out agreeable legisla tion from the different findings of the Lampert Committee and the Morrow Board, while also fending off proposals from congressmen who favored air independence. The Air Service needed to walk a fine line. Fresh from seeing Billy Mitchell's fall, the flyers understood that they had to demonstrate a new attitude of cooperation with the War Department and the administration. Major General Mason Patrick led the way, fighting for his vision of the Air Corps concept and working with the War Department and Congress to hammer out an agreeable settlement. When the remaining air rebels tried again to influence Congress, Patrick acted swiftly and sent a message that Mitchell-style tactics were no longer tolerated and would be met with swift punishment.

Temptations remained, as Representatives Charles Curry, Fiorello La-Guardia, and Frank Reid introduced air-friendly legislation and held hearings that could cause rebellious air elements to reemerge. Some of the bills ignored both politically charged reports of late 1925 and pushed for an independent air force and a unified department of defense, but no officer rose to replace Billy Mitchell as another martyr. Major Henry "Hap" Arnold remained Mitchell's primary and most influential supporter within the Army. He kept in close contact with Mitchell and engineered one last scheme to lobby support for a more autonomous air force, but the time for firebrands had abated. Arnold's effort failed and he avoided dismissal, but he now understood the pressure for moderation. After the passage of the Air Corps Act, Representative Frank James even worried that the new attitude within the service would result in officers' unwillingness to come before Congress and report truthfully on their needs.[1] Patrick was almost singularly responsible for the change in mind-set. Four and a half years after his appointment to the top air position, and in less than six months after Mitchell's court-martial, the general had fulfilled his mandate to bring order, discipline, and a new attitude to the Army's air arm.

POLITICS AND THE LEGISLATIVE FLURRY

In the words of one publication, the rash of aviation-related matters that came to a head in December 1925 left a "howling baby" on the congressional doorstep when it reconvened for the winter session and found the "aviation squabble again on its hands."[2] More than ever before, arguments over air power became enmeshed in politics and political alliances. Even the service's own publication doubted if any air legislation would pass during the first session of the Sixty-ninth Congress, and members themselves doubted the feasibility of action. The new Speaker of the House, Nicholas Longworth (an Ohio Republican), reflected on the amount of serious study the air situation required, while his fellow Buckeye James T. Begg, a GOP legislator, commented more frankly, "the committees are now busy with other legislative matters and aviation legislation will be intricate and difficult."[3] Political infighting and a congressional Republican insurgency further complicated the prospects for producing meaningful legislation from the different aviation proposals that lined up before both chambers.

The power struggle within Republican ranks dated back to Theodore Roosevelt, Senator Robert M. La Follette, and the Progressive movement. Although the movement reached its high tide in 1912, Progressive elements remained strong within the Republican Party into the 1920s.[4] As late as the Sixty-eighth Congress, the Progressives from the Farm Bloc held the balance of power in the Senate.[5] La Follette, a Republican from Wisconsin, led the insurgency, which included other senators from the upper Midwest. Beginning with the winter 1925 session of the Sixty-ninth Congress (and coinciding with the Mitchell court-martial), Coolidge tried to close the party ranks and made conciliatory gestures to the radical wing, including Robert M. La Follette Jr., elected to fill his father's seat in a September 1925 special election and the youngest member of the upper chamber.[6] With a narrow majority in both houses and the next election only a year away, the party needed to reach out and at least give the appearance of harmony.[7] The GOP did not actively court the House radicals, who renewed their mutiny by refusing to vote for Longworth as Speaker. Still, since the House gap was larger and the House's rules allowed the majority to stifle any rebellious acts, Republicans concentrated on keeping party senators firmly behind the president.[8]

President Coolidge also concentrated on the Senate to fulfill his air legislative program. While remaining in frequent contact with key congressional leaders of both houses, he and others recognized that his wishes regarding aviation would hold sway in the upper chamber but not necessarily in the lower. The president also made it very clear that he desired air legislation along the lines of the Morrow Board recommendations. In his annual message to Congress he announced his intention to reduce the services further and said that he did not support any great changes in the national defense structure.[9]

Yet all who followed the machinations of the aviation investigations and pending legislation knew that the House would support the Lampert Committee and urge creation of a Department of Defense with three services, while the Senate would support the president, the Morrow Board, and the more limited fixes to the aviation problem. [10] The membership of the House Military Affairs Committee provided another indication that the House would follow the Lampert recommendations. Air-friendly congressman John Morin, an influential Pennsylvania Republican, chaired the group, and other Republican members included W. Frank James of Michigan, John P. Hill of Maryland, and Mayhew Wainwright of New York. [11] Both Morin and James were friends of the Army's "air radicals," which included Mitchell. In 1919 Morin had sponsored legislation for a separate Department of Aeronautics, and he and James had introduced a total of three aviation bills in the Sixty-ninth Congress along the lines of the recommendations by the Lassiter and Lampert Committees. Hill would also offer his version of an air corps bill, and he and Curry started the legislative parade. [12]

Wasting no time, the air-friendly congressmen actually introduced the first pieces of air legislation on the first day of the congressional session, which came prior to the announcement of the findings of either air investigation. On 7 December 1925, at the height of the Mitchell court-martial, Curry and Hill introduced similar measures, both of which called for the creation of a Department of Defense with an air armed service equal to those of the Army and Navy. [13] Curry also introduced a second bill the next day, which, instead of a united Department of Defense, created a separate Department of Air and a United States Air Force. [14]

During the first five weeks of 1926, three congressmen introduced four more aviation bills. [15] Frank James, acting chairman of the House Military Affairs Committee during Morin's absences, derided the first bill sent over by the War Department, saying it either "was not drafted in good faith" or ignored the recommendations of the Morrow Board. When Secretary of War Dwight Davis appeared before the committee, James chided him and said he should have been ashamed to submit such shoddy legislation. [16] Morin, under whose name the bill entered Congress, unabashedly washed his hands of the bill, saying, "It was left on the doorstep." Morin's comments on the bill came during Mitchell's testimony, and he assured the airman, "Do not hesitate to express yourself, because I do not claim it as mine." [17]

Congressional rules required the chairman of the Military Affairs Committee (in either chamber) to introduce bills drafted by the War Department, which was a perfectly acceptable practice and did not violate civil-military divisions of duties. War Department officers often assisted in drafting the legislation, thus lending their expertise to the civilian secretary. However, the Secretary of War rightfully retained the authority and the responsibility to shape all department bills sent to Congress. The War Department also often sought military officers' input on legislation in order to better prepare for hearings and ensure it met the service's needs. An internal Air Service report on the War Department bill noted that only those congressmen "who normally are behind all Administration measures" approved of the bill. The Air Service believed the Navy, Post Office, and other departments "who fear enactment" of any separate air or national defense bill gave the measure silent approval. The report went on to list others who would not support the War Department bill, including "a great many Congressmen, especially members of the House Military Committee," the Democrats, and Army fliers.[18] The Air Service clearly understood the political coalitions that both supported and opposed air autonomy. Discussion of all of the aviation bills occurred within an eight-week span and often blurred together. The debate even spurred the introduction of a new bill.

During Patrick's testimony, James peppered the general with questions designed to expose the lack of War Department responsiveness to past congressional suggestions, as well as the department's recent history of stalling on aviation matters and legislation. James specifically mentioned the defunct Lassiter Board's recommendations. Under pointed questioning, Patrick admitted that he was neither consulted on any legislation nor called upon for recommendations. Regarding the War Department–sponsored Morin bill, Patrick concluded, "The basic recommendations that I made are not in the bill at all."[19] James therefore suggested that Patrick sketch out his own legislation that would accomplish the Morrow Board's objectives. Patrick did so, and the next day he submitted the bill through Secretary Davis to Congressman Wainwright, chairman of the subcommittee on aviation.[20] Since Congress had specifically requested that Patrick draw up the bill, Davis could not refuse its submission. However, Secretary Davis made it clear that he supported the bill from the War Department, which codified the Morrow recommendations against independence. Although the House Military Affairs Committee limited the amount of testimony on the different bills, the air-friendly congressmen made their positions known and took Davis to task at every opportunity.

After a few questions from the chairman, James asked Davis why his proposed legislation did not more closely follow the Morrow Board's recommendations and why it did not include recommendations from the Lampert report. James also asked Davis for his opinion on allowing Mitchell to testify.[21] McSwain pushed Davis on a number of points and even had him admit that the "principal part of this agitation and unrest" in the Air Service originated with "the discontent of the junior officers."[22] Congressman Hill, sponsor of rival bill H.R. 46, rebuked Davis for writing a bill that did not completely fulfill the requirements of the Morrow Board and implied that

Davis did not seek the advice and expertise of his aviation experts—notably General Patrick.[23]

Despite these conflicts, the hearings on the various aviation bills occurred without the fireworks of their predecessors. Patrick testified for the Air Service and supported his air corps idea as a stepping-stone to his ideal solution: a unified department of defense.[24] Mitchell testified four days after his resignation from the Army, and even he seemed more subdued. He did not make outlandish remarks or attack anyone in uniform in a position of authority.[25] Mitchell did assert his usual position that the services remained tied to traditional means of warfare, but his testimony lacked the derision and outright slander of his previous appearances, when still in uniform, before friendly committees. He did comment that the Army and Navy continued to worship infantry and battleships, and he likened the latter as praying to Buddha.[26] As for the various bills, Mitchell predictably backed those of Hill and Curry, both of which called for a unified national defense and an independent air arm.[27]

Compared to the previous seven years, the testimony on the myriad of proposed legislation passed quickly and without demonstrative tactics or serious attacks on the administration. Three elements contributed to the more subdued hearings. First, the bills began to resemble one another (and bills from previous Congresses) and fell largely into two distinct categories: those supporting the Morrow Board's limited recommendations for minimal changes, and those supporting wider changes in the national defense structure. Second, the histrionics had already occurred during the two investigations of the previous year and during Mitchell's court-martial. Finally, one cannot discount the lasting effect of the Mitchell trial and the publicity surrounding it, as well as and the public's perception of his actions. The previous years clearly demonstrated how the tactics of confrontation and exaggeration had failed to bring about change and had perhaps hindered more than helped aviation's cause. Arnold believed that in the immediate post-Mitchell era, the War Department "set their mouths tighter, drew more into their shell, and, if anything, [took] even a narrower point of view of aviation." He contrasted that with the Navy, which undertook an introspective and far-reaching study and became more "air-minded."[28] Only one instance of questionable conduct by Army officers occurred, and it was dealt with quickly and decisively.

While the Military Affairs Committee debated the merits of the various bills, two Army aviators from the Office of the Chief of the Air Service printed an inflammatory circular in an effort to sway congressmen. The circular urged all Air Service officers—Regular as well as Reserves—to support the bill Patrick had drafted (H.R. 8533) and asked them to contact their legislators and "people of prominence within [their] state who can communicate with the Senators and Representatives; people whose communication will be given more than casual consideration."[29] The circular predicted that the bill would likely pass the entire House if the Military Affairs Committee favorably reported it, and unabashedly announced:

> We have tried to put across the idea of reorganization, in which the Air Service can be developed and operated so that it will be able to give its maximum efficiency and effectiveness.
>
> This educational work is as much yours as it is ours, and now is the psychological moment for you to get busy. There is more interest in aviation throughout the United States now than we can hope to attain again for many years to come, so that there will never be a better opportunity than right now to try and get recognition commensurate with our actual offensive power within the scheme of national defense. . . . This is your party as much as it is ours. We must all get busy and do it now. Next month will be too late. We are relying on you to do your share of the work. Do not throw us down.[30]

"Hap" Arnold, a major and then the Chief of the Information Division in the Office of the Chief of the Air Service, masterminded the campaign. He wrote the circular and then asked his longtime friend Major Herbert A. "Bert" Dargue, also stationed in the chief's office, to help with the publication. They also enlisted a reservist, Captain Don Montgomery, to help cut the stencils and run the mimeograph machine. They attached the circular to an outline of the general concepts of the Patrick/Wainwright bill (hereafter referred to as the "Comments") and made sure that no other marks on the paper divulged its origin. Copies somehow made their way to Capitol Hill.[31]

After consulting with Secretary Davis, Patrick and Major General Eli A. Helmick, the Army's Inspector General, ordered an investigation. They assigned two officers to lead it and made Patrick's assistant, Brigadier General James Fechet, the director.[32] Fechet aimed to find out who had "surreptitiously" prepared a document intended to influence military legislation, especially since the bill the circular supported "had been definitely and distinctly disapproved by the War Department."[33] He started his investigation on 5 February 1926 and issued the final report eight days later. During that time Fechet interviewed thirty-six officers and civilian employees, all but one assigned to the Air Service or in Office of the Chief.[34] Arnold quickly became the target of the investigation.

Arnold remained angry and frustrated by recent events, especially the outcome of the Mitchell trial. He even recalled how he and Dargue took action as "the first ones to try to keep the battle going."[35] Arnold later admitted that he had many friends and contacts with the press and in Congress due to his long and varied service in the District and that he had visited Mitchell's ranch in Middleburg, Virginia, to plan strategy.[36] Initially, though, Arnold admitted to nothing specific, and he even denied his participation when confronted by Fechet and Lieutenant Colonel Thorne Strayer, the Inspector General's representative for the investigation. Arnold lied to the investigators (he was not under oath in this initial phase) and denied any knowledge of the documents, saying he first heard of them when he received a copy, and he offered his belief that the circulars had originated outside Air Service's offices. Fechet noted, "His bearing and demeanor were indicative of mental stress and perturbance."[37]

Arnold's guilt surfaced during the investigation's second phase, when Fechet, accompanied by Arnold, interviewed the civilian clerk in charge of the mimeograph rooms. Also appearing stressed, Thomas J. Rowe looked to Arnold for an indication of how he should answer Fechet's questions. Rowe initially denied all knowledge, but when Fechet informed him of the consequences of false statements he confessed to printing six hundred of the "Comments" section under Arnold's direction, and he then produced the master stencil for Fechet's inspection.[38] Arnold then admitted his role in this action, but he and Rowe continued to deny responsibility for the more controversial letter—even after Fechet used a machine to cut a stencil and proved that the letter had been reproduced on those machines.[39] Fechet found that after preparing the "Comments," Arnold, Dargue, Montgomery, and First Lieutenant Burnie R. Dallas decided to attach the lobbying letter and send it out without official markings or in official envelopes, targeting the Senate especially. Arnold drafted the letter, with Montgomery's assistance, and Rowe reproduced it and agreed with the officers that the letters should be distributed without markings or indications of point of origin.[40]

The investigators recommended that Arnold be tried before a general court-martial for breaking General Order No. 25, which barred officers from attempting to influence legislation, and for lying during the investigation.[41] Fechet concluded that Dargue should only be reprimanded and reassigned, because of his lesser role, because his initial statements were not intentionally meant to deceive, and because under questioning confusion arose between his answers on the "Comments" and on the circular. Strayer wanted a stiffer punishment for Dargue, but Fechet believed that Dargue was only responsible for the "Comments" section and not the letter or the intended distribution.[42] Since Fechet's opinion carried more weight, Dargue received a reprimand due to being "less culpable," while Patrick threatened Arnold with a court-martial before "exiling" him to Fort Riley, Kansas. Patrick bluntly told the newspapers: "Both of them will be reprimanded, and one of them, no longer wanted in my office, will be sent to another station."[43] Initially, Patrick gave Arnold a choice of court-martial or resignation, and twenty-four hours to decide. Arnold supposedly demanded the court-martial, and Patrick, when reminded that the accused would reference similar lobbying by the chief, relented and sent him to the "worst post in the Army."[44] According to Patrick's biographer, Arnold never forgot or forgave Patrick for the "exile" and thus never fully grasped the impropriety of his lobbying or his lying during the investigation.[45] Patrick took no action against Montgomery.

During the affair, Dargue and Dallas reflected upon proper conduct of officers and influencing legislation. Dargue, more involved with drafting legislation with Representative James and with Patrick, took a more liberal view. He avoided an answer when Strayer specifically asked him, "Did you consider it a proper document to send out?" Dargue replied that he did not want to incriminate himself, but he believed that the proper methods to achieve an independent air arm remained a matter of opinion. He acknowledged, however, that military customs dictated the use of political influence only through proper channels, adding, "My service has taught me that, and I appreciate that."[46]

Lieutenant Dallas, in spite of his junior rank, showed an even deeper insight into proper conduct. He took fifty to seventy-five copies home, with general instructions to distribute the letter and attached "Comments" from a list given to him by Arnold. So as to hide the source, he was instructed not to use War Department (franked) envelopes. Arnold instructed him to send them "where it would do the most good," including to members of the Senate.[47] According to his sworn testimony, Dallas did not send out any of the letters but rather burned them at home. When asked why he did not mail them (using blank envelopes and stamps at his own expense), Dallas replied that the more he thought about it, the less he thought he should be involved in such activity. Of primary importance, he reasoned that if the letters contained information that could not be distributed officially and in franked envelopes, he should play no part in sending it out covertly.[48]

The whole affair infuriated Patrick for a variety of reasons. First, the press reported how Secretary Davis had implicated Patrick himself in the circular's publication, linking it with Patrick's recent testimony before the Military Affairs Committee and the drafting of his own legislation.[49] Originally, Davis ordered the Inspector General to investigate the entire affair, suggesting that Patrick had violated the General Orders and the orders of the president.[50] Coolidge also wanted to enforce discipline within the Air Service and did not want to allow a violation of this type to go unpunished, coming so close on the heels of the Mitchell trial. The president made his views clear to Davis, and they both used the Coolidge memorandum upholding Mitchell's sentence as a policy statement on civil-military relations.[51]

Second, the affair allowed Mitchell another opening to inject himself into Air Service matters, even though the deposed assistant chief supported Patrick in this instance. Mitchell publicly accused Davis and the War Department of being on a vendetta and claimed that Patrick was "a victim of an espionage system within the Department" by a "War Department clique."[52] Mitchell further stated that the War Department targeted Patrick for his views that were counter to the administration's and that Patrick would be "bull-dozed by this bureaucracy."[53] Finally, the whole affair undercut the positive attitude and direction Patrick had worked for during the past few years. Instead of concentrating on constructive legislation, the incidents produced more unfavorable press and gave the appearance that rebellious elements within his branch remained outside his control.

Patrick did not need to worry about his public reputation, as most newspaper editors supported him. The *New York Times* echoed most national papers, writing, "No one could suspect General Patrick. He is a stickler for the regular order and a high-minded officer who works through 'regular channels.'"[54] The papers did not look upon the actions of the air officers in question so kindly and supported the president's crackdown on discipline and proper behavior. "The point does not need to be labored," wrote the *New York Times,* "that propaganda by officers of the army Air Service to bring about legislation that they want and that the President as Commander-in-Chief regards as unwise must be subversive of discipline."[55]

One of Patrick's unspoken fears did come true, though, as the commotion over the circular added fuel to the fire in the debates over air legislation and the political infighting between the president and Congress. On the heels of ordering the investigation, Davis blasted Patrick's bill in a letter to the chairman of the House Military Affairs Committee. In the letter he offered a point-by-point refutation of virtually all of Patrick's suggestions and ridiculed Patrick's recommendations as "far from being the logical solution" and "unsound and uneconomical."[56]

Davis's vehement opposition, combined with his recently announced investigation of the Air Service, prompted the air-friendly House to act. The same day of Davis's letter, the Military Affairs Committee met in executive session and decided to investigate Davis's investigation. Some members believed that Davis targeted Patrick due to his recent testimony and bill. Fiorello LaGuardia led the attack on Davis and the War Department. He denounced the department for its "tyranny and oppression" and reiterated the charges of muzzling and coercion that had dominated the Lampert and Morrow investigations.[57] Trying to put out the fire, Davis met personally with key congressional members and assured them that he bore no animosity toward Patrick and had even appointed him to the committee investigating the propaganda scheme.[58]

Patrick and Davis also publicly patched the rift between them in written statements. The air chief said that a misunderstanding took place due to the almost simultaneous announcement of Davis's investigation and the disagreement with Patrick's testimony. In accordance with established policies and General Orders, Patrick had indeed informed Davis of his forthcoming testimony before the Military Affairs Committee. For his part, Davis agreed that Patrick had acted in a "perfectly proper manner" in his relations with Congress and had fully participated in the investigation with the Inspector General.[59] Despite these pleasantries, Davis did not apologize for vilifying Patrick's legislation.

In his press release, Patrick noted that although people in the War Department might differ with him, he respected the sincerity of their convictions and their honesty of opinions and hoped they held him in the same regard.[60] The air chief may not have acted in a "perfectly proper" manner, as he remained in peripheral contact with several congressmen and still argued for a policy that conflicted with the president's, but his actions were a vast improvement over those of Mitchell and the other, more radical, air officers. Not only did Patrick inform Davis of his testimony, but when he disagreed with official War Department policy on the air bills he would almost invariably reaffirm, in his reply to the many questions, that those ideas reflected his own opinions.[61] Also, when asked to submit his own version of air legislation, which competed with the administration's bill, Patrick submitted it through proper channels and then backed away from it when events proved that Congress would not pass it.

The discord between Patrick and Davis undoubtedly carried into the House committee's consideration of the various aviation bills. Coolidge, like Davis, did not hide his dislike for Patrick's proposal and applied as much pressure as he could in the House to pass legislation along the lines of the Morrow Board. Davis and Coolidge probably made their positions even more clear to Patrick, because the air chief never did defend his proposal stoutly. Although many members liked Patrick's plan, they swept it aside when it came to a vote.

Apparently Patrick had made a deal, or had some sort of understanding with his superiors, for he agreed to accept legislation that improved the Air Service's position incrementally instead of making great leaps toward independence. The *New York Times* recounted the committee's voting and debating as a "stormy" session but noted how they easily defeated the Patrick/Wainwright measure, which many members had earlier strongly supported (in fact, less than a week earlier the *Times* expected an early favorable report on the Patrick bill to the House).[62] According to the paper's congressional sources, the committee rejected the proposal because of "the indifference of its sponsor, General Patrick."[63] The sources further stated that Patrick and the War Department "reached a peaceful agreement" and the former decided not to press his case for his bill. In an almost total reversal, Patrick concurred with a bill that Davis had drafted (along the lines of the Morrow Board) in response to the Patrick proposal.[64] With the Chief of

Staff, General Hines, also in agreement, Davis achieved his goal—which had been the goal of his predecessor as well—of having the War Department speak with one voice.

No such harmony was heard within the House Military Affairs Committee as the legislators debated and voted on the different aviation bills. In yet another session described as "stormy," they waged an internecine battle that basically represented in microcosm the aviation arguments extant since the end of World War I. Election-year politics intensified the struggle. The bill proposing a consolidation of the services into a Department of National Defense with three separate branches actually came the closest to passing. On 2 March 1926 unofficial polling showed the measure would pass by one vote, eleven to ten, but overnight one unidentified member changed his mind and the committee declined the measure by the same count.[65] The measure retained solid Democratic support, and Representative McSwain promised that his party would continue to fight for such a department—a blatant nod to the upcoming elections and probable campaign issues.[66] After rejecting a new national defense structure, the committee then rejected the bills representing the Morrow and Patrick plans, both by one-vote margins. The plan for a unified air service (Mitchell's desire) suffered the worst defeat, failing by a tally of sixteen to five.

With all of the proposed bills defeated, James encouraged the members not to give up but rather to establish a goal of producing a bill that would pass their committee unanimously and win congressional approval.[67] He not only desired to settle the matter while aviation matters held the public's attention, but the Republicans now wanted it settled before the Democrats could make a unified defense structure a key election issue. Yet the Republicans were divided. Those friendly to aviators remained incensed at the War Department for its recent treatment of Mitchell, Patrick, and Arnold as well as the many problems over the previous seven years. Finally, James, as acting chairman during Morin's many absences, set a goal for the committee to draft its own bill that would satisfy all committee members. They ultimately succeeded, and H.R. 10827 emerged from the committee unanimously reported to the House on 29 March 1926.

James called the measure the ultimate compromise, as it was satisfactory to both the Air Service and the War Department on issues that had caused key disagreement.[68] To ensure a smooth and rapid passage by the House, James and a small staff worked on an exhaustive report for the legislators that clearly identified the arguments and answered many questions. Notably, James used Major Dargue and two other army officers to prepare the report and collect data from the War Department. James also met with LaGuardia and Mitchell to make the proposal as complete as possible. On the floor before a sparse House that contained only thirty-eight members present, James read a letter that he had solicited from Patrick. The air chief reiterated his desire for a department of national defense as the "ultimate solution," but he supported this incremental step.[69] James's work paid off, and the bill passed the House in under forty minutes.[70]

Coolidge, in the meantime, did not want such a quick resolution in the House. His strong hand remained in the Senate, and he wanted to keep the air forces small and the budget low. The president also feared that if the air radicals got their way the proper balance of civil-military relations would be adversely affected. As the *New York Times* explained, "If the efforts of air officers, advocating an expansion program beyond the Administration's recommendations were successful the President sees that it would be only a step further for military men to make demands calculated to make this a military nation."[71] Coolidge, a proponent of disarmament and a fiscal conservative, did not want to spend money on armaments. He disagreed with any proposed legislation that called for a five-year plan to fund Air Service expansion. Coolidge opposed any set plan that tied his hands on the budget. He also publicly worried about a proposed deficit for fiscal year 1927 and used that as ammunition against military increases. During the House debates, Coolidge announced that he would not increase the military budget and that any increases in personnel or funding for the air arm would come at the expense of the other branches.[72] He continued to hope that the Senate could derail any unwanted legislation; in case it did not, he kept his veto pen close at hand.

Predictably, the Senate endorsed measures agreeing with the wishes of Coolidge and the Morrow Board. When H.R. 10827 finally reached the Senate, they amended it to the point that it became unrecognizable to its authors.[73] Senators Hiram Bingham and Joseph T. Robinson led the debate over the House measure. Bingham, usually a friend of the aviators, in this case supported Coolidge and backed a version closer to the Morrow recommendations. Two factors likely influenced his decision: the pressures of his party and the president and a proprietary interest due to his membership on that board. Robinson, a Democratic senator from Arkansas, battled Bingham on a few points and supported the views of many Air Service officers.[74] During the debate, Senator James W. Wadsworth Jr., chairman of the Military Affairs Committee, vented his frustration over air officers "lobbying" to assure themselves of promotion: "Since the beginning of our consideration of this bill it has been perfectly apparent that there is a special little group of Air Service officers, of the rank of colonel and lieutenant colonel, who would like to have Congress legislate in such a fashion that they and only they shall be eligible for these extraordinary promotions. They have been coming to Senators and members of the House asking that amendments of this sort be put in so as to freeze into the law the certainty of their future promotions."[75]

Robinson, who had offered the promotion amendment, called it an "injustice" to say that Air Service officers pressured him into presenting the changes.[76] Still, there remained little doubt that Robinson and others stood with the Air Service officers, and some flyers may have taken their wishes to individual congressmen. While the direct involvement of airmen in political matters lessened after Mitchell's demise, and officers undoubtedly understood the provisions of the relevant General Orders, they continued to find a way to make their views heard. As in the House, the Senate vote fell along party lines, with the Republicans supporting the president's program.[77] A conference committee worked out the final details of the Air Corps Act over a period of ten days, and after many years and political struggles, Army aviators finally obtained an advance in their status. The president signed the Air Corps Act, which substituted sections of the existing National Defense Act, into law on 2 July 1926.[78] The major provisions of the Air Corps Act recognized the air arm's coequal status with the other combat branches of the Army (rather than being merely an auxiliary or support service), mandated an expansion program (in personnel and equipment), directed that only air officers would command air units, and provided for the addition of an Assistant Secretary of War for Air.

During the political fray over the air bills, the airmen themselves remained restrained and inactive compared to the previous years. With Mitchell gone and his protégé Arnold exiled, no other firebrand or ringleader emerged from the aviators. Patrick, even amidst the investigations of Arnold's unprofessional behavior and the activity surrounding the air chief's own legislative proposals, publicly called for calm within the service. At a speech in Chicago, he asked extremists at both ends of the debate to "kindly take a seat" and let Congress work it out. He postulated that moderation would prevail and Congress would find a solution "somewhere between the extremes of the enthusiasts and those who call themselves conservative."[79] Mitchell tried to stir activity and began a nationwide lecture tour to promote aviation interests, but he met disappointingly small audiences, and those who did come wanted more to see the famous airman than to support his positions.[80] No public upheaval occurred, and unlike with past debates, the rank and file of the Air Service officers did not parade before Congress to further inflame the debate.

Congressman James, perhaps sensing the danger of *not* having officers speak up when it became necessary, warned the service's senior leadership in a speech before the Army War College. He cautioned them against becoming too conservative—too silent—if they believed something was wrong with the conduct of national defense. "There is no way that we can pass legislation for the national defense," he instructed the soldiers, "unless we know all the facts; we do not consider a man disloyal to the military establishment if he can give us something good for the national defense." James went on to assure the officers that Congress would protect them from retribution for expressing their own views and would attempt to punish those who would reprimand any officer for expressing his views, even personal ones, before the legislature.[81]

James's concern was not without substance, as it seemed the pendulum had begun to swing the other way. Finally, the Air Corps had achieved some of its goals, and the rebellious elements had seemingly been subdued. For those in the House who liked to needle the War Department, the loss of a thorn in the department's side gave them less leverage and further strengthened the president's power.

Within a few months of Mitchell's resignation, and after six years of turmoil since the creation of the Air Service in the 1920 legislation, the Army's air arm had obtained a small victory. Yet this ending of the outwardly radical stage of the air officers came with still-disturbing behavior by high-ranking military members. Aviation leaders allowed themselves to become political pawns to the anti-administration elements in Congress. Officers lobbied Congress and used a favorable press and key pro-aviation congressmen in an effort to gain concessions for a national military policy not supported by the commander in chief. Patrick's actions indeed set a more moderate tone and alternate approach, but his contacts with key congressmen and his agreement to write a bill opposing Davis's submission (even though he did so by request and properly submitted the legislation) bordered on the improper and on violating the apolitical behavior expected of the nation's military leadership.

The air officers had to walk a fine line between advocating what they thought was best for the national defense and pursuing policies that opposed the administration's wishes. The General Staff and Davis wanted total control over Army aviation, while Coolidge did not want to radically alter the military structure, purchase additional weaponry, or take any actions that increased the national budget. As the nation's experts in Army aviation, Patrick, Mitchell, and Fechet should have expressed their opinions as to how they believed air power should be organized and employed, but this advice should have been given as their professional opinion only, and in the proper forums. Therefore, when the aviators openly pressed their case beyond providing judgments and expert advice, they crossed over into party politics, since anti-administration elements in Congress and the public used the aviators' cause for partisan ends and sometimes with the active participation of the air officers themselves.

Still, as the air force had taken an incremental step toward the eventual goal of independence, its officers had likewise begun to change their tactics. Patrick teetered on the thin line between proper and improper civil-military behavior, but it was a vast improvement over Mitchell. Fechet would improve on this technique further, and both he and Patrick would soon have the political "buffer" of a civilian Assistant Secretary of War for Air who could appropriately partake in political maneuvering. The era of confrontation and insurrection had come to a close, and even the Mitchellites who remained in the service and remained passionate for future independence recognized that the new flight toward their goal would rise on the wings of moderation.

CHAPTER FOUR

The Impact of an
"Air-Minded" Civilian, 1926–1932

The 1926 Air Corps Act did not give the air officers all they wanted, but they took a step toward autonomy with a new name and greater recognition of their unique mission. However, the inclusion of a civilian representative in the War Department provided the single most important change for the next seven years. The new political clout in the form of an Assistant Secretary of War for Air gave the service a civilian buffer—and advocate—between the Secretary of War and the air arm, someone not bound by General Orders and limits on actions by professional officers. Not only could the new assistant secretary play the political game, but he was expected to do so. As a political appointee, though, he was also expected to support the president's programs and policies.

By mid-1926 the new Air Corps found itself in a better position and looked ahead to the promise of expansion and appropriations, yet the promise remained unfulfilled. The vaunted Five-Year Plan continued to vie with Coolidge's program of economy and a Congress unwilling to provide the required appropriations. War Department appropriations, slim enough during the boom years of the 1920s, faced even dimmer prospects during the Great Depression. When the stock market crashed in October 1929, the Air Corps expansion program, the most promising proposal for the service emerging from the Air Corps Act, had yet to reach its halfway mark. The public, which had never fully supported Billy Mitchell's pleas for increased air power, now focused on personal economic survival.

Amidst this turmoil the top Air Corps leaders and F. Trubee Davison, the new Assistant Secretary of War for Air, battled Congress for more money to fill the holes left open by the Air Corps Act. Yet despite the chaos and broken promises, the flying service did not fall back upon the tactics of desperation. Instead, the steadying hand of Davison and the moderate approach of its top officers kept the ax of severe budget cuts from falling entirely on the Air Corps. Some observers even commented that the newly named branch received favored status for funding in the Army. Emerging from the Mitchell experience of inappropriate civil-military conduct, and aware that the presidents and war secretaries would not tolerate further rebellious behavior, Major Generals Mason Patrick and James Fechet advocated—and enforced—discipline and a attitude of compromise within the service. The two chiefs also encouraged others to follow their lead and publicly support War Department policies. Davison helped them. The air secretary represented the service in political matters, which left the military leaders free to concentrate on operational matters. This teamwork and division of duties allowed the Air Corps to avoid confrontation and keep the service on a steady path.

THE VICTORY OF THE AIR CORPS ACT

Aviation scholars have debated whether the Air Corps Act actually advanced the Air Corps or if it represented only a false progression. The statute did not immediately fix the service's promotion problems or grant the branch a separate budget. However, it mandated a review of the relative promotion of air officers versus ground officers, and the Air Corps budget, although not separate, always received individual attention from the president and Congress. The 1926 law provided incremental steps toward autonomy and independence sought by its moderate leaders. The Air Corps rose in status to a combatant branch, and the legislation mandated numbers of both aircraft and personnel.[1] Given the spectacular conflicts and politics, the Air Corps Act probably represented the best compromise the flying officers could have obtained.

Yet the addition of an Assistant Secretary of War for Air, often described as a minor victory, actually was a significant gain. It gave the air branch its own political representation—the only branch to have that—and thus a way to circumvent the Army General Staff, which had previously stalled many aviation initiatives. In addition, it provided virtual direct access to the Secretary of War.[2] More importantly from the perspective of civil-military relations, the civilian air secretary acted as the political representative of Army aviation. Having an active and informed person in this position went a long way to alleviate the need for the Army aviators to involve themselves in political matters they should have more properly avoided.

The post also expanded the ability of the Air Corps to coordinate with other government branches, a job that was also better suited to a civilian.[3] As in the traditional Army branches, the Chief of the Air Corps and his officers would still provide expert advice to Congress, which, as Representative Frank James so rightly noted, was essential for that body to make informed decisions. More importantly, though, a civilian air secretary could legally and properly participate in the backroom political wrangling that was essential to obtaining the appropriations for the newly authorized Five-Year Plan.

Just as significant, the air secretary helped to delineate the parameters for civil-military relations by eliminating the need for the air chief to advocate legislation.

F. TRUBEE DAVISON AND THE FIVE-YEAR PLAN

Interestingly, Coolidge picked a former Navy aviator as the Army's first Assistant Secretary of War for Air, and he selected a former Army aviator for the same position in the Navy Department. F. Trubee Davison became the Army's first aviation secretary on the morning of 16 July 1926. Although Davison was only thirty years old, his background in politics and aviation reflected a seemingly perfect pedigree for such a position. Born and raised in New York City, he graduated from Yale in 1918 and from Columbia Law School in 1922. During the war he volunteered as an ambulance driver in the American Field Service, and while at Yale he and some classmates took flying lessons on their own, starting the "First Yale Unit" of flyers. His aviation opportunities in the service diminished after a 1917 crash broke his back. In 1921 he won election to the New York Assembly, and his constituents re-elected him subsequently until he left to take the War Department position.[4] The wealth of Davison's family and the connections to Coolidge's confidant Dwight Morrow undoubtedly helped the young politician gain this visible office.[5] However, he would soon prove that his abilities were equal to the task; he had experience in politics, important connections, high regard in the Republican Party, and a thorough understanding of aviation.

Davison quickly realized that the Air Corps Act did not specify his duties. He requested written confirmation of his specific responsibilities from Secretary of War Dwight Davis, asking especially to be included in all Air Corps procurement decisions.[6] Davis provided the following description for the new air secretariat: "to aid the Secretary of War in fostering military aeronautics" (a paraphrase of the short provision in the Air Corps Act authorizing the additional secretary); to execute the provisions of the Air Corps Act; to supervise experimental aeronautics and development of military aviation with the aid of the Chief of the Air Corps; to coordinate with the other Army branches; to act as Secretary of War during the absence of the secretary and the other assistant secretary; and to be the point man for carrying out the provisions of the Five-Year Plan and for defending estimates submitted to Congress.[7] Thus Davison's primary concern became the Five-Year Plan. Even though the Air Corps Act mandated the expansion program, he needed to use his political skills to wrangle with the Army, the president, and Congress for the budget needed to complete that program. Davison would not publicly counter the president's desire for economy, but he sought every dollar he could for the Air Corps.

Davison quickly realized that he could maneuver in political circles in ways officers could not. He could attend political rallies, engage in partisan politics (when necessary), and work closely with legislators, even lobbying them. A political dinner was another prime example of what Davison could do that officers could not. He and another assistant secretary, Charles B. Robbins, held a dinner for members of the Senate Military Affairs Committee at the Mayflower Hotel in Washington on Valentine's Day in 1929. Members of the House Military Affairs Committee also attended. General Summerall was the lone military officer on the guest list.[8] No known record exists of the dinner's specific political purpose, but it demonstrated the clout and abilities of a civilian secretary.

Davison did even more; when he disagreed with the administration, as he would on funding of the Five-Year Plan, he readily entered the legislative arena. He took the lead when appearing before congressional appropriations committees, while Patrick, and later Fechet, sat close by and answered questions when asked. Not only did this arrangement more appropriately avoid the air chief's political load in the budget battles, but Davison's presence gave the Air Corps more political weight; congressmen seemed to respect him and his views. On one occasion, Davison made his opening statement and then answered different senators' questions. At a point later in the questioning, probably following a silence after a question from Senator Bingham, Fechet spoke up and asked, "Are you asking me, Senator?" Due to Davison's primacy in the dealings with the committee, Fechet had seemingly lost interest and attentiveness.[9] On a number of issues, Davison worked closely with Representative Frank James and Senator Hiram Bingham.[10] Davison believed that his position allowed him to work out differences within the War Department and promote a better understanding of aviation. Congress required Davison to provide updates on the progress of the Five-Year Plan. Once when Davison appeared before the House Military Affairs Committee, several congressmen inquired as to his opinion on the value of his position. Davison commented that the office offered a way to iron out different opinions on aviation within the War Department and allowed better understanding between the top military and civilian leaders.[11] In other words, he became the focus for civil-military relations and helped to soothe them.

The civilian secretary got along well with his Air Corps subordinates. They shared the same ideas for moderation and a positive use of the press, and Davison and Fechet enjoyed the same passion for hunting, occasionally taking trips together.[12] Davison urged the air officers to be seen in public, to promote aviation and "air-mindedness," and to provide good role models for the country. Like his military officers, Davison made speeches and trumpeted aviation and its promises and possibilities, although the officers' speeches did not have the same political tone. In the midst of limited appropriations and funding, Davison urged that all speeches be positive and stress how aviation would help national defense.[13] But Davison also participated in party politics, which the Air Corps leaders could not.

Davison remained active in New York, visiting there quite often. He spearheaded the effort to have New York City fund and upgrade its airport, and he used his position and contacts to drum up support. Although he held a national office, Davison also remained the permanent chairman of the New York Republican state convention, and he aided the nomination of Albert Ottinger (who ran against Franklin D. Roosevelt for governor in 1928).[14] Davison also campaigned nationally for the GOP and urged women to turn out and vote Republican.[15] On other occasions he urged veterans groups and other civilian organizations that lobbied for military causes to promote "air-mindedness" and support aviation.[16] By toeing the party line he proved himself a reliable partisan who would not use his position within the War Department and among the formerly rebellious aviation branch to cause trouble. On the contrary, two Republican presidents counted on his party allegiances and dependability.

Davison's position would tap all of his political acumen and charisma, for the Five-Year Plan required him to seek money for a program the president would not fund. Even before final passage of the Air Corps Act, observers raised doubts about the availability of money for the program. Just before the Senate debated the bill, the president met with Senator James Wadsworth and relayed his fear of a budget deficit.[17] Coolidge did not want a compulsory program included in the final bill, but he still signed the act, which mandated that at the end of five years the Air Corps must contain 1,800 serviceable aircraft, 1,500 officers, and 16,000 enlisted men. Congress decreed the program should begin with the fiscal year starting 1 July 1926 and be equally distributed over the next five years. Consequently, lawmakers requested submission of a supplemental increase for the budget cycle that was only days old.[18] Coolidge took advantage of the permissive wording of the supplemental section, which noted that additional estimates "*may* be submitted," and deferred the start of the Five-Year Plan until the following year (fiscal year 1928, which began on 1 July 1927).[19]

Congress showed no inclination to override the president, and neither wanted to raise the total amount given to the military. Thus, any increases in Air Corps manpower and budgets would come at the expense of other branches and areas in the Army. Such arguments would dominate budget and manpower discussions within the service and with Congress for the remainder of the interwar period, constantly frustrating those in the other Army branches.

While the Air Corps often portrayed itself as a mistreated stepchild, Army leaders pointed to the percentages of the budget and called the air arm a favored son. These arguments reappeared within a month of the Air Corps

Act's passage. Major General Fox Conner, the Deputy Chief of Staff and departmental spokesman on Army financial matters, proclaimed that unless the entire War Department budget increased, the Air Corps expansion plan would cause a reduction in other branches of 20,000 to 25,000 men, ruining the Army "beyond the hope of recovery."[20] Coolidge would not significantly raise the Army's budget or its overall manpower, so the Five-Year Plan would come at the expense of other Army branches.[21] Although many in the non-flying Army wailed about impending doom if the overall budget suffered under the Air Corps expansion program, in the long run the remainder of the Army fared about the same as it would have without funding the air arm, given the limits placed on all military spending. The budget battles placed the Army in a bad position as it wrangled with the Navy and its always-powerful congressional lobby. As much as the Army may have wanted to limit the Air Corps in order to provide for other areas, the War Department could not afford to overly restrict its air arm's growth for fear of losing the coastal defense mission (and budgets) to the Navy, especially as the latter prepared to increase its aircraft carrier force.[22] The success of the Air Corps in obtaining the money it did came from the political intervention of Davison and the continuing compromising attitude of the Air Corps leaders.

Throughout the early part of the expansion program, the Air Corps maintained its tone of moderation, and at public events Air Corps leaders even talked of teamwork and cooperation with their Army counterparts. Both the Army and its Air Corps seemed to declare a public truce and tried to change public perception of their animosity.[23] During the expansion plan (1928–33) the Air Corps never realized fulfillment of its mandated number of officers, men, or aircraft. Aircraft could not be built quickly, and their purchase required sizable allocations in the face of economic constraints. The limited budgets affected personnel issues tangentially: since appropriations limited the Army's numbers, the Air Corps grew at the expense of the other branches.[24] The fight for the expansion program for men and planes centered on budgets and transfers, but Air Corps leaders knew they would continue to make gains, though slowly, because of the mandatory legal status enshrined in the Air Corps Act.

Although the Five-Year Plan never met its goals, its presence gave the Air Corps a legally mandated target and political leverage in the budget battles, and Davison used both effectively. He worked with Congress to keep the program on track as much as possible while still supporting Coolidge's economy program. Congress wanted to buy cheaper aircraft in higher numbers—thus legally fulfilling the program and saving money.[25] As the expansion program fell behind each year, the Air Corps needed to request more in order to purchase the current year's requirements and make up for the previous year's shortfalls. Davison warned Congress of that very fact during questioning, and he received no disagreement.[26] Not only did the Air Corps need to request more money to restore past shortfalls, but the service needed higher appropriations to replace grossly outdated aircraft and equipment, as the Air Corps continued to reduce its stock of surplus World War I equipment.[27] Even during the first few years of the depression, Davison and the air chiefs were able to persuade Congress to increase the flying service's budget, which climbed every year from 1928 to 1931. Aircraft numbers also increased every year except for fiscal year 1930. Likewise, when Congress reduced the War Department budget severely, from over $509 million in 1931 (of which $58 million represented special funds allocated for depression relief operations) to just over $322 million in 1932, the Air Corps did not suffer proportionately. While the overall War Department lost 37 percent of its budget, the Air Corps budget dropped only 12 percent.[28] Still, Davison articulately argued that these figures represented an unfair cut for the Air Corps. In his 1932 annual report he contended that the Air Corps cuts, by percentage, represented the single largest cut in the Army budget—far greater than cuts that other branches of its size experienced.[29]

Davison also continued to press for a closer adherence to the mandated expansion program. When the program neared its end and was falling short of its goal, even Congress began to look at the expansion plan differently, realizing it could not provide the funding to meet its own mandate. A report on the 1931 budget noted: "The 5-year program is not a hard and fast schedule which must be adhered to rigidly. It is nothing more than an authorization to appropriate, subject, of course, to considerations and eventualities that could not be foreseen when the program was adopted. It would be a mistake to expect or require strict adherence to a procurement program for a product so unstable that obsolescence occurs between order and delivery dates. Procurements are so nearly in accordance with the program that there is no room for complaint."[30] Air Corps leaders did not like the delays in completing the expansion program, but they understood that, given the economic times, their budget represented a liberal share. Overall, the Air Corps made solid gains toward its program. Between 1930 and 1932 the service added 550 aircraft and stood less than 200 short of the 1,800-aircraft goal. The improved condition of the Air Corps occurred largely because Davison could fight for the air arm before Congress and around the General Staff. The air service was the only branch to have its own political representation, and its depression-era budgets proved the effectiveness of Davison's influence.

Air Corps leaders also used their cooperative approach and Davison's influence to try to correct the promotion system, a major problem in the service that had been unchanged by the Air Corps Act. The Army promoted officers on seniority; the highest-ranking officer (in terms of time in service) of a grade advanced when a vacancy became available, regardless of branch.

However, due to the influx of officers in the service during World War I, a "hump" of 5,800 officers below the rank of lieutenant colonel existed. The hump represented more than half of the entire promotion list, and the men on the list all had about the same length of service.[31] Due to its relative youth as a combatant arm, the Air Corps faced the worst of the problem. The youth of its officers in age and time served meant that the air branch did not have enough officers in higher ranks to fill Air Corps positions. Congress, unable to solve the promotion woes during the early 1926 legislative tussles, decided not to make specific recommendations within the Air Corps Act but instead required the Secretary of War to submit a study within six months.[32]

The study confirmed the "war hump" problem, but Congress failed to act and fix the problem. Even as late as 1930, roughly 400 of the Air Corps' 494 first lieutenants had World War I experience yet could not advance in rank. A woeful shortage of Air Corps field-grade officers resulted. A not so unusual example revealed that one air squadron contained 19 second lieutenants, 129 enlisted men, and 15 aircraft, all commanded by a second lieutenant (in a major's position). Also indicative of the lack of promotion possibilities, and creating increased future problems, a record number of officers resigned during fiscal year 1930 (twenty-seven) — even amidst the troubled economic times. Of the fifty-three total Air Corps squadrons, majors properly commanded only five.[33] As Davison eloquently summarized the shortages and their impacts in his 1930 annual report: "Sluggish promotion raises havoc with the morale of Air Corps personnel; is responsible for the resignation of valuable officers whose services the Government can ill afford to lose; places burdens of responsibility upon junior officers entirely out of proportion to their grades; and finally, tends to undermine the efficiency of the entire service by paving the way for a spirit of hopelessness which is bound to develop into apathy."[34] Davison and the air chiefs often stated that only the enactment of new legislation, requiring a separate Air Corps promotion list, would correct the imbalances and save the service from further loss of men, wasting of money, and drop in morale.[35]

Air leaders made their wishes known constantly and supported different legislative efforts to obtain promotion equity, but their tactics differed from Billy Mitchell's: no rants before Congress, no newspaper propaganda efforts, no "information" circulars, no embarrassing statements about Congress or the administration. Public statements usually came via their published annual reports and congressional testimony, and neither sought to shame Congress or the president. Davison's 1929 report was typical: "I regret, however, to observe that the most pressing question affecting the Air Corps — namely, the promotion problem — still exists despite repeated efforts to obtain legislation which would eliminate injustices and handicaps imposed upon our flying personnel due to sluggish promotion schedules. . . . Legislation is the only remedy."[36] Several different legislative efforts occurred, and the Air Corps backed those offered by Representative Allen J. Furlow, a Minnesota Republican.[37] His oratory on the House floor mirrored the comments of Davison's annual reports: unless someone fixed the Army promotion system, the Air Corps would lose pilots and cost the government additional money, time, and effort.[38]

The ordeal concerning promotion differed dramatically from the fights of the previous years, again a testament to the new leadership style and the presence of Davison. When Furlow's bill reached the Senate, different senators offered changes and amendments. However, contrary to past air legislation, as Senator Joseph T. Robinson pointed out, the Secretary of War, Secretary Davison, the Army Chief of Staff, and the Chief of the Air Corps all supported the same bill (the original one as passed by the House). Unfortunately, a filibuster that had no relation to the promotion legislation stopped the Senate from passing the Furlow bill, and the end of the congressional session tabled the measure. Although Furlow and Senator David A. Reed (a Pennsylvania Republican) reintroduced similar legislation during the next session, the worsening economic situation sidetracked all efforts to improve Army promotions, and neither chamber considered any such legislation until 1933.[39] The failure of promotion legislation set the Air Corps and the Army overall back. More significantly, however, the events clearly demonstrated the new dynamics in Washington regarding aviation matters and a proper division of civilian and military responsibilities.

POLITICAL CHANGES, IDEOLOGICAL CONTINUITY

Despite a new president in the middle of the Five-Year Plan in 1929, the changes were limited to faces and personalities and not political ideology or attitudes toward Army aviation. Republicans retained the Oval Office, and changes in the leadership of the War Department and the Army Chief of Staff did not significantly alter the Army's attitude toward aviation. Most importantly for the Air Corps, the new administration retained the services of Trubee Davison.

When Coolidge decided not to seek another term, Herbert Hoover emerged as the clear-cut choice as successor. He had earned the reputation of a great humanitarian for his relief efforts in Belgium during World War I and had served both Harding and Coolidge well as commerce secretary. Even Franklin D. Roosevelt, lauding Hoover's wartime efforts, announced, "I wish we could make him President of the United States. There could not be a better one."[40] Although Hoover did not fit the "mossback Conservative" mold of his "normalcy"-era predecessors, his fiscal policies toward the military did not drastically differ from those of Harding and Coolidge. Budgets would remain tight.[41]

As secretary of the Commerce Department, Hoover had overseen the expansion of commercial aviation, which also benefited military aeronautics, and had praised its positive contribution to America. He later claimed responsibility for recommending the formation of the Morrow Board to Coolidge and said that for civil aviation "the Morrow commission adopted my recommendations in full." The accomplishments of civil aviation under Hoover's Commerce Department can be stated with more certainty. William P. MacCracken Jr. became chief of the aviation division, and together they led the efforts that ended with dramatic increases in the number of airways, the lighting and electronic marking of those routes, and the number of civilian airports and licensed aircraft. These advances improved public confidence in aviation and spurred manufacturing output (both with carry-over effects on military aviation).[42]

During his campaign, several aviation leaders publicly supported Hoover, including Charles Lindbergh and aircraft designer Igor Sikorsky. Hoover's campaign even used Lindbergh's and Sikorsky's letters of support in press releases.[43] In his campaign, Hoover mentioned his past support for aviation and promises of its bright future. Although most of his statements advocated expanding commercial aviation, he also noted how the commercial side, with government purchases and assistance, could aid the efficacy of providing for the national defense.[44] Unfortunately for the Air Corps, Hoover's previous air-mindedness and aviation insight did not carry over into his presidency, and Army aviation never became one of his priorities.

Not only did Hoover not support military aviation in his speeches or budgets, but he even backed away from his earlier overt support of aviation as being good for the country. As president he repeatedly turned down offers to speak at the annual convention of the National Aeronautical Association (NAA). During this time, Senator Bingham served as NAA president and always invited Hoover to address the organization's conventions. Bingham even tried to entice Hoover to appear by touting the positive political benefits for the 1932 campaign of having such a forum available.[45] Hoover refused, although he sent letters to the conventions supporting aviation. Yet his actions never lived up to his words. In a proposed letter to the Aeronautical Chamber of Commerce of America, Inc., Hoover struck out a sentence that read, "The people everywhere should be educated to appreciate the greater command of time and space which can be theirs through the use of aircraft."[46] He also turned down the invitation to attend the public dedication of the Wright memorial at Kill Devil Hill, North Carolina, although he touted his own earlier efforts on the committee that initiated the memorial project.[47] Hoover likewise turned down the chance to publicize aeronautics. The National Broadcasting Company and the Aeronautical Chamber of Commerce jointly prepared a long series of weekly "air education" radio broadcasts for the public, and they invited the president to speak first. Davison wrote a letter urging the president to accept, but Hoover declined.[48] Although he had earlier flaunted his efforts in the field of aviation, as president Hoover did not promote the "air-mindedness" sought by Davison and Fechet.

Hoover's cabinet comprised different personalities, but its views on military aviation did not differ substantially from those of Coolidge's. The president picked James W. Good, a close friend and fellow Iowan, as his Secretary of War. Good, known as a shrewd political tactician who stressed party discipline (especially during his congressional career, 1909–21), did not get a chance to make his mark in the War Department due to his death in November 1929.[49] Army veteran Patrick J. Hurley, who had been the Assistant Secretary, became Secretary of War. A self-made millionaire and the youngest member of the cabinet, Hurley never lost his rough Oklahoma edge. He attacked problems directly, took shortcuts whenever possible, and displayed an energetic attitude.[50] In a letter probably indicative of most air officers' opinions, Captain Ira Eaker believed Hurley's appointment a solid one for aviation. Eaker called Hurley a "sincere friend of aviation" who could guide the service through "the darkest story of our period." He also noted that Hurley and Fechet were close friends.[51]

In deference to Davison's accomplishments and abilities, Hoover kept him as Assistant Secretary of War for Air. As a former cabinet officer, Hoover undoubtedly knew the successes Davison had achieved and how he had helped control a formerly troublesome military branch. Davison submitted the customary letter of resignation on Hoover's Inauguration Day, which Hoover did not accept, and the New Yorker served the entirety of Hoover's presidency. Hoover retained only three members of the "Little Cabinet," and all three held aviation posts: Davison; William P. MacCracken Jr., the Assistant Secretary of Commerce for Aeronautics; and Second Assistant Postmaster Warren J. Glover (responsible for air mail).[52] The new president was likely pleased with the jobs they had done during Coolidge's tenure and with how these men had helped the aviation situation, so volatile earlier in the decade, regress into the political background.

Summerall remained Army Chief of Staff for the first twenty months of Hoover's term. Still unfriendly to aviation, the general ranked among those who considered the Air Corps a political "favored son." In his public speeches on strengthening national defense, he regularly ignored air power.[53] Summerall was also not particularly astute about civil-military relations. In a speech on 27 June 1927 to the Army War College, he lambasted political leaders for always destroying the peace gained by the Army. "The political phase of war must eventually be controlled by military men," he commented. "In war the political phase has often nullified the skill of arms and the mastering of economic difficulties."[54] When Summerall's term ended, Hoover replaced him with a general who proved to be more politically perceptive. In November 1930 the president bucked the trend of selecting the Chief of Staff based on seniority, "searched the Army for younger blood," and tapped Major General Douglas MacArthur to lead the service. Hoover did not want a general who would have to retire before completing his term, and thus he justified the skipping over of officers senior to MacArthur.[55] By doing so, he bolstered his claim of infusing young blood into the position.[56]

MacArthur's relative youth did not translate into radical views on air power and the place of Army aviation, even though he had voted to acquit Billy Mitchell. His main goal remained protecting the Army from the budget crunch spurred by the depression. As far as aviation was concerned, he sought to minimize the reductions in other branches to achieve the mandated Air Corps expansion program. MacArthur's ideas dovetailed nicely with Hoover's, as both wanted to eliminate Army deadwood and maintain economy.[57]

Fechet served as Chief of the Air Corps for three-fourths of Hoover's presidency. He had replaced Patrick upon the latter's mandatory retirement in December 1927, thereby establishing the tradition of the Assistant Chief's rising to Army aviation's highest post. Brigadier General Benjamin D. Foulois then became Fechet's assistant. Upon his selection as chief, Fechet announced he would only serve one term, believing it better to rotate in new personalities and ideas. At that time, Secretary of War Davis accepted Fechet's announcement that he would not consent to reappointment, saying that a continuing rotation of leaders would benefit the service during peacetime.[58]

Although Fechet was an ardent supporter of increased funding for the expansion plan, which contradicted Hoover's minimal defense spending, he refused to create open controversy and agreed to focus on leading the Air Corps. In his annual reports he covered controversial matters delicately, such as the need for promotion legislation or the shortage of aircraft, and stressed positive gains in the service.[59] In an example of his optimism, Fechet called the third increment of the expansion program "successful," even though the service fell 500 aircraft behind its authorized strength for the year and remained more than 800 behind the 1,800 goal.[60] Fechet relied on Davison to fight the political battles. From the beginning of their relationship, the two men agreed that the military officer would administer the Air Corps, while Davison would handle all relations with the civilian elements of the government. Davison took the lead when the press wanted information.[61] Their rapport worked wonderfully, and they guided the service through a difficult period of decreased government spending.

The nation's mood toward military spending did not change during the Hoover years, and as budgets became leaner after 1929 defense spending receded even further as a priority. Much like Coolidge, the public saw little need to spend much money on defense without an immediately visible threat. Immediately after the death of Secretary of War Good, letters and telegrams poured into Hoover's offices requesting that the War Department's name be changed to "Department of Peace" to reflect America's peaceful and defensive nature regarding military force.[62] Hoover understood this mood and agreed with it to a point. He wanted world disarmament while retaining enough force to defend the United States and the Western Hemisphere.[63] Three ideas shaped his relations with the military: an abhorrence for war, a desire for economy in government, and a belief that preparedness served as a deterrent.[64] Since the United States floated the world's second-largest navy, while the Army contained fewer men in proportion to population than the Versailles Treaty allowed Germany, the Navy would obviously take the largest cut under Hoover's plans. These factors combined to cause the Navy to conserve its ships and the Army to salvage what it could of a skeleton force. Leaders of both services looked at the Five-Year Plan and Air Corps budgets with different levels of jealousy and contempt.

With the worsening economic crisis, Hoover asked the Army to propose ways to save money. The president wanted to cut taxes, and he thought that the world situation, with no visible war on the horizon, allowed cuts in the Army budget. Hoover wanted "constructive and not destructive" reductions within his stated goal of preserving "a completely adequate national defense."[65] Summerall led the effort and surveyed the military establishment. He wanted a critical reappraisal from his field commanders and hoped the survey would serve as a vehicle to begin a restructuring of the Army. However, many of the replies from the field demonstrated the disparity between field commanders and the General Staff. Very few commanders wanted cuts or radical alternatives. For the Air Corps, Fechet actually recommended the opposite of economy: he wanted to jettison the current expansion plan and institute an even larger one.[66]

Summerall's report to Hoover and Secretary of War Good demonstrated the Chief of Staff's continued hostility toward aviation. He called current air doctrine an unproven hypothesis but conceded that the airplane "has captured the popular imagination and the five-year Air Program has been the result."[67] Summerall supported a separate Air Corps budget for the explicit purpose of making clear to Congress and the nation the high costs of military aviation, which he believed might assuage any public and legislative objection to cutting aviation's appropriations. He further reasoned that a stronger civilian aircraft production base made the arguments for aircraft expansion over the last five years outmoded. He believed that, should hostilities occur, factories could quickly produce aircraft. Summerall concluded that the budget could obtain substantial savings if the aircraft program could be extended to ten years. He further calculated that the immediate savings amounted to

$4 million but that over the final eight years of the extended program the government might add as much as $32 million to the budget. Summerall did not elaborate the reasons for the increased costs, but they undoubtedly included the higher unit costs of aircraft due to longer acquisition times. [68] Yet, reiterating the expansion program's popularity, he cautioned Hoover: "The five-year Program is the product of a civilian board, and received the support of the people and of Congress. It would not be wise for a revision downward to be undertaken except by a similar agency." [69]

Although Summerall had surveyed field and branch commanders for recommendations, he did not share his conclusions with others. He annotated his personal copy with a penciled note: "Seen only by President Hoover and Secretary Good. They agreed on its conclusions." [70] Fechet and Davison did not know that Summerall had recommended to the president policies detrimental to the Air Corps. Soon, though, another problem reemerged that placed the Army in a quandary. The Army leadership needed to support the Air Corps and the need for aircraft in the coastal defense role to counter incursions by the Navy for that mission and its associated budget.

In the meantime, the Navy simultaneously fought for aircraft, aircraft carriers, and larger budgets. Controversy began in 1925, when Navy leaders revealed that their aviation expansion program would base planes ashore, touching off a legal and interservice battle lasting for five years. [71] In early 1930, Hurley directly challenged Navy leaders, who had taken the matter to Senator Bingham and the attorney general, and requested arbitration from Hoover. Hurley presented the case to Hoover as one of "unnecessary duplication of effort" pertaining to land-based aircraft and argued that the Navy had circumvented the law. [72] Hoover, seeing both the military and economic ramifications, refused to decide and sent the issue back to the secretaries.

Three months later, in May 1930, Hurley again pressed the president for a decision, along his own pro-Army recommendations, calling the issue "the gravest moment" for the Army. [73] The issue stirred unrest between the two departments and continued until MacArthur and his naval counterpart, Admiral William V. Pratt, issued a one-paragraph statement in January 1932. The MacArthur-Pratt Agreement was only informal and between the two chiefs, but it lasted until Pratt left office in 1934. The two leaders agreed that the Navy's air assets would be based with the fleet and perform fleet missions, while the Air Corps would remain land-based and support the Army, to include the coastal defense role. The agreement undoubtedly hurt Hurley, considered one of Hoover's favorites. Due to a close relationship with the president, and Hoover's desire to reduce the Navy, Hurley hoped the president would decide in the Army's favor. [74]

In yet another sign of aviation's new moderation and Davison's influence, the Air Corps leaders stayed out of the argument. Although Air Corps roles and budgets were at stake, air leaders let the Army leadership fight the public battles and relied on Davison to take the lead on the political front. During the lengthy controversy, Davison worked closely with Bingham, who not only sat on the Senate Military Affairs Committee but also chaired the Joint Committee on Aerial Coastal Defense.

Davison and David S. Ingalls, the Assistant Secretary of the Navy for Air, appeared together before Bingham's committee. In his usual accommodating style, Davison noted that he believed the two services could work out the details, obviating the need for Congress to legislate a radius of action for Army and Navy shore-based aircraft. Davison reiterated his belief that the Army and Navy generally agreed on broad policy matters and that existing bodies such as the Joint Board could work out the minor details and disagreements.[75] Davison also provided Bingham with pro–Air Corps "facts" on the coastal defense controversy and likely urged him not to further incite the controversy with congressional proceedings.[76] Davison soothed the controversy and kept Air Corps leaders from once again becoming embroiled in a political fray. While the fight for control of coastal defense missions and budgets occurred, the Army leadership sought to reduce aviation budgets within the Army to keep other branches from further depression-induced belt tightening by President Hoover.

With the economic situation worsening, and faced with a projected deficit of $900 million for 1931 and more for 1932, Hoover needed additional cuts. Recent congressional changes increased his urgency for economy. For the first two years of his presidency, Hoover benefited from having a Republican Congress. Riding the Iowan's coattails, the GOP had gained eight seats in the Senate during the 1928 elections, extending what was previously a two-vote majority over the Democrats to seventeen seats. The same pattern occurred in the House, with the Republicans more than doubling their majority. However, the depression spelled the end of Republican dominance. After the 1930 midterm elections, the GOP held advantages of only one seat in the Senate and two in the House.[77] Hoover believed the Senate would not support him anyway, as he counted only forty sure votes for his programs. He discounted the support of the party's "left wing," which included Senators William Borah and George Norris.[78] The situation got even worse for the president. By the time the Seventy-second Congress convened in December 1931, thirteen elected members had died—most of them Republicans. The Democrats then controlled the House, and two men in particular—Speaker of the House John Nance Garner (Texas) and chairman of the Subcommittee on Military Appropriations Ross Collins (Mississippi)—stressed economy and the need for the Army to become more amenable to reorganization and change.[79] Faced with this tough situation, Hoover asked the Army to help him propose spending cuts without gutting the force.

Hoover held a conference at his Rapidan Camp (a log-cabin retreat built in the Shenandoah Mountains at the headwaters of the Rapidan River) in May 1931. Only non-flying Army officers attended, accompanied by Secretary Hurley.[80] This meeting—and Hoover's negotiations with the Army on how to make depression-era cuts—purposely excluded the Air Corps leadership. Davison's omission from some of the high-level talks was especially surprising. Hoover liked to surround himself with experts, and he gave significant weight to their opinions, so his decision to keep Davison in his administration demonstrated his trust in and respect for the air secretary. But now Hoover needed deeper reductions in the military budgets, and the expensive Air Corps armaments represented a prime target. He probably wanted to minimize the number of people involved and to obtain solidarity with the overall Army leadership on the budget reductions.

The president opened the meeting by announcing the budget shortfalls and stating that he could not increase taxes to obtain the money. He added that although he did not want to "jeopardize national defense," he needed some concessions until employment improved. "Aviation is a very large item," he concluded.[81] MacArthur, who had taken over as Chief of Staff from Summerall the previous November, agreed, declaring that "the air is the most expensive element war has ever known." Viewing the air budgets as being too large, MacArthur argued that reducing air expenses would support Hoover's commitment to reduce the military budget while preserving defense. Perhaps with a tinge of parochialism, MacArthur noted that Army infrastructure outlasted the expensive armaments for the Air Corps and the Navy, which he called fragile and soon obsolete.[82]

When MacArthur finished, a rapid exchange immediately occurred between the president and the assembled War Department staff. All agreed that no European nation could threaten the United States from the air, that the nation led the world in aircraft production capacity, and that the Navy led the world in aircraft assigned to the fleet. The president then asked a series of questions: "Would we be justified in allowing that air fleet to run down? Have we any justification for further expansion? Can we make a study for the financial viewpoint toward letting that air force deplete itself in the next two years?"[83] In the end, all present agreed to study the amount of air strength that the nation needed, but they concurred that the expansion program could withstand some cuts. When the Army's Deputy Chief of Staff later remarked, "Public sentiment has caused abnormal development in our air forces," Hoover replied, "Yes, they get the public to believing you can win a war with nothing but airplanes, while we all know you couldn't win a war that way."[84]

Having been in the cabinet during the height of the air insurgency led by Mitchell, Hoover surely remembered the airmen's publicity campaigns and assertions of the predominance of air power. Although the air leadership had disavowed Mitchell's tactics and polemics, the president clearly understood the political turmoil caused over Army aviation and how the public and key members of Congress supported the air arm. Hoover believed that to save the country and his own job, he needed to save government money, but he also realized that he needed to tread lightly with the Air Corps. During previous years and different economic times, he had touted his support for aviation, but when it came to a depression budget and closed negotiations, he thought differently. Likewise, Hurley, although liked by the aviators and a friend of Fechet, did not object to his boss's support for Air Corps cuts. To keep their proceedings closely held—especially from the Air Corps leadership— the meeting notes were marked "Confidential" and only three copies were made. [85]

Although MacArthur publicly supported a strong Air Corps—going so far as to participate publicly in the nationwide air maneuvers for 1931—privately he did his best to limit its budget. [86] To modernize the Army overall, he needed to transfer money to other branches. Following the May 1931 Rapidan Conference, MacArthur undertook a secret study of defense needs. Presented to Secretary Hurley, the report argued that the optimum total of aircraft for the Air Corps to carry out all its missions was 2,950 (over 1,000 more than the Air Corps Act and Morrow Board recommended). However, MacArthur noted that the geographic situation of the United States and the current period of relative peace required only 1,800 total aircraft. He believed the Five-Year Plan had significantly reduced the Army's effectiveness by reducing the strength of the other arms and that no revision upwards of aircraft numbers or budgets should occur. [87] Later that same year, MacArthur ordered a reduction of the 1933 budget request by over $15 million, of which more than one-third came from the Air Corps. [88] Whether intentionally or not, MacArthur's views coincided with those of the president.

Despite the discord with the Navy and the apparent collusion of the administration and the Army to limit aviation funding, the Air Corps expansion program plodded slowly along, bolstered by public and key congressional support. Hoover was likely a factor as well. On the one hand, the nation's economic turmoil made him reluctant to support increased funding for military aviation. On the other hand, he was reluctant to curtail the air budget more severely because it would hurt the nation's aircraft manufacturers. Hoover had already received warnings that reducing aircraft orders and money for procurement could harm the manufacturing sector and increase unemployment. The Air Corps published warnings that the aviation industry would lose vast sums of money for 1930, and the outlook forecast no pending improvement. [89]

Davison also played a key role in spurring the expansion program. During these times he appeared before the administration and Congress to keep the program as close to its goals as possible, fighting for the best and most feasible appropriations, and keeping Army aviation in a positive light before the country. As the depression worsened, Davison's understanding and flexibility undoubtedly pleased Hoover and Congress. Following fiscal year 1931, Davison noted that the Air Corps still needed legislation on pay and promotion but that he would defer any recommendations "until improved general conditions warrant" action; he further mentioned how the officers and men of the Air Corps would make such sacrifices for the nation's benefit, "confident that they [the airmen] will not be unjustly or disproportionately imposed upon." In eloquent political language, Davison pointed out that he would still recommend those items needing attention but that "the present is not a propitious time for climbing costs" and that until better economic times, efficiency would be maintained unless it affected vital areas of national defense.[90] Davison even economized by curtailing flight training so that the Air Corps could return money to the Treasury.[91]

Davison's role during the Coolidge and Hoover presidencies repeatedly benefited the Air Corps while cooperating with the Army. Although some conflict remained (Summerall's attitudes and jealousy over Air Corps budgets), the general relationship between the branches and within the entire Army had changed. In 1930, Major Hugh Knerr, attending the Army War College, noted a "growing appreciation" among Army officers for air power.[92] Patrick had begun the transformation, and Fechet continued it, staying even more in the background. Fechet kept a lower profile than any of his predecessors. He concentrated on running the aviation branch and let Davison take the lead with the press and in politics. Davison knew when and how to push for budgets, using his political skills and lobbying, actions forbidden to air officers by General Orders and the unwritten rules of proper civil-military conduct. Years later, Davison agreed that Hoover did not do all he could for aviation, but he thought that Hurley and Davis had tried to help the Air Corps. On tactics, Davison noted that he could only "push" against the different war secretaries. "I wasn't fighting them," he recalled, "because they were my bosses. . . . All the time we [the Air Corps] were just trying to get ahead, and we got something when they thought they were justified."[93] He also urged the "air-mindedness" for the country, and his air officers responded.

"Air-mindedness" included not only Air Corps actions and positive coverage but reaching out to the nation, especially its young boys. Promotions included one where Babe Ruth tried to catch baseballs dropped from aircraft. Arnold and Major Ira Eaker wrote books on flying, and the former wrote a six-volume series with a young aviator hero, Bill Bruce. Arnold also helped influence cinema, as he rubbed elbows with some of the stars during his monthly air shows at March Field, California. Officers also assisted aviation clubs such as the Junior Birdmen and the Jimmie Allen Flying Club.[94] Positive coverage of aviation feats, joint maneuvers, and new aircraft kept air advances before the people (and the politicians in charge of the purse strings) without resorting to whining about inadequacy or forecasting doom as leaders had in the 1920s.

One later evaluation of the Hoover administration credited Davison with much of the progress made by the Air Corps after 1926.[95] Davison had also kept himself visible to the airmen and promoted "air-mindedness" by flying hundreds of hours and keeping in touch with aviation advances. Hoover lauded Davison as he left office, saying, "Your fine public service in building up the Aviation section of the Army is manifest and I need not extol it."[96] As Davison's tenure ended with Hoover's defeat in the 1932 elections, perhaps the most telling gauge of how much the Air Corps had changed came with the early resignation of its chief.

During his long tenure in the top two positions of the air arm, General Fechet had followed the General Orders regulating civil-military relations and had acted carefully and conservatively with his civilian counterparts and superiors. During his term, and due to Davison's taking the lead before Congress, Fechet, unlike Mitchell and Patrick, never became embroiled in controversies over congressional testimony or public statements. Because Davison handled all of the legislative matters, Fechet did not have to draft legislation or provide information directly to Congress. Instead, he concentrated on his military duties of organizing, maintaining, and administering the Air Corps. When he disagreed with his superiors, he did so through proper channels and in a deferential manner.[97] Fechet likewise did not seek personal publicity, preferring to stay out of the newspapers.

With only a few months left in his self-imposed single term, Fechet left office. In September 1931 he took a three-month leave of absence, which left Benjamin Foulois as the acting chief.[98] When his four-year period expired in early December, Fechet retired (ten years earlier than his mandated age-retirement date) and notified Davison of the decision.[99] The president then appointed former firebrand Foulois as the new chief.[100] Fechet retired primarily because his term as the Army's highest-ranking airman was soon to expire, and he could not rise any higher in the Army.[101] He had already held the two highest jobs in the Air Corps, and he had no chance to become the Army's Chief of Staff. However, during the new year the nation discovered another reason why Fechet left the service early.

Saying that the nation needed an "awakening" from believing in the adequacy of its air defenses, Fechet explained that in a civilian capacity he could push publicly for improvements in military aviation. He announced that he would open a Washington bureau of the *Aero Digest*, owned by his lifelong friend Frank A. Tichenor, and begin writing for that publication and otherwise promoting aviation and the national defense.[102] He summed up his motivations for resigning: "I have come to the end of my military career and pass the control stick on to younger hands. I go in sorrow at the state of our armed forces, in sadness at the attitude of our country towards its defenders and because of the false sense of security my people seem to feel. I want my last act to be this word of timely warning. We are the most hated nation in the world. . . . Unless there is immediately a national consciousness of impending trouble and ample preparation to meet it, your fool's paradise will be lost."[103] Perhaps still feeling the tie of loyalty so central to thirty-three years in uniform, Fechet tried to avoid criticizing Hoover: "I have no fuss with the administration. Existing conditions are what should have been expected to follow the World War." But he did not absolve administration leaders from all blame, closing, "they have gone too far."[104]

Aero Digest published Fechet's first article in its January 1932 edition, and Fechet contributed articles monthly for the next year and a half. The first three articles discussed the shortfalls of American defenses overall but concentrated on the lack of an adequate air force.[105] Fechet's articles ranged from general comments on the national defense and the state of civil and military aviation to the more technical and specific examinations of different type of aircraft. He did not limit his coverage to the Army but also commented on the need to expand naval aviation and add more aircraft carriers. Fechet never named anyone in the administration or Congress, but he alluded to the overall situation and its development as shortsighted and dangerous.[106] When he published a book in 1933, he used the same approach—urging without disparaging, and commenting positively on aviation's benefits and the need for more attention.[107]

Fechet also wrote a series of articles for the *New York American* in early 1932. Each contained language similar to his initial announcement, although the articles never singled out the president, members of his administration, or individual congressmen. Like Mitchell, Fechet seemed to want to arouse the ire of the country regarding the state of its military aviation, but without the personal or organizational attacks that had been Mitchell's hallmarks. Now that he was out of the service, Fechet was free to disparage the government's handling of aviation. Following the Japanese invasion of Shanghai, his article opened, "China today is paying the price of unpreparedness. America, hamstrung by pacifistic propaganda and misled by the prophets of false economy, is slowly but surely drifting toward the same folly."[108] He went on to show that if Japan attacked American bases in the Pacific (including Hawaii), a quick and heavy hammer blow could leave the West Coast open to invasion or additional attacks from Japanese aircraft carriers. Arguing that the current number of aircraft available (1,591) represented only half of what was needed, he urged Congress to appropriate the money immediately to make up the difference.

Although he used some of the same rhetorical devices that Mitchell had used ten years earlier (casting current state of aviation as being far behind other countries, conjuring up simulated attacks and their effects), Fechet did not denigrate the character or assign malicious intent to government agencies or people.[109] His post-service writing career did not seem to raise any ire with the administration (as opposed to Leonard Wood's controversies with President Wilson a decade earlier, for example). During his time as Chief of the Air Corps, Fechet had never undermined Hoover's policies. He did not become a professional lobbyist or use his access to encourage those still in the military to act against administration wishes, nor did his articles encourage a civil-military schism. His writings seemed more directed to encouraging the population to better understand what was going on around the world and attempting to shake the country out of an isolationist mood.

Fechet was no Mitchell. Even in retirement he retained the same conservative style he had exhibited during his eight years in Washington. Although he proved to be more vocal and visible in the print media after leaving the service than he had been as air chief, he still remained well within the boundaries of normative civil-military behavior in the United States at the time. His post-military career activities fit into a pattern pioneered by such retired advocates as General of the Armies John J. Pershing and Major General Leonard Wood (though Fechet was not nearly as active or vocal as these two men) and followed by many military leaders for the remainder of the twentieth century.[110]

The years of the Hoover presidency were tumultuous ones for the Air Corps and the country. Military leaders needed to fight for budgets and missions, but they also understood the need to limit military spending. Hoover wanted an adequate military and understood aviation's role in national defense, but he also believed that without a visible and credible threat to the country, he could curtail military spending. He knew that the root problems of the depression were the nation's top priorities and that if he did not get the country on the road to recovery, he would be out of a job.

Despite turbulent times, the Air Corps continued the moderate policies that Patrick had established during the post-Mitchell years. Fechet reminded civilian leaders (primarily through his annual reports) of the need to support aviation and the expansion program, but he did not publicly rant when the appropriations fell short or the expansion program fell further behind each year. Not until he left the service did Fechet more explicitly push for aviation advances, and even then he did so without personal attacks. He skillfully led the service through four difficult years, but the efforts of Trubee Davison offered the most value to the flying service.

Davison provided political protection for the service and interceded in matters and manners where military officers did not appropriately belong. When economic times toughened, he continued to press for improvement in Army aviation while also finding ways to economize and return money to the Treasury. He helped avoid congressional intervention in the coastal defense debate, and he took the lead in appearances before appropriations committees. He smoothed relations between the Air Corps and the Secretary of War and the General Staff, and he let the air chief more appropriately direct his attention to commanding the service.

Despite the turmoil of Hoover's term, the initial six years of the Five-Year Plan would seem placid indeed compared to those of Franklin Roosevelt's first administration. With no replacement for Trubee Davison, a seemingly reformed firebrand leading the Air Corps, and the darkest days of the depression looming ahead, the political forecast did not seem to favor smooth flying for the Army Air Corps.

CHAPTER FIVE

Conflict Within and Without, 1933–1934

The first six years since the passage of the Air Corps Act had gone relatively smoothly for the Air Corps, primarily owing to the steady hands of Generals Mason Patrick and James Fechet and the presence of an intelligent and effective political intermediary in Trubee Davison. The Air Corps' budgets, although not enough to allow the service to complete the Five-Year Plan, represented a fair share compared to overall War Department appropriations constrained by the depression and the mood of the American public. Because its civilian and military leadership took a more restrained approach with Congress and the administration and not agitating for independence, the Army's flying service conducted its civil-military and inter- and intraservice relations more within contemporary practice than at any other time since World War I.

All of this relative serenity came crashing down during Franklin Roosevelt's first administration. The new president and his New Deal initially limited defense spending, and Army leaders began to fight more aggressively to divert the flow of money to more neglected branches of the Army.[1] Riding the New Yorker's coattails, the Democrats now controlled both houses of Congress and thus limited the power of aviation's proponents, who were mostly Republican. However, the most important changes occurred inside the Air Corps' leadership.

Until 1941, Roosevelt left vacant the position of Assistant Secretary of War for Air, which Davison had used to mend civil-military relations (aided by Generals Mason Patrick and James Fechet). This vacancy, coupled with former firebrand Benjamin D. Foulois' command of the Air Corps, made the flying service vulnerable to setbacks in its status, its budgets, and its proper interaction with civilian elements of government. Foulois abandoned the moderate line and began behaving more like Billy Mitchell, even though the two hated each other. Foulois derided the Navy, wrote and submitted legislation for air-friendly congressmen (circumventing the War Department), openly chided the General Staff and the War Department, and pursued programs unsupported by the president. Civil-military relations plummeted in

1934, when the air mail fiasco combined with a congressional investigation of Foulois to send the Air Corps reeling. Congressional allies abandoned Foulois and rejected air legislation and programs, while the president fumed at being let down by Foulois' promises. Suddenly, Air Corps civil-military relations resembled the nadir of the Mitchell years.

THE PRESIDENTIAL FOX AND HIS HENHOUSES

The onset of the Great Depression dominated Hoover's presidency and led to his landslide defeat. While the 1930 elections narrowed the Republican majority in Congress, the presidential elections of 1932 overwhelmed the GOP. Franklin D. Roosevelt carried all but six northeastern states; he even took Hoover's home state of Iowa.[2] As Roosevelt prepared his cabinet, no distinguishable process guided the selections except the requirement of loyalty to the chief executive.[3] The new strong-handed leadership would also affect how the Army conducted its affairs. Unlike Hoover, Roosevelt did not arrive on the scene with a record as either a steadfast friend or a virulent foe of aviation. He had taken the unprecedented step of flying to Chicago to accept the Democratic nomination, but he did not use the trip as a statement for aviation. The services also well remembered his attacks on Mitchell and his opposition to removing naval aviation from the fleet, but those events had occurred thirteen years earlier when he was Assistant Secretary of the Navy.

Now he was president, with a lot more pressing problems than air power. Foulois recalled not knowing the new president's position on the Air Corps, but he believed FDR would continue to favor the Navy and remain unsympathetic to the Air Corps' desire for technical improvements.[4] After the election, a minor squabble ensued when both services approached Roosevelt about transporting him on their planes. Roosevelt shunned both and traveled by car and train.[5] If the aviation elements of either service, but especially the Army (due to the president's prior affiliation with the Navy), expected the new chief to support their aims, they sadly misread his priority on the economy. The military more clearly understood FDR's fiscal policy when he appointed Lewis W. Douglas as director of the Bureau of the Budget. The man through whom the services would submit their spending plans arrived with a reputation as a fiscal conservative and bottom-line budget balancer.[6] And Roosevelt himself brought a new style of leadership as commander in chief.

Whereas Hoover had often leaned upon the War Department for advice on military decisions, Army leaders soon realized that FDR would not operate the same way. Army chiefs saw how Congress quickly responded to Roosevelt's economic programs, and they correctly assumed that the new president would take a larger role in defense matters.[7] Even the service's unofficial publication foresaw the change only weeks after the inauguration:

"Nothing is more evident than the fact that the destinies of the services are in the hands of President Roosevelt."[8] For his Secretary of War FDR chose Utah governor George H. Dern, called a "pillar of strength in the West." Roosevelt wanted Dern somewhere in the cabinet, and after initially slotting him for Secretary of the Interior he assigned him to the War Department. Although Dern occupied a cabinet position, Roosevelt did not consider him a close adviser, and Dern did not count among the inner circle of the Brain Trust or New Dealers.[9] One Washington insider, *Army and Navy Journal* editor John Callan O'Laughlin, called Dern a "weakling" who would not stand up to the president.[10] Dern admitted to not entering the office with any preconceived plans or ideas, and he specifically mentioned the air component in this remark. However, unlike Patrick Hurley, Dern operated with understated modesty.[11]

Perhaps due to lack of experience in military affairs and being a Washington outsider, Dern quickly came under the sway of Chief of Staff Douglas MacArthur. Their conservative views on military policy helped them forge a good relationship, but the charismatic general undoubtedly affected the newcomer. Learning the job with MacArthur at his side, Dern would strongly promote the views of the General Staff—long the bane of any aviator, and especially those with radical ideas.[12] MacArthur also shared a good relationship with Dern's boss. Roosevelt seemed to like and respect MacArthur and would often bounce ideas off of him.[13] With MacArthur so trusted and relied upon by the two most powerful civilians overseeing the Army, the General Staff became all the more unsympathetic to the ideas or actions of air enthusiasts. The General Staff's retrenchment became especially evident with the assignment of Major Generals Hugh Drum (Deputy Chief of Staff) and Charles Kilbourne (Chief of the War Plans Division). Both of these officers resented the money and power given to the Air Corps over the past few years and almost devoutly viewed aviation as a support unit for ground operations. Kilbourne wrote articles (some published) countering public statements by Foulois and Mitchell. He also prepared War Department rebuttals and notes to congressmen opposing any proposals for increased money or influence for the Air Corps. In one instance, Kilbourne suggested that the War Department object to any legislation that would meet the goals of the Air Corps.[14]

Congress also turned over in 1932. With the Democrats now solidly in control of the House by a 313–117 margin, and the Senate by 23 seats, those Republican supporters of Army aviation who survived the Democratic tidal wave lost their ability to control the Military Affairs and Appropriations Committees.[15] Luckily for the Air Corps, Randolph Perkins and Frank Reid retained their House seats, and Michigan reelected Frank James in 1932 (but not in 1934). Senator Hiram Bingham and Representative Fiorello LaGuardia were the most painful losses, both failing in their 1932 reelection campaigns.[16] On the positive side in the Senate, Joseph Robinson, one of the few Democratic supporters of Army aviation, became the Senate majority leader due to the reversal of political power. Robert La Follette Jr., the insurgent who often supported the aviators (usually to be a thorn in Coolidge's and Hoover's sides), still roamed the upper chamber due to his 1928 campaign win. He remained in the GOP but would win his next two elections as a Progressive.[17] Two bright spots for aviators in the post-election period came with the ascent of two supporters as chairmen in their respective Military Affairs Committees. Morris Sheppard, the long-serving Texan, became the chairman of the Senate Military Affairs Committee, and South Carolina's John McSwain rose to the same position in the House.[18]

Sheppard, known as the "Dean of the Congress" because he had the longest continuous service, could not be counted among the air enthusiasts, although he likewise declined to ally himself with the traditional Army group that would defend the General Staff. His long voting record and many speeches revealed his desire for an adequate and balanced national defense. In one speech, citing the theories of Giulio Douhet, Sheppard argued nevertheless that "we must have a proper balance of our forces in the air, in the sea and on the land."[19] He favored no service above the other and seemed content as long as Texas continued to benefit from an inflow of defense dollars. During FDR's first administration Sheppard supported efforts to bolster Army enlisted strength and correct the promotion situation.[20] His foremost objective, however, became the New Deal, and he was among the staunchest supporters of Roosevelt's program.[21]

In the House, McSwain's voting record demonstrated his affinity for the flyers and their cause—but not to the extreme degree of Frank James. McSwain once announced his motto as "Where national defense begins, partisan politics must end."[22] The South Carolinian's actions supported that statement, and unlike some other congressmen, McSwain would not support an aviation bill just to make a political statement. He also did not support aviation only for the Air Corps' sake. "I have not been the partisan of the army as against the navy, nor of the Air Corps against the ground troops or the sea forces," he announced. "I am as much interested in one instrumentality for insuring the national defense as I am in another."[23]

In the swirl of uncertainty and gloom infecting the nation in 1933, Army aviators could look forward to reasoned support from McSwain, who announced his support immediately upon taking over as chairman of the House Military Affairs Committee. "I would place the highest emphasis upon the power of aviation," he pronounced boldly, and he questioned the need for funding obsolete items such as cavalry units and their horses.[24] Soon after FDR took office, McSwain demonstrated that he would not hesitate to take aviation causes directly to the president. He pleaded with Roosevelt for the allocation of public works money for airplanes and other aviation equipment and construction, and he also wrote, in more specifics, to Secretary Dern.[25] Although McSwain continued to support aviation's advancement, he did not work closely with Foulois, and according to the airman's biographer, McSwain "cooled" in his feelings toward Foulois.[26] Foulois lost the full support of this influential politician, and he would soon be without the services of another.

One vacancy stood out when the new administration took the reins of government in March 1933. The office of the Assistant Secretary of War for Air sat empty, despite its previous seven-year success of vastly improving civil-military relations between the Air Corps and two administrations. Billy Mitchell, who backed FDR and actively assisted in securing his nomination at the 1932 Democratic National Convention, hoped for a job within the administration, and some of his congressional friends pushed the retired airman for the air secretary post.[27] *Aero Digest* recommended John Dwight Sullivan, a World War I veteran who had served in various aviation posts in New York, including being appointed by Roosevelt to the New York State Aviation Commission. Nevertheless, the journal asserted, "somebody must fill the shoes now occupied by Trubee Davison."[28] One historian gave MacArthur, whom Roosevelt kept as Chief of Staff despite his conservative views and reputation, the credit for eliminating the office simply for economy.[29]

The saving of this minor salary, especially when compared to the overall defense budget, could not have possibly been the primary reason, but MacArthur did make the decision. When Lewis Douglas asked the War Department for recommendations to save money within the Army, Dern put the matter in MacArthur's lap. MacArthur's sole recommendation called for the elimination of the air secretary position.[30] Perhaps due to Roosevelt's power and the lack of enough of the old air supporters in Congress, no cry rose up questioning the executive branch's leaving vacant an office created by statute. The economic explanation—and the lone MacArthur recommendation for saving War Department money—belied the true motivation for the recommendation: increasing the power of the General Staff and reducing the power of the Air Corps.

For years, the ground Army leaders had decried the excess amount of money (in their opinion) given to the flying branch, and the group resented the political power Davison wielded. Davison himself agreed that his old office became vacant not due to economy but because the General Staff wanted to regain control over Army aviation and "reassert their influence" throughout the War Department and "downrate air power." He staunchly brushed aside the economic explanation and declared that MacArthur and his deputy, Major General George Van Horn Moseley, disliked the fact that Army aviation had a civilian secretary and had worked to eliminate the office and its political power.[31] The *New York Times* agreed that the vacancy revealed a power play to eliminate the office it termed "a 'buffer block' between the services, Congress, and the public."[32] The actions taken by the General Staff after the decision provided insight on the group's push not to have it filled.

The General Staff revealed its true intentions in several studies and memoranda designed to divvy up the duties of the then empty air secretariat. In the view of the G-4 (Supply Division), the civilian air office infringed upon the statutory duties of the Assistant Secretary of War, the General Staff, and the Chief of the Air Corps. The G-4 concluded that the air position created "a War Department within the War Department."[33] The memo went on to accuse the "improper assignment of duties" to the air secretary as causing procurement problems. The study recommended that the General Staff take over the majority of the old air secretary's functions, especially procurement and budget planning. G-4 closed by giving the Air Corps notice that its semi-autonomous days were over: "The Air Corps cannot be made an exception to the policies governing the other Arms and Services in these matters."[34] Lower-ranking members of the Air Corps realized the loss of power and the significance of the air secretary to their position. As Major Carl Spaatz told Hap Arnold, "The War Department is hot on the trail of the Air Corps since the buffer has been removed."[35]

Only General Foulois' reaction was more surprising than the General Staff's attack on a congressionally created position. In replying to the General Staff's proposal, Foulois did not react vehemently to the elimination of the office or the reduction in power for the Air Corps. In fact, he supported the proposal. His only disagreement came with wording of the original memorandum that disparaged him for supposedly failing to submit a requirements study on time, and other minor items, which seemed to impeach his leadership and officers.[36] Even more surprising, Foulois correctly envisioned the trouble the Army might bring upon itself by working to eliminate the office of the air secretary, and he suggested improvements in language to assuage the military takeover of duties previously apportioned to the civilian staff. He believed such an aggressive action and language could bring congressional ire on the General Staff: "After all, the law still provides for an additional Assistant Secretary of War (Air) . . . [and] it is my understanding that the President has not, as yet, taken definite [action] toward the abolishment of the Office."[37]

To allay the fears of such an action, Foulois recommended altering the language to present the reorganization as "simply delegating *temporary*" control of the air duties to the other Assistant Secretary of War, Harry H. Woodring, and to himself as Chief of the Air Corps. Foulois also suggested issuing less-binding letters of instruction to make the changes, rather than the amendment of Army regulations, as suggested by the G-4. Foulois moved to strengthen his own position by recommending that Secretary Dern formally ask for the air chief's recommendations as to the reallocation of the nonstatutory duties of the air secretary, and that the air chief temporarily, pending a final decision on the abolition of the office, be given the responsibility for making recommendations to the Secretary of War.[38] Interestingly, Foulois did not mention the loss of the air secretary office at all in his memoirs. Obviously, he sought to increase his own power, but he also recognized the General Staff's power play and probably made the latter recommendations to ensure that the Air Corps received more than the General Staff when it came to the division of Davison's old duties.

Perhaps Foulois simply tried to make the best of a deteriorating situation, but he only assisted in weakening the Air Corps' position. His help in rewording the General Staff document aided in a virtual coup over the civilian air office so cherished by the aviators. No evidence exists that Foulois stood his ground and fought against MacArthur or Dern. Foulois also failed to highlight how the War Department virtually ignored U.S. statutes and left such a valuable post vacant. Instead of writing memoranda supporting the office and the need for it, Foulois fought only to ensure that he obtained some of the office's power—and he did not highlight the office's vacancy at all in his annual reports. The press had already pressured Roosevelt to fill the office. The *New York Times* reminded its readers of the history of the office's creation, congressional support for retaining the office it had created, the low cost, its usefulness over its seven-year existence, and how the unsettled world situation—more than at any other time since 1926—actually necessitated an air secretary. The article, entitled "Ask for Air Secretaries," closed, "It is felt by many who have studied the problem that true economy and efficiency in the air arms can best be served by the retention of such secretarial appointees who act not only as buffers between Congress and the Services, but frequently saved the services from well meant but restrictive administration by the general staff in the army and the general board in the navy."[39]

Foulois could have highlighted the need for an air secretary in his annual reports, or at least requested a meeting with Dern and Roosevelt to discuss the post's importance and the national and congressional support for the office. He would not have been inappropriately opposing the administration's wishes and going outside proper bounds, and perhaps he could have averted the General Staff's coup and ensured that the War Department obeyed Congress's wishes and laws. At the time, Roosevelt seemed undecided on filling the office. His delay allowed MacArthur and the General Staff to win the fight and thus weaken aviation's voice in national defense.

Foulois would come to regret the loss of a civilian advocate and buffer. Less than a year later, a member of his staff drafted a document (in case Foulois or his deputy, Oscar Westover, needed it) complaining about the loss of the air secretary. The statement, not known ever to have been used or distributed, called the loss of a civilian secretary "unsatisfactory," causing delays in procurement and "deplorable" efficiency. The document recommended an immediate appointment.[40] After his retirement, Foulois lamented, "had the Office of the Assistant Secretary of War for Air continued to function, it is further my sincere opinion, that the procurement of the 1800 airplanes authorized under the Act of July 2, 1926 would have continued to show upward progress."[41] Clearly, he understood the office's significance, but he made no real effort to retain the secretariat in 1933 or 1934, nor did he ever explain his assisting the General Staff actively and his inaction.

Neither FDR nor the War Department could formally abolish the air secretary office, but by July 1933 the *New York Times* reported that the president had decided to "abandon permanently" the post (by not filling it) and had transferred all duties to Assistant Secretary of War Woodring. Woodring announced that although he remained sympathetic to the aims of the Air Corps and wanted to continue the expansion program, he believed that the "moral and general fitness" of the air service outweighed buying new equipment.[42] At the end of fiscal year 1934, Dern mentioned the abandonment of the post and gave the reason as bringing the air service in line with other Army branches, reporting that the office remained unfilled "because the Air Corps, like the other branches of the Army, now functions directly under the Chief of Staff, to the mutual benefit of the Air Corps and the Army as a whole."[43] The most important part of the now empty office's duties concerned the recommendations for budgets and procurement, and all realized that with Roosevelt's economic plans, the situation of the last two Hoover years would only get tighter.

Roosevelt's primary focus was the economy. While Hoover did not want a large federally directed program, FDR wanted action, even though he did not really know what would work. Each attacked the problem in a different way, but Roosevelt gave the country a far different impression: he looked like a decisive leader who sought change and as someone who cared about the people.[44] The different approaches, however, still relegated military affairs to the background and provided military leaders with little hope for increased budgets for a rapidly deteriorating infrastructure and upgrades to weapons and other equipment.[45] In a quick departure from Hoover's policies, Roosevelt included the military in the 11 percent government pay reduction and signaled support for a smaller Navy.[46]

Military leaders and their political allies did not stop trying to gain what they thought they needed. They understood the need to support the president and help the country recover, but they were also obliged to ensure that the military would be able to carry out its orders and duties. Yet FDR showed he held tight the reins of the military and the budget, and when he could not use his "vaunted Roosevelt charm," which one historian ranked as the president's primary political asset, he applied the power of his office and his firm control over Congress.[47] Without an intermediary like Trubee Davison, the Air Corps began the Roosevelt era with little political protection within the War Department or between the service and the president. Even with an air enthusiast leading the House Military Affairs Committee, the Air Corps would make little gains, and Foulois' ineffectiveness did not help.

OUTSIDE HIS MILIEU: FOULOIS AND
THE POLITICS OF BEING THE AIR CHIEF

Foulois' position as Chief of the Air Corps was obviously not as secure as MacArthur's as Chief of Staff. Both had known Roosevelt since World War I because of his prominence in the Navy Department, but Foulois and Roosevelt had crossed swords in the immediate postwar fight for air independence, and the airman had aggressively opposed Roosevelt's negativism toward aviation and air independence.[48] Additionally, Foulois' recent requests for more Air Corps funding challenged Hoover's economy program, which seemed shallow compared to Roosevelt's, and signaled yet another contentious issue between the airman and his new commander. Just prior to the election, Foulois asked for more funds for Air Corps expansion. While announcing his understanding of tough economic times, he still stressed the need for defense and touted the air forces as America's first line of defense.[49] In an unpublished article, Foulois actually railed against the Democratic platform on defense. He believed the Democrats would not spend enough money and that, due to Roosevelt's background, the Navy would receive the lion's share. In Foulois' opinion, the Air Corps could better utilize the money and more efficiently protect the nation.[50] Even without the publication of this article, Roosevelt knew where Foulois stood given their histories.

Roosevelt kept the former firebrand as the head of the flying service, but Foulois did not sense the trouble ahead. He possessed neither the tact of Fechet and Patrick nor the charisma of MacArthur or Mitchell, nor did he have "top cover" protection, as the flyers called it, from an Assistant Secretary of War for Air. With the aviation allies in Congress whittled down due to death and election defeat, the Democrats in charge of the government, and the General Staff asserting more power over the entire Army's functions, Foulois was in trouble. Instead of being cooperative, moderate, and showing a willingness to work within the Army (as his two predecessors did so well), Foulois reverted to the more aggressive tactics reminiscent of his early years fighting for air independence. By his actions he significantly impeded, if not reversed, the progress of both intraservice and civil-military comity. Within months of the Democratic assumption of political power, Foulois' troubles began, and they would not abate for the next two years.

In March 1933, Foulois supported McSwain's new measure for a national department of defense and a coequal air service. McSwain actually fired three volleys for aviation. First, he ordered a hearing before his committee on the status of the Air Corps and summoned Fechet, Foulois, and Mitchell to testify. Then he introduced a bill to add one hundred pilots to the service and further refine the requirements for the selection of the Air Corps chief to ensure that a long-serving and qualified aviator held the slot. Finally, he reintroduced a bill (H.R. 4318) to create a unified department of defense with an air service equal in status to the Army and Navy. McSwain deftly allowed the last measure to enter the Committee on Expenditures in the Executive Departments, thus presenting the measure as affording economy; he still held hearings on the bill in the Military Affairs Committee, but with only two witnesses. The bill once again attempted to form a national defense department in which one secretary would run the department, with three assistant secretaries overseeing the three services.[51]

During the hearings in his committee, McSwain called only Foulois and Lieutenant Colonel James E. Chaney of the Air Corps to testify.[52] Instead of using moderation and supporting the structure as better for all services and good for the national defense, Foulois returned to his earlier rebellious ways—a disastrous approach that negated the gains the Air Corps had made in recent years. Foulois agreed that the Army and Navy could keep some inherent support and observation aviation, but he wanted the air service to control most of the combat aircraft. Reminiscent of his archenemy Billy Mitchell, Foulois also took the chance to strike a blow at the Navy. He asserted that the Air Corps had replaced the Navy as the nation's front line of defense and thus deserved priority for expenditures.[53] Foulois also appeared in the related hearings McSwain called to determine the status of the Air Corps, and one day after Mitchell appeared, Foulois' comments mirrored those of Mitchell over a decade earlier. Foulois asked the committee, "Why build battleships when they can be put out of commission by cheaper airplanes?" He also played to the nation's defensive mood and employed the tactic of the past decade of presenting air power as defensive. "We need no more battleships," he announced, "unless we intend to invade a foreign country."[54]

Foulois also used an old Mitchell tactic of presenting the aircraft situation in an unfavorable (and inaccurate) light. He told the committee that the service had only 924 serviceable combat planes. Only six months earlier, Davison had reported the number as 1,604 serviceable aircraft (1,041 of these bombers and fighters), with 67 more on order.[55] Foulois used the old rhetoric and methods to make the situation seem worse than it actually was in order to spur appropriations.[56] However, his bold statements not only kindled mistrust between these services but also demonstrated once again his lack of political savvy. Foulois asked for more appropriations in a time of economic difficulty, and at the expense of the Navy, with a president who in his first days in office openly wanted defense cuts and admired the sea service. Foulois thus differed with the president's policies less than a month after the inauguration.

The air chief's comments also caused problems within the War Department. Foulois argued for a program counter to presidential and War Department policies and forced the General Staff, already intent on consolidating power, to parry Air Corps moves and radical statements. He had also acted against the existing General Orders, as he did not clear his remarks with the War Department prior to testifying, and the War Department took immediate action. With Foulois having left Washington on other business, his assistant chief, Brigadier General Oscar Westover, remained to take the heat. Westover informed the General Staff that Foulois had received the call to testify before Congress only the previous night. Foulois had supposedly contacted MacArthur to inform him of the appearance. Perhaps sensing another attempt to circumvent the General Staff, Deputy Chief of Staff Drum distributed a memo and reiterated to all officers, "No presentation of evidence will be made without prior approval of the Chief of Staff."[57]

For a short time, the memo reignited the internal War Department debate concerning officers providing testimony to Congress. The General Staff wanted everyone to follow the "party line," since the Army enjoyed the full support of the president and the Secretary of War in matters regarding aviation and defense department reorganizations. The Air Corps wanted the opportunity at least to present an alternate view to Congress. If the air officers presented their views as personal opinions to allow Congress to make better decisions, and if they kept the War Department informed of their testimony beforehand, theoretically there should be no ire. Drum immediately informed Westover that the memo might have caused confusion. The deputy chief clarified: the directive was not to restrict any congressional testimony but merely to ensure that MacArthur approved the Plans Division's official position before that division testified. Drum's letter to Westover may have indeed been for clarification, or he may have realized his mistake and, remembering the mid-1920 charges of "muzzling" of air officers by congressmen, wanted to avoid those dangerous political arenas. No political fight emerged, as McSwain's defense bill never reached the House floor, and a contentious debate was averted for the time being.[58]

With the air independence legislation dead in committee, Foulois did not have a chance to follow up his attacks on the Navy and recommendations for department reorganization. In fact, he remained publicly silent on the issue. However, he continued to work to improve his relationships with several congressmen. In May 1933 he sent English translations of Giulio Douhet's article on air power to McSwain. He also wrote to California Republican Henry E. Barbour, after the congressman's 1932 election failure, to express his both his appreciation for Barbour's support of aviation while on the House

Appropriations Committee and his hope that Barbour would soon return to Congress.[59] Foulois also remained on good terms with Iowa Democrat Alfred C. Willford, but he rightly refused to accept a gift of Iowa bacon from the congressman.[60] While he did not accept this gift, in certain instances Foulois was fond of giving privileged insight. Most notably, he provided information to Mississippi Democrat and firebrand Ross A. Collins.

Collins supported aviation, and he reportedly admired Billy Mitchell; historians have documented his adversarial relationship with MacArthur, which probably counted as a motivation for supporting the aviators.[61] Foulois provided Collins with confidential information on the needs of the Air Corps and its importance in the national defense structure, and in April 1933 Collins publicly reasserted his support for separate air service. Whether his support derived from an honest belief in aviation's capabilities and promise, from Foulois' influence, or from a dislike for MacArthur and the General Staff— or (more likely) a combination of the three—cannot be known for certain.[62] But Collins and McSwain could not gather enough support for the aviation bills, nor could Foulois' political maneuvering assist them.

Foulois tried to play some of the political games, but his efforts could not match the contacts of airmen such as Mitchell and Arnold. Undeniably, Foulois' lack of deep political connections and influence came from not being the caliber of leader of those officers and from not being broadly respected among the air officers. He did not draw men to him and his ideas as did the charismatic air leaders Mitchell, Arnold, and Frank Andrews. Elwood R. Quesada, then a captain and one of the Secretary of War's pilots, later noted how Foulois did not have a dynamic personality and was not a forceful leader and that Air Corps officers "liked [him] more because he was decent than because he was effective."[63] Many years later, Davison called Foulois "a small minded man" deficient in both imagination and leadership.[64] Lacking the personality of other leaders, Foulois could not take advantage of party politics, as could earlier air leaders in the 1920s, due to the changing times and the new political dominance of congressional Democrats. During the previous three Republican presidencies, air chiefs understood—and played to—partisan politics. They could use black-sheep Republicans and the Democrats against the three Republican presidents. But with FDR in firm control, especially in the first year of his administration at a time of national economic crisis, and with a solid Democratic Congress, Foulois' options remained limited to McSwain and an occasional rebellious congressman such as Ross Collins or Frank James. Thus weakened, Foulois would endure many trials in the next two years. The ordeal began immediately with Roosevelt's economic program.

When Roosevelt first arrived in the White House, military leaders clearly understood that Hoover's economic measures would be light compared to the cuts the new executive would implement. This news struck the Air Corps hard. In January 1933 the Hoover measures had reduced the Air Corps budget by 19 percent, compared to 10 percent for the War Department overall and 9 percent for the entire government.[65] With the Navy Five-Year Plan nearing completion, the Air Corps remained short of the planned 1,800 aircraft by about 200, and it returned almost $400,000 to the Treasury in 1933 due to procurement curtailments.[66] During the first months of his administration, FDR, through the advice of Budget Director Douglas, proposed a further slashing of the military. Roosevelt wanted a reduction in the 1934 War Department budget of $90 million out of the already-approved $277 million. In the new plan, he reduced Air Corps spending from $26.3 million to $11.6 million, a 56 percent reduction that would have eliminated all planned aircraft purchases.[67] Roosevelt's other "economy measures" included furloughing officers at half pay, cutting veterans' pensions, and reducing all federal salaries (including military pay) 15 percent.

MacArthur, of course, fought these efforts. The *Army and Navy Journal* helped to arouse veterans and civic groups. In light of the events in Germany and Japan, even the nation's press seemed unwilling to support such a reduction. The *New York Times* and *Washington Post* both compared the move to reducing a city's police force during dangerous times, and other papers also condemned the cuts.[68] Dern, caught between protecting his department and the national defense and remaining loyal to his boss, confronted Roosevelt about the cuts, accompanied by MacArthur. The president resisted. Dern received a tongue-lashing; MacArthur then declared that when American men died in the next battle, he wanted the soldiers' dying words to condemn "Roosevelt," not "MacArthur." The president became livid. Realizing his indiscretion, MacArthur immediately offered his resignation, but FDR soothed over the situation as MacArthur and Dern rose to leave. MacArthur was so upset by the confrontation that he vomited on the White House steps.[69] Although MacArthur had momentarily challenged the president, he had done so in private. Roosevelt understood the general's passions and, given the circumstances, let the matter pass.

In the end the budget reduction exceeded $51 million, half of what Roosevelt originally planned. MacArthur helped save the Army, including the Air Corps, from deep cuts, but the cuts that occurred created even more intraservice tension. Foulois did not back off, still supported air independence, and pushed for additional aircraft purchases, thus inciting even more animosity from the General Staff. Rekindling the push for air independence broke the tenuous peace the airmen had established with the other branches of the Army and the General Staff during the Fechet and Davison years. Foulois showed no intention of accepting the status quo or reconciling his differences, and thus he further isolated the Air Corps.

At the heart of the Air Corps' plans was the development of a long-range heavy bomber, the key to fulfilling the prophecies of Mitchell and Douhet and the rationale for independence. In the early 1930s, the officers of the Air Corps Tactical School (ACTS) refined their ideas into a strategic bombardment doctrine. Although the doctrine remained ahead of capabilities, Air Corps officers would henceforth consider the heavy bomber their primary *raison d'être* and pinned their hopes of eventual independence on the bomber's fulfilling its potential.[70] In 1933 the Air Corps began a design competition for a multi-engine long-range bomber, and Woodring, very early into his tenure as Assistant Secretary of War, signed the development order for what would eventually become the Boeing B-17 Flying Fortress.[71] The airmen justified development of the heavy bomber to the War Department and the nation as a defensive and humane instrument of war—defensive in that it could protect the nation and its possessions (the Panama Canal, Hawaii, and the Philippines, primarily) from enemy attack, and humane because it could decide wars without trench warfare stalemates.[72] The development of the heavy bomber and the Air Corps' ideas for its use clashed with the War Department at a crucial time.

During this same period, MacArthur continued the implementation of his Army organizational reform program, the Four Army Plan. The reorganization provided a plan for defense of the United States and a framework for mobilization and expansion.[73] Reassessing the air needs for the Four Army Plan, the General Staff asked the Air Corps to submit recommendations on how the air arm would support those war plans, based on an air strength of 1,800 aircraft. In mid-July 1933, General Westover sent the completed study, which he entitled "Air Plan for the Defense of the United States," up the chain of command. The study diverged from the General Staff's guidelines and instead outlined autonomous air operations. Westover called for 4,459 aircraft and postulated that "the plan for the use of air power initially will bear little relation to the details of any of the existing . . . war plans . . . initially it is necessary to consider a plan for the phase in which air power is applied either alone or in conjunction with the Navy."[74] Predictably, this tactic caused a stir within the General Staff, especially in the War Plans Division.

In a seven-page memorandum, Kilbourne, Chief of the War Plans Division, rejected the plan as incorrect in its evidence and conclusions. Since Westover had prepared the report, Kilbourne asked Foulois to confirm his agreement with the plan, then admonished the air chief for submitting so faulty a document as to "cause grave doubt of your military judgment" and leaving the impression "that you are more concerned about securing an increase for your arm than you are about the national defense." Kilbourne closed by recommending to MacArthur that Foulois be ordered to refrain from advocating the air proposals in front of Congress, in the press, or in any publication, and to instruct all Air Corps schools to conform to published doctrines.[75] Foulois responded that although he was out of town during the report's preparation, he kept in constant contact with his office and remained "in thorough accord with the principles contained therein."[76] To resolve the dispute, Dern directed a board to work out the differences. General Drum chaired the board, which subsequently took his name, and Foulois represented the Air Corps.[77]

The Drum Board submitted its report in October 1933 and did not recommend any radical departures from previous War Department policies. As it had on many occasions since before World War I, the Army opposed air independence and asserted the necessity for unity of command.[78] The officers recommended a total air strength of 2,320 aircraft. This number represented an interpretation of the Air Corps Act's authorization of 1,800 serviceable aircraft by adding an excess number for aircraft undergoing maintenance and overhaul and providing for a 25 percent war reserve. However, the board noted that the Air Corps should not be allowed to procure more than 1,800 aircraft if the additional expense would impede funding of other branches and services. The one minor victory for the Air Corps came with the recommendation for the formation of a General Headquarters (GHQ) Air Force, which would unify all air forces under the control of an air officer for the purpose of supporting coastal defense and ground forces. The Drum Board avoided any interpretations that hinted at the independence of air operations.[79] The report, classified as secret and filed away in the War Department, did not reach Congress or the public until another investigation the next year revealed its findings. Nor were its recommendations immediately acted upon.[80]

Foulois' part in the Drum Board's proceedings remained clouded. His stated goals and desires for the Air Corps did not correspond to the board's findings, yet he did not file any dissenting report or mention the proceedings in his memoirs. Perhaps he knew the board's membership doomed any of his proposals and decided to go along and get what he could: the GHQ Air Force.[81] One year later, during a separate investigation of his activities on a variety of fronts, he informed Congress that the Drum Board acted unfairly toward him, did not allow him to participate fully, and used procedural rules to outmaneuver his attempts to make his views known. One board member, Major General George Simonds, when informed of Foulois' recollection of obstruction, told Congress, "I am astounded to hear that." Two other Drum Board members, Kilbourne and Major General John Gulick, emphatically contradicted Foulois' testimony, calling the latter's sworn statement a lie. Drum also countered the allegations that the board blocked Foulois.[82] During the Drum Board, Foulois probably went along with the General Staff, being outnumbered by nonflying Army officers. But he had now set himself up for a mighty fall. Now the Navy and the General Staff were his open enemies, and he had no real allies within the administration or the War Department (especially without an air secretary). His limited contacts in Congress could little affect the situation. It would get worse in the coming year, and Foulois dug himself a deeper hole, soured his only political contacts in Congress, and found himself besieged by Congress and the administration.

FOULOIS UNDER FIRE: LEGISLATION,
AIR MAIL, AND INVESTIGATIONS

The pivotal year of 1934 began with Foulois still in the good graces of the key air-minded congressmen, notably Collins and McSwain. McSwain ushered in the year calling for further studies of national defense, once again expressing his desire to create an independent air force. Even though members of the Military Affairs Committee may have been able to get McSwain's pending bill to the floor, the eventual success of any such legislation depended on Roosevelt's attitude on air independence, and the president remained against any such proposal.[83] McSwain himself admitted that he did not know the president's attitude toward air independence, but "I do expect the support of the common-sense people of the country." He continued by noting that the lack of appointment of Assistant Secretaries of War for Air for both services necessitated air independence, insinuating that those vacancies left aviators with little voice in national defense policy.[84]

In response to the pressure of McSwain's bold pro-aviation announcements, the War Department took action to appease aviators and their congressional supporters. Using the Drum Board's findings, the War Department submitted a bill (H.R. 7553) to create the GHQ Air Force. Dern called the changes necessary because recent advances in military aviation enhanced aviation's value to the Army and required a new organizational structure. McSwain countered the War Department by introducing his own bill (H.R. 7601) providing for some Air Corps autonomy by placing Army aviation directly under the Secretary of War, expanding the Air Corps to 4,832 aircraft, and stipulating a separate budget and promotion list.[85] Unknown at the time, Foulois' staff had drafted H.R. 7601 for McSwain at the congressman's request. Foulois later gave MacArthur the lame explanation that while his office prepared the bill "under the personal instructions of Mr. McSwain. . . . [n]one of the above bills have ever received my approval."[86] Foulois even tried to persuade the General Staff that he was unaware of the measure until McSwain introduced it in Congress. In what would soon become a controversial and failing tactic, Foulois would make unsubstantiated statements to cover earlier statements or lies. When caught in his lie regarding knowing about and supporting the bill, Foulois equivocated: he never saw the bill in its printed form.[87]

In fact, Foulois had ordered Lieutenant Colonel Chaney to prepare the bill in concert with other officers in the Office of the Chief of the Air Corps. Foulois also ordered a committee of at least seven air officers to draft Mc-Swain's other bill, H.R. 7872 (a bill to resolve the promotion system that discriminated against airmen). Evidence existed that Foulois changed the wording of the bills to make him eligible for reappointment after his term expired. The air chief also directed the preparation of a third bill, which went further toward air independence than either of the other two, and he prevented the members of his drafting committee from working on other important Air Corps tasks so that they might finish before Congress adjourned. Chaney presented this bill to McSwain personally, in the presence of another officer (Colonel W. R. Weaver), along with a statement by Foulois approving their product.[88] An Air Corps officer also later revealed that several other congressmen knew of Foulois' activities, as well as defense-interested citizens and lobbying groups. Foulois appointed Major Follett Bradley as the liaison officer between the Air Corps and the Air Defense League, a civilian lobbying organization. Bradley furnished the league with inside information on the bill, and any other assistance the league wanted, in order to lobby for Air Corps desires.[89] Foulois' lies would betray him during congressional and Inspector General investigations, but in January 1934 his coordination with McSwain remained secret, and Foulois testified in firebrand style during the hearings on H.R. 7601, denouncing the General Staff and pleading for independence. His antics infuriated the General Staff and caused Secretary Dern to enter the fray, much as Secretaries Weeks and Davis had done during Mitchell's heyday.

Foulois incensed the War Department, and the General Staff felt betrayed. Although MacArthur wanted to retain Army control over aviation, and requested more money for the Army overall, he still supported Air Corps expansion and funding to a point. The General Staff, which thought its support of a GHQ Air Force and the requests for a new Five-Year Plan demonstrated support for the air arm, felt deceived by Foulois. General Kilbourne led the charge.

Kilbourne wrote letters to congressmen, composed memoranda for the General Staff and MacArthur, and conducted a rearguard action to counter Foulois' actions, especially the air chief's assertions of constant harassment and obstructionism by the General Staff. Kilbourne spearheaded efforts to defeat any pro-aviation legislation and often wrote the War Department's responses and position papers.[90] He also wrote information for release to the public to defend the General Staff, and even asked congressmen to release it in their names.[91] In the wake of the pro-aviation bills introduced in early 1934, Kilbourne wrote letters (under the authority of Drum) to every member of the House and Senate Military Affairs Committees and offered his services to discuss his ideas on the matter and to counter the claims of the aviators. He mentioned the aviation legislation and asserted that "air power has been accentuated by presentation of unproven claims of effectiveness and by press articles and public addresses prepared by those whose enthusiasm has led them to make recommendations without first examining all factors."[92] In preparation for the upcoming hearings, Kilbourne, also under Drum's direction, ordered all elements of the General Staff to make a thorough report on all of their contacts and work on aviation subjects over the previous four years, and especially on their work with aviation since Davison's departure. This information, he stated, would be used to support the General Staff and the Secretary of War to show how aviation officers and programs had received fair, if not preferential, treatment.[93]

Kilbourne appropriately bristled under Foulois' attacks, the first from such a high-ranking air officer since Mitchell in 1925. However, the General Staff had no reason to throw stones. The Air Corps had received budgets exceeding those of other branches, but that fact merely reflected the costs of its war implements. The General Staff sought not necessarily to reduce the Air Corps budget and counter air expansion plans but rather to obtain money to improve the more neglected parts of the Army.[94] The air service still remained behind the Five-Year Plan/Morrow Board number eight years later, and some National Guard aircraft still used World War I Liberty motors.[95] Additionally, under the immediate cuts by the Roosevelt administration, the Air Corps took the highest-percentage cut in the Army. The General Staff remained rightly concerned over the deplorable state of the rest of the Army, but Kilbourne's reaction to Foulois' inappropriate actions and comments only made a bad situation worse. Kilbourne should have left the matter entirely to Dern, since the secretary did not ignore Foulois and the Air Corps' activities, but took the case immediately to Congress and the president.

In a long, tersely worded letter to McSwain, Dern chided the chairman for allegedly agreeing to support the War Department measure (and the GHQ Air Force) and then changing positions. Dern upheld the War Department's goal as being "devoted entirely to the constructive problem of increasing" the Air Corps, and he alluded to McSwain's support of air autonomy as a contentious issue of "patronage."[96] Dern also cautioned McSwain that air autonomy, combined with the influence of powerful civilians in the aircraft industry, could create privilege, favoritism, and "a specialized officer corps de elite," which could destroy the Army's effectiveness.[97] In a power play of his own, Dern informed McSwain that he would impede congressional progress on any Air Corps bill, including his own. "To these two measures," he wrote, "I am unalterably opposed—opposed to such an extent that I will not attempt to advance the constructive thought involved in the simple increase of the Air Corps, if it is your intention to couple it with these other issues."[98] Dern attached to his letter an eighteen-page statement to the entire Military Affairs

Committee and sent copies of both to the president.[99] By the first week in March the press reported on Dern's statement, and dirty laundry involving the Air Corps was once again spread out before the country.[100]

Dern's statement repeated the War Department's traditional arguments for proper balance, unity of command, and the indecisiveness of the airplane in war. In a swipe against the press and Air Corps propaganda, Dern noted that "due to an unremitting, though distorted publicity, many Americans are predisposed to the belief that the airplane will dominate future war," a notion he dismissed as "romantic."[101] Further, he attacked the press's support for aviation, calling the study of war and reliance on the airplane "[not] so simple as might appear to the layman after perusal of the front page of his newspaper."[102] In a thinly veiled rebuttal to Foulois' dismissal of the Navy and the Air Corps' assertions of defending the coasts, Dern observed that only a "zealot" would agree that air power alone could defend the coasts.[103] He conceded the value of Army aviation, but only as an integral part of the Army and when controlled through proper channels (the General Staff).

Dern's represented the most complete and strongly worded statement by a Secretary of War to congressional air supporters and, by extension, the Air Corps leadership, in ten years. The secretary strongly denounced the tactics aviators and their supporters used, firmly backed the existing War Department programs and proposals, and threatened to stonewall Congress unless it backed the administration's proposals. McSwain, obviously upset by Dern's statement, refused to drop the autonomy bills and vowed to fight the War Department.[104] However, the coming air mail and procurement controversies overtook those bills, and once again, as in 1925 with the *Shenandoah* disaster, air deaths would cause a full reassessment of the Air Corps.

The development of post office routes and air mail contracts had received intense scrutiny, including investigations by three different government departments, at varying times, since Hoover appointed Walter F. Brown as Postmaster General in 1929. Brown radically restructured the way the government awarded contracts and paid airlines to carry the nation's mail. These investigations intensified when the Democrats took control of the government; a Senate investigation, headed by Alabama Democrat Hugh L. Black, took on a decidedly partisan air. The Democrats wanted to find wrongdoing and indict "responsible" Republicans for a system they believed was monopolistic, one in which big companies profited and then filled Republican coffers.

With these investigations still incomplete, Roosevelt had seen enough by February 1934 and wanted action. The Senate moved to charge Hoover's Secretary of Commerce, William P. MacCracken Jr., with contempt for not cooperating with their investigation. The Justice Department wanted to prosecute Postmaster Brown, and Roosevelt reportedly wanted to blame all of the air mail problems on the Hoover administration and show collusion and fraud, thus allowing him to cancel the contracts. Roosevelt's cabinet did not want an immediate cancellation, so FDR promised to contemplate his actions and respond to the cabinet. Roosevelt never resubmitted the question to his cabinet. Instead, two days later, he decided to cancel the air mail contracts and turn over delivery of the air mail to the Army Air Corps.[105] Foulois' positive assurances to Roosevelt's staff—assurances given without proper study and consultation—influenced the president to take the bold action.

Roosevelt did not contact Foulois directly. Instead, the morning after Roosevelt conferred with Postmaster General James A. Farley (who gave FDR information on the ongoing Post Office investigation), the president sent Second Assistant Postmaster General Harllee Branch to see Foulois. Roosevelt already knew he wanted the Air Corps to carry the mail, but he wanted Foulois' assurances that the service could do the job. Branch asked Foulois whether the Air Corps could carry the mail, to which the latter confidently replied, "Yes, sir. If you want us to carry the mail, we'll do it." Foulois believed the service could ready itself within nine days for the project. Roosevelt, immediately informed of Foulois' reply, canceled the air mail contracts and issued an executive order for the Air Corps to do the job, beginning 19 February. Neither Roosevelt nor Foulois consulted with MacArthur (the former's military adviser and the latter's boss).[106]

Foulois took the job because he knew the pressure to do so came from the president himself and because he saw the chance to increase the Air Corps' visibility, and thus an opportunity for positive publicity, larger budgets, and an expanded force structure. Prior to carrying the first bag, Foulois testified to Congress that "this operation is going to be a great benefit to our pilots and personnel" and an aid in building an organization that could react to emergencies.[107] In his zeal to better the Air Corps' position, he failed to examine the matter completely. Foulois did not contemplate or warn his superiors or Congress that pilot crashes and deaths (inevitable in such a project) could follow.

The Air Corps lacked the equipment necessary to carry the mail in the same capacity and with the same degree of safety as the civilian carriers. Even with all his experience, Foulois never informed others that the Army planes lacked enough up-to-date navigation instruments (especially essential due to the amount of night flying required) or that civilians flying the mail had very different piloting skills than Air Corps pilots training for combat.[108] One insightful congressman, Author Lamneck of Ohio, asked Foulois about the types of Air Corps planes and if they had open or closed cockpits. Foulois admitted that the planes carrying the mail (which would operate at night, in winter, and in northern areas at high altitude) had open cockpits but that his pilots were acclimated to such conditions and would have no problems.[109]

He also misled the congressmen by assuring them that he had assigned his most experienced pilots to fly the mail. Instead, the majority of pilots (140 of approximately 262) possessed less than two years' experience, and many were one-year Reserve pilots. More telling, their readiness in night flying (only 31 pilots had more than fifty hours' experience) and weather flying (214 had less than twenty-five hours' experience or simulated training) showed quite the opposite of Foulois' assertions. [110]

The pilots and staff officers realized that the Air Corps did not have the equipment, yet Foulois pressed ahead. [111] He accepted the "calculated political risk" in the hopes of advancing the Air Corps (and perhaps his own reputation and position). [112] Before the Air Corps delivered one bag of mail, three pilots died preparing for the missions, the victims of inclement weather and inadequate navigation equipment. The respected aviator Eddie Rickenbacker, a World War I ace who was preparing to fly the last commercial air mail flight, called the deaths "legalized murder." [113] Foulois assured Dern that those accidents required no policy changes, but the Air Corps rushed to install navigational equipment, and Foulois urged his flyers to be especially cautious in the early flights. [114]

The "legalized murder" claim emerged again from a Republican after the death toll rose to six officers by 25 February 1934. Representative Clarence J. McLeod of Michigan called for an immediate investigation into the carrying of the air mail by aircraft reportedly ill-equipped to the task. The Army disputed that faulty equipment caused the deaths and pointed to mechanical causes and the extremely bad weather. Democrat and pro–Air Corps congressman Joseph Hill called the "legalized murder" talk "a lot of political claptrap." [115] With a little better weather, a Foulois speech publicizing the safety record, and some accident-free deliveries, the press coverage improved and congressional ire waned slightly. The short-lived respite ended on 9 March, when four more airmen died. With the problems again front-page news, Roosevelt called a halt to the air deliveries, exclaiming, "the continuation of deaths in the Army Air Corps must stop." [116] Foulois set in place new restrictions and safety measures, but the political damage had been done and FDR began looking for scapegoats.

For the first time, public dissention appeared in the Roosevelt administration. In the words of Arthur Schlesinger, the air mail fiasco "dented the myth of Roosevelt's invulnerability." [117] Republicans predictably offered the more impassioned calls for explanations, as they finally saw an issue in front of the American people to use against the Roosevelt administration. With fissures appearing in the New Deal, Republicans could also exact some revenge for their business leader constituency, especially those affected by the cancellation of contracts. Republican Hamilton Fish of New York led the early charge on the House floor, and Harold McGugin, from Woodring's own Kansas, called for a nonpartisan review of the "cloud" over the War Department. McGugin defended the assistant secretary, and hoped the investigation would uncover the truth, but without partisan character assassination.[118]

Joining McSwain, congressional Democrats also questioned the handling of the air mail situation—none more prominent than Speaker of the House Henry T. Rainey of Illinois. Rainey questioned the Air Corps' readiness and training, but he also took a jab at the Republicans and attempted to deflect criticism away from Roosevelt. Rainey called the situation a mess left by three previous Republican administrations for Democrats to clean up.[119] According to insider John Callan O'Laughlin, "The administration realizes it made a terrible mistake, and is anxious to correct it while at the same time saving the President's face."[120]

Roosevelt felt let down by his military staff. In his letter to Dern, FDR noted that he assigned the Air Corps the mission after receiving what he called "definite assurance given me that the Army Air Corps could carry the mail."[121] The president then called MacArthur and Foulois to the White House. Roosevelt may have received Foulois' assurances of being able to fly the mail through others, but he took out his frustration in person. The two officers entered the president's bedroom, and MacArthur introduced Foulois. In these circumstances, after a year in office, Foulois finally met his commander in chief.

Roosevelt demanded to know when the deaths would stop, and Foulois responded with more frankness than he had while sitting before Congress a month earlier: "Only when airplanes stop flying, Mr. President."[122] Foulois admitted to receiving the worst "tongue-lashing" of his career, but it obviously failed to change his attitude. The career military man viewed the president as "one of those politicians who did not understand air power, airplanes, or any of the problems of flying and apparently did not want to learn."[123] He did not grasp his own failure to inform the president, when asked beforehand, of the inherent dangers and probability of accidents and fatalities. A total of twelve officers died during the air mail preparations and operations. After Foulois' meeting with the president and a mission stand-down, the Air Corps implemented different procedures, added more instrument-equipped aircraft, and reduced the number of routes. They also benefited from better weather. The Air Corps delivered its last bag on 1 June, and only two men died after Roosevelt's scolding of Foulois and the related changes. According to Foulois, MacArthur, who understood politics much better than the airman, encouraged Foulois to be ready for upcoming hearings: "The Republicans will give us the chance to fire a few rounds for effect. Make sure you have your ammunition ready when you get the chance to fire [on the president]."[124] Foulois, however, would be on the receiving end.

According to O'Laughlin, a decidedly pro-military Republican, Roosevelt had also attempted to get one of his officers to take the fall. O'Laughlin reported to Hoover that the president called the War Department the day after his berating of Foulois and MacArthur and asked the latter if he remembered giving the president assurances the Army could fly the mail. MacArthur supposedly replied, "Mr. President, I dislike intensely saying what I am going to say to you to your face, but I never telephoned you. I knew nothing about your plan to have the Air Corps carry the mails." Roosevelt persisted, "But you are mistaken, Douglas. You phoned me as I have said." MacArthur again brushed aside the president's suggestion, and the president put his secretary, Marvin McIntyre, on the phone to urge MacArthur that such a conversation took place. McIntyre reportedly told MacArthur that another secretary, Early, had taken the call. When MacArthur failed to take responsibility, Roosevelt supposedly switched to pressure Foulois to admit the "assurances." Exposing his partisan leanings, O'Laughlin called the event confirmation of his beliefs that Roosevelt bent to public opinion and that he would unload blame for unpopular decisions onto subordinates. Nothing else could confirm the telephonic confrontation, but O'Laughlin's belief that someone would take the fall came true. [125]

The air mail disaster coincided with the beginning of congressional interest in the misuse of funds for the Air Corps and alleged illegal procedures for purchasing aircraft. As the airmen readied to fly the mail, Foulois testified before the House Appropriations Committee on the 1935 budget. In circuitous questioning, initiated by queries concerning $7.5 million of Public Works Administration (PWA) funds provided for the Air Corps, congressmen realized that the Air Corps still used negotiated contracts with aircraft manufacturers as the primary means of procurement. Although section 10 of the Air Corps Act required competitive bidding, unclear language also allowed leeway to buy certain items, if it best served the government, at a negotiated price. [126] However, the clear intent of the law was to foster competition, and Air Corps leaders understood this. But they believed the negotiation system gave them better aircraft, and they had used that method since 1926. Additionally, an Adjutant General ruling in 1927 offered them the loophole of buying with negotiated contracts. [127]

Assistant Secretary of War Woodring, who remained in charge of procurement by statutes, wanted to use the competitive bidding with the PWA allotment for purchases of aircraft. The Air Corps wanted to pick the specific aircraft, with the performance requirements they desired, and negotiate with the manufacturer. According to Foulois' testimony, Woodring's insistence on competitive bidding would result in inferior aircraft. [128] Woodring actually ordered Foulois to stop negotiated bids, hurting the Air Corps by slowing the receipt of the aircraft and possibly obtaining aircraft inferior in performance. [129] These revelations coincided with Senator Black's investigations on government collusion and other ongoing investigations into government wrongdoing in sales of other military equipment and surplus. The swirl of events only angered Congress more, and Foulois lost more congressional allies.

Chairman McSwain, a proponent of competitive bidding, became especially incensed upon learning of the reliance on negotiated bids. He dropped his earlier pro–Air Corps bills and launched an investigation into Air Corps and War Department procurement practices. Other committee members, especially Frank James, also became upset by the revelations.[130] The initial investigation conducted by the Military Affairs Committee only confused matters more. Foulois spoke at times in meandering fashion and often contradicted himself. Woodring stuck by his decision to move to competitive bidding, but he also agreed that he had earlier authorized negotiated contracts, as the Air Corps had operated that way for the past few years. The only lucid and clear explanation came from the Air Corps' Chief of Material Command at Dayton, Brigadier General Henry C. Pratt, who often cued Foulois to correct or amend his comments to reflect the Air Corps' actions and beliefs.[131] Operating under the strain of the air mail operations, budget testimony, and now separate investigations, Foulois began to make mistakes or to offer intentionally vague testimony. McSwain wanted a broader investigation into many different aspects of War Department procurement and financial practices, but the investigation's primary focus remained the Air Corps purchasing system. The House approved the investigation with bipartisan support.[132]

The air procurement part of the investigation came before the Military Affairs Subcommittee on Aviation, known as the Rogers Subcommittee for its chairman, New Hampshire Democrat William N. Rogers.[133] The subcommittee wanted a scapegoat to present the American people (and Roosevelt) to placate the problems exposed by the air mail problems and to explain how millions of dollars spent had not resulted in a viable Air Corps. Due to Foulois' testimony before the House Appropriations Committee and the subcommittee's predisposition to support aviation, the Rogers Subcommittee initially focused on Woodring.

Assistant Secretary Woodring faced the Rogers group on the first day of testimony (7 March 1934), and the pro-air committee attacked him constantly. Woodring admitted to wanting to change the procurement to more closely follow the Air Corps Act, but he denied responsibility or intent to lower aircraft performance characteristics. He patiently withstood the attacks and tried to sidestep the subcommittee's finger of guilt. Woodring placed the blame upon the Air Corps, since the service's practices predated his recent arrival in the War Department. Woodring admitted that he did not support competitive bidding in all instances but that he tried to convert procurement over to this proper policy.[134] Foulois then made a mistake that would focus the subcommittee's attention upon the Air Corps Chief for the remainder of the investigation.

The subcommittee's questions to Woodring turned to Foulois' accusation that Woodring had forced the Air Corps to change procurement methods and thus obtain, in some instances, lower-quality aircraft. Three weeks after Foulois had made the comments (on 14 February), the *Washington Post* published extracts.[135] Suddenly Foulois, perhaps feeling the heat and wanting to get in Woodring's good graces, placed himself in the subcommittee's bull's-eye and sent the Assistant Secretary an apology. Woodring then brandished that message (which he claimed to have just received), in which Foulois admitted that the paper's comments took him grossly out of context and distorted his testimony. To the subcommittee, and to anyone examining the statements, it seemed as though Foulois had changed his testimony, or perhaps recanted. A now infuriated group of congressmen voted to call Foulois immediately to testify and sent a clerk to find him.

Rogers led Foulois through his previous testimony point by point and sought new and clarified answers. Foulois seemed even more confused and rattled, but he called his previous testimony a misinterpretation and misunderstanding of Collins's questions before the House Appropriations Committee. Now he agreed with Woodring and absolved the Assistant Secretary of lowering aircraft standards. Knowing the fury that would come following these revelations, the subcommittee closed the hearings to the public on the first day of testimony.[136] Foulois steamed at Rogers's prosecutorial tone and the comparing of the chief's past testimony to his current assertions.[137] Foulois' actions so upset the committee that even longtime aviation proponent Frank James showed wrath.

James took Foulois to task for ignoring the will of Congress and violating the Air Corps Act. When Foulois remarked that the Air Corps had continued to use negotiated bids since 1925, "as Mr. James undoubtedly knows," James interrupted, "No; I did not know that. I had not the slightest idea that went on. I do not think a man on this Committee knew it was going on, or we would have had hearings a long time ago and found out why the act of Congress in 1926 was being ignored."[138]

McSwain had provided Rogers with a Committee Print of Foulois' statements, given in secret executive session on 1 February 1934, after promising Foulois, in front of the full Military Affairs Committee, that the remarks would remain "absolutely protected" and considered a personal opinion. The remarks allowed Rogers to compare notes on other Foulois testimony, exposing the air chief's blasting of the War Department and the General Staff.[139] Foulois, by his own words and actions, lost the trust and backing of air-friendly congressmen—even in the committee most dedicated to supporting the Air Corps. At one point the subcommittee considered pressing not only for Foulois' court-martial but also for that of a dozen of his assistants, as well as the civil prosecution of F. Trubee Davison for collusion and fraud.[140]

The Rogers Subcommittee did not issue its preliminary report until 1 May, but in the meantime congressmen began to back away from the political hot potato of the Air Corps. Many began to doubt the legitimacy of all previous Air Corps claims, especially those of a War Department and a General Staff subverting air needs and endangering the national defense. Committee members began to report that they no longer supported any legislation for air independence or autonomy. "These members," the *Army and Navy Journal* reported, "include many who were champions of separation, and who held the belief that the Air Corps was being ham-strung by the rest of the Army."[141] Representative Goss, an opponent of independence, reported, "The military committee, you know, has always been very pro–Air Corps. Now, however, I believe that any such proposal . . . would not have a chance. Many of the members who have been sitting in on the Rogers subcommittee have completely turned around in the matter."[142] The final report would further damage the Air Corps and lead to the end of Foulois' long career.

Issued 15 June 1934, the Rogers Subcommittee's final report accused Foulois on a number of counts. On procurement, it charged "deliberate, willful, and intentional violations of law by the Chief of the Air Corps, aided and abetted by his assistants in charge of procurement."[143] The report presented Foulois' testimony in column format, showing alterations that indicated an intent to deceive. On the air mail, they called his actions "a glaring example of mismanagement and inefficiency" in planning, preparation, and execution. They especially berated Foulois for not seeking any advice or input from his officers, in particular his assistant, General Westover, whom he put in charge of the overall operation.[144] In closing, the subcommittee assured Congress and the nation of respect for the rank and file of the Air Corps but unanimously recommended Foulois' removal.[145] Foulois lost the battle with Congress, and specifically those in Congress who favored the service, and he measurably hindered relations with the Air Corps' most vocal supporters in government at a time when budgets were strained and the service was losing personnel and could not speedily promote those who stayed. The Air Corps needed political support to rectify problems in the service at a crucial time, but Foulois' actions caused all legislation to stop for an inquiry into his wrongdoings.

Yet, despite losing Congress, at this stage Foulois still retained the support of the press. Much as with Mitchell in 1925, it seemed that as long as Foulois played the "underdog" role, did not make open statements denouncing the administration or Congress, and avoided being found guilty by a military court, he could keep his press support and thus affect American popular opinion. The *New York Times* supported him and the air service overall during the air mail crisis, as did other papers. Most newspapers portrayed the Air Corps as gallantly performing a mission against bad weather in spite of outdated equipment and inadequate training, both the product of miserly budgets.[146] As in the Mitchell days, the Hearst syndicate backed the aviators. Hearst not only supported Foulois initially but also disparaged McSwain.

McSwain and Hearst both supported aviation and independence, but from different angles. McSwain vowed to support measures that he believed would aid the national defense, while Hearst and his papers usually supported aviation and backed statements of the radicals, like Mitchell and Foulois, and took the air insurgency causes to the public. Mitchell had played to this support, but McSwain did not like sensationalist journalism even in the cause of aviation, because such an approach naturally derided the other elements of the military. After McSwain asked for a congressional investigation, Hearst's newspapers ran stories implicating the Democrats for politicizing the air mail situation to attack the Air Corps and Foulois. McSwain immediately sent Hearst a telegram, followed by a long letter in which he recounted his own many years of support for aviation and asked Hearst not to make judgments before reviewing all of the facts.[147] McSwain's reasoning did not shake Hearst's commitment to the more radical positions and the publisher's desire for air independence legislation. Despite McSwain's letter, Hearst's papers soon criticized the chairman for supporting the Foulois investigations. Hearst publications specifically fingered McSwain for holding up air legislation and "squandering the time of his committee in various and sundry investigations for political purposes."[148] The Hearst syndicate remained firmly in the corner of the airmen.

Once the Rogers Subcommittee announced its findings, opinions began to diverge. Some newspapers called Foulois a scapegoat for the Army and the administration, believing that he alone could not bear all responsibility. Other publications derided the subcommittee's legal handling of the matter and for not allowing Foulois to present a proper defense. The *Christian Science Monitor* astutely noted how the Rogers group came to a unanimous decision, even with three different political parties represented.[149] Soon, however, a sampling of newspapers demonstrated a consensus that Foulois should retire early. Some editors agreed that he should have his day in a more formal and legally appropriate venue (such as a military court of inquiry), but many agreed that the stain of impropriety could harm the service. For the sake of the Air Corps, Foulois should step down and then fight his battle.[150] None mentioned a more harmful loss—the loss of prestige, power, and support by a committee heretofore dedicated to Air Corps growth and autonomy. Perhaps for that reason alone, Foulois should have stepped down and let the Air Corps repair its relationship with both Congress and the administration.

Meanwhile, Foulois tried to save his career and reputation. He released statements to newspapers and service journals, calling the subcommittee's actions illegal and without authority, as he could not defend himself. He challenged the group to allow him to appear "in open court," and he defended his procurement actions as in the best interest of the service and nation.[151] Foulois also attacked McSwain and the Military Affairs Committee for releasing his February testimony after McSwain had assured Foulois that it would not be printed. Foulois called the release of the remarks unethical and the members of the committee untrustworthy.[152]

Foulois also took McSwain to task personally and began a battle to have the secret transcripts and "evidence" from the subcommittee released so he could prepare a defense.[153] The subcommittee became very stubborn toward Foulois and his request for records. McSwain would not help, and he told Foulois that he must go directly to Rogers. Rogers responded that Foulois could see some of the records, with advance written notice, but many related to "matters vital to national defense and much of it is not germane to the charges made against you, and is not open to public inspection at this time."[154] Foulois also wrote to McSwain and provided copies to the members of the Military Affairs Committee, expressing outrage that they printed his testimony and then did not help him obtain materials needed for his defense against the charges. Most committee members backed McSwain, and only Illinois Democrat Chester Thompson replied that he supported the release of records.[155] The committee's replies demonstrated the loss of congressional support for Foulois and for his career. Yet in a stunning reversal of the political support for Foulois, Secretary of War Dern came to his defense.

Dern would not agree to the subcommittee's request for Foulois' removal, and he asserted the air chief's right to a military court of inquiry. After Rogers wrote directly to Roosevelt and requested Foulois' removal, Dern took up the airman's defense not on the specific charges but his on legal rights for a fair hearing and proper defense.[156] Dern criticized Rogers for the subcommittee's acting as judge, jury, and executioner, saying, "The report is not limited to an indictment, but in effect finds the accused guilty, fixes the sentence, and calls on the Secretary of War to execute it."[157]

The Secretary of War did not fight for Foulois' retention as such but only for a fair trial. Many papers lauded Dern's fairness, and one, the *Boston Transcript,* called his approach "admirable diplomacy" with Congress.[158] Perhaps because of the heat of the air mail problems and Foulois' failure to take full responsibility, the administration's backing of Foulois (in the person of Dern) did not extend any further. The investigation and controversy relieved the political pressure and press coverage of the administration's air mail problems, which had since faded with restoration of contracts, undoubtedly to Roosevelt's relief. Roosevelt also appreciated the subcommittee's placement of responsibility for the Air Corps deaths squarely on Foulois, not on the administration.[159] It was the Air Corps that was wounded far worse than the administration, which actually benefited politically from the Rogers investigation.

Dern eventually began an internal inquiry, ordering the Inspector General to undertake a full and fast probe into Foulois' actions and statements of the previous spring. This investigation, and the subcommittee's agreement to turn over records, came after MacArthur negotiated an internal investigation in exchange for the records.[160] Colonel Walter Reed led the inquiry, which began with a review of all the subcommittee records and then further interviews.[161] Reed re-interviewed the four General Staff officers who testified against Foulois in the subcommittee, and they further stated their claims of his unfitness for duty as Chief of the Air Corps. The generals did not believe Foulois acted illegally in his procurement activities, but they did not like his statements against the General Staff or his handling of the air mail crisis. They asserted that Foulois' actions demonstrated his inability to execute a high and important command.[162] Due to his statements against the General Staff and the long-standing animosity between that group and the Air Corps, those assertions surprised no one. The statements of other Air Corps officers, however, provided more insight into the service's view of their leader.

Brigadier General James E. Chaney, who worked in Foulois' office, did not think the chief violated any procurement orders or mismanaged the Air Corps. Chaney believed the Air Corps did the best it could flying the mail, but he agreed that Foulois probably erred by not studying the matter further and talking to his staff.[163] Chaney consented that Foulois probably made errors in judgment and statements that seemed misleading. Of interest, Chaney did note that the presence of an Assistant Secretary of War for Air would probably have avoided the procurement confusion or taken the political brunt of the investigations, thus avoiding Foulois' problems.[164] Westover and Pratt agreed, and so probably did a majority of Air Corps officers. Westover, a voice of moderation in the same vein as Patrick, went even further, criticizing Foulois for unsubstantiated assertions and statements deriding the General Staff.[165] Reed also interviewed Foulois, who did not offer any substantially different evidence and only tried to clarify previous statements. Foulois did give Reed a compilation of solicited character references from a variety of congressmen, including Speaker of the House Joseph Byrns, a Tennessee Democrat, and Senator Sheppard, chairman of the Senate Military Affairs Committee. Even Trubee Davison provided the beleaguered Air Corps chief only a lukewarm letter affirming Foulois' loyalty and technical competence.[166] None of the letters defended Foulois' remarks or misrepresentations. In the end, no one provided any political weight to swing the case in Foulois' direction.

The entire affair lasted well into 1935. Reed did not conclude his investigation speedily, and Congress gave the Inspector General a 1 May 1935 deadline. If Reed failed to submit a report by then, Congress promised to stall all pending War Department legislation. Dern sent a partial report to placate Congress in April, and then a full report on 14 June.[167] The evidence showed that Foulois had not violated any procurement laws, but he made "incorrect, unfair, and misleading statements to a congressional committee," which "violated the ethics and standards of the military service."[168] Dern and MacArthur reprimanded Foulois—what the accused and his congressional nemesis Rogers called a "slap on the wrist."[169] Rogers also called the action a virtual acquittal and demanded to carry on the fight. In a reversal of earlier times, the Military Affairs Committee held up aviation and other War Department legislation, including the budget, until they gained satisfaction from the War Department. Foulois, casting an eye toward Congress and another toward the growing threat of war in Europe, resigned for the good of the Air Corps.[170]

As it had ten years before, the air service bid farewell to a leader who probably knew the most about Army aviation and all its intricacies, from war to budget to procurement to personnel. Each leader, however, failed to temper his desire for air independence with an understanding of the political and public landscape and, more importantly, failed to work within the programs of the presidents who drove those policies. Roosevelt, like Harding and Coolidge before him, did not want an independent air service and did not like high-ranking air officers pushing for this end against presidential wishes. Mitchell had the advantage of political clout and strong partisanship in Congress over the issue. Foulois' political connections came not from his clout or public abilities but only because his wants coincided with those of some influential congressmen. When Foulois fell out of favor, through his own words and hasty actions, no political support came to save him or soften his fall. However, as happened with Mitchell ten years earlier, Foulois' problems and publicity highlighted the need for changes in the Air Corps and spurred the War Department and the civilians to act. It would take the return of moderation to the Office of the Chief of the Air Corps to reinvigorate political support and advance the changes in the air arm, but the controversy did create a favorable atmosphere for improvements to the Army Air Corps.

POLITICAL FALLOUT AND A NEW BEGINNING

Like Coolidge nine years earlier, Roosevelt created a civilian board to calm the controversies involving the Air Corps. Seeing the muddle created by the air mail contract cancellation and the threats of a political quagmire stirred by reinvigorated Republicans, FDR moved to placate all parties with a study of the Air Corps situation. Dern and Roosevelt decided on another air board in March 1934, during the height of the air mail controversy. Roosevelt, who also formed a civil board (the Howell Commission) to examine the aviation industry, supported the boards as a way to study policy before taking any action.[171] O'Laughlin may have reflected the sentiment of many when he reported to Hoover, "It will be surprising if, when the government is through there will be any confidence at all in our aviation and if we have not been set back years in the development of this art."[172]

Originally the board consisted of the members of the War Department's Drum Board (including Foulois), Orville Wright, Charles Lindbergh, and noted pilot Clarence Chamberlin.[173] However, Wright (poor health) and Lindbergh (political reasons) refused to serve. Lindbergh's refusal created quite a stir, and a slight embarrassment for the administration.[174] With his two high-profile civilians declining, Dern wired the president's secretary, Marvin McIntyre, and requested that McIntyre contact Newton D. Baker, the World War I Secretary of War and veteran of some of the early political aviation battles. Dern noted his need for an "outstanding national figure as chairman."[175] Baker accepted, Dern added more civilians, and they began work on 17 April 1935 with the purpose of reporting on Air Corps operations and the "adequacy and efficiency" of the service.[176]

The Baker Board mirrored the Morrow Board in many ways. Although it was not as politically predisposed to a position against air independence, the Baker Board would not go against the administration's wishes and recommend such a solution. The choice of Baker as chairman signaled that fact, as he had opposed air autonomy during the Wilson administration.[177] Like the previous board, Baker's used many earlier air studies and findings and worked as quickly as possible to provide an acceptable political solution to the issues of independence, but not so fast as to make the report seem hasty and biased (Baker issued the report three months after the board first met). Baker also solicited testimony and input from aviators. Air Corps officers not officially called as witnesses were invited to submit their own testimony and suggestions, and the board received a flood of advice urging more air autonomy as the only means of improvement for the Air Corps.[178] As with the Morrow Board, the pleas fell on deaf ears. The board questioned 105 witnesses over a period of three months, recorded more than four thousand pages of testimony, and visited Air Corps bases and civilian aircraft factories. Even with such attempts at an in-depth and comprehensive survey, the findings changed little from the past sixteen years.

When the Baker Board issued its report, it only strengthened the case of the General Staff and the War Department and did nothing to contradict the president's desire for keeping the current national defense structure.[179] Like the fifteen similar studies in the preceding sixteen years, the board's report recognized that the country remained safe from a massive air attack, that air power's weaknesses limited its ability to control territory and provide an independent decisive outcome, and that air power remained too expensive to justify independence given those shortcomings.[180] It thus rejected the two main arguments for independence: economy and unique mission. On the first, it found that any structure besides the Army organization would not produce economy and would only jeopardize national defense.[181] However, in rejecting the Air Corps' ability to perform a unique mission, the board inadvertently identified for the aviators the conditions they must meet to argue in the future for autonomy. The report noted that current aircraft technology could not fly long distances and deliver an appreciable amount of munitions. "To date," it read, "no type of airplane has been developed capable of crossing the Atlantic or Pacific with an effective military load, attacking successfully our vital areas, and returning to base."[182] While currently rejecting the Air Corps' ability to undertake strategic bombing, the board invigorated the Air Corps' desire to obtain such a capability already under development. The Air Corps Tactical School continued to refine the doctrine, and the Air Corps now had increased motivation to purchase a four-engine heavy bomber in mass quantities.

Foulois' testimony before the Baker Board came as he was still under attack by the Rogers Subcommittee and in bad odor with the administration after the air mail problems. Foulois handled the meetings quietly, though he later asserted that he fought "many verbal battles during closed-door deliberations."[183] Although he could have, like Mitchell, tried to exit with a bang, Foulois decided not to confront the General Staff more than he had already done in other venues, and he worked to secure aircraft increases and obtain further support for a GHQ Air Force. Foulois even refused to comment publicly when Air Corps officers argued a case for independence.[184] Thus, in a forum where he could have openly and appropriately expressed his opinions, he did not, because of his weakened authority and loss of political support. Having destroyed his relationships with Congress and the president, Foulois presented only a token and ineffective presence on the Baker Board.

Despite Foulois' inaction, the Air Corps did obtain something positive from the board. The service officially received recognition that its shortcomings, epitomized by problems delivering the mail, derived from not receiving its required appropriations and budgets, and the board heartily advocated the Air Corps' case for increased funding. As a by-product, the board noted that funding for military aviation would also reap benefits for the aircraft industry overall, which Roosevelt would undoubtedly support due to its boost for the economy. The findings also walked a fine line on the issue of contracts. The report found that there existed in the Air Corps Act sufficient leeway for different procurement methods. As such, it found that the government could best benefit by using all three methods (purchasing design competition aircraft, negotiated contracts, and competitive bidding) while also assisting the industry by bearing some of the costs of research and development.[185] Finally, the Air Corps received support for a GHQ Air Force. Arnold summed up the feelings of many Air Corps officers when he said these findings were "the first real step ever taken toward an independent United States Air Force."[186]

The Baker Board reassured the American public of the administration's and the Army's support for aviation as a combatant branch within the Army and not as an independent force. Politically, the recommendations supported the War Department's and administration's positions and took some of the impact out of the Rogers Subcommittee indictments. The report also provided enough to the Air Corps, in the form of the GHQ Air Force, to placate both the service and public opinion. As with previous boards and commissions, putting the findings into action required political support and congressional funding. Roosevelt announced his support of the plan and authorized new spending to bring the Air Corps up to strength.[187] Yet Baker's efforts and the support demonstrated by all for the board's findings did not quickly or easily repair the relationships that Foulois had damaged. For the Air Corps to take advantage of the changes and the administration's support, new Air Corps leadership and the return of moderation would be required. Much as during the immediate post-Mitchell era, the Office of the Chief of the Air Corps needed to improve its public image and show itself as a team player in the Army and the administration.

Foulois' actions harmed his reputation and career. His inappropriate behavior probably impeded the progress of the Air Corps in the short term. The General Staff, though never fully supportive of any actions it believed unfairly favored the flying service, actually hardened its stance against the Air Corps during Foulois' time as chief. The Secretary of War clearly supported the General Staff and further hindered any pro-aviation actions arising from within or without the War Department—and the absence of an air secretary limited aviation's voice. Most importantly, Foulois alienated the last bastion of air power supporters in Congress. With the country's desperate economic situation and Roosevelt's focus upon the New Deal, the air service could probably have done only marginally better, if at all, on its budgets. However, during this same time the Navy had begun a strong building program, and the Navy's air service had completed its five-year building program on schedule while the Air Corps still lagged behind.[188]

Still other areas of pressing concern to the Air Corps required congressional support, especially the issue of officer promotions. The service desperately needed promotion reform, and a flurry of proposals came before Congress during Foulois' tenure.[189] Between being distracted by the Foulois investigations and perturbed by the chief's words and deeds, the pro-aviation congressmen did not, and perhaps could not, aggressively push any Air Corps legislation (of which the highest needs remained the "war hump" promotion problems and personnel and equipment shortages). The service settled for temporary fixes until after Foulois' departure. Not until 1936 did the Air Corps obtain most of the promotion benefits it had long desired.[190] Promotion fixes and all the other needs of the service stalled until the Air Corps changed its leadership and returned to more proper relations with the Army and civilian officials.

Although Foulois had hated Billy Mitchell, he ranked close behind the court-martialed airman in disrupting American civil-military relations during the interwar years. Foulois did not directly challenge a presidential administration in the same manner or over as long a period as Mitchell, but he had just as surely acted outside the accepted boundaries of behavior for military officers in a democratic republic devoted to civilian control of the military.

Foulois had abandoned the conservative approach instituted by Patrick and continued by Fechet, and he did so at a time when the service lacked the political buffer and advocate in the person of an Assistant Secretary of War for Air. Foulois testified before Congress for aviation policies unsupported by the administration and the General Staff. His testimony actually contradicted their policies and pushed an agenda rejected by five presidential administrations, from Wilson's through Roosevelt's. But with the American public and the president focused on the crisis of the Great Depression, Foulois' occasional early advocacy for changes in military aviation had received little notice. Only with the air mail controversy and the death of Army flyers did aviation suddenly burst into public view once again and require the attention of the administration and Congress. In the resulting inquiries, Foulois was exposed as lying to Congress and misleading the president. By the time the investigations had ended, Foulois was ready for retirement and had concluded his term. That timing saved both him and the president from an embarrassing sacking. Thus the Secretary of War and the commander in chief probably allowed Foulois to retire instead of taking further actions that would have received even more unwanted public attention. Instead, the civilians quietly appointed a new air chief whom they knew would restore moderation, discipline, and a sense of propriety for civil-military relations to the Air Corps.

CHAPTER SIX

Moderates and Money, 1935–1938

Beginning in 1936 with the German remilitarization of the Rhineland, Hitler's saber rattling spurred President Roosevelt to begin limited rearmament and revitalization of the services. Two years later, with the Munich Pact giving Hitler parts of Czechoslovakia only six months after the Austrian Anschluß, FDR set the country on the path toward rearmament. The president emphasized the need for air power, especially strategic bombers, as a deterrent to Hitler, but his stated motivations and unstated intentions (whether it be rearming Europe or economic pump priming at home) never matched those of his Secretaries of War or Air Corps leaders. Thus the fight over Army aviation, and the civil-military conflict it produced, differed from the pattern of the previous sixteen years. Instead of battling directly for independence, the Air Corps and its congressional allies concentrated on the appropriations to purchase bombers — bombers that would demonstrate why the Air Corps deserved to be an independent service.

From 1936 to 1938, Congress played a lesser role in major Air Corps issues than it had before. With a virtual absence of air independence legislation, and air leaders cooperating with the War Department and the administration, occasions for open disagreements between aviation leaders and civilian policy makers diminished. Air Corps legislation passed, authorizing a new strength of 2,230 aircraft, and additional bills increasing personnel numbers became law, usually with full War Department support.[1]

Thus the civil-military rifts that occurred over air power during this period arose *within* the War Department and centered upon the Air Corps' desire for the heavy bomber and how these purchases affected Roosevelt's overall economic plans. Even after the civilian leaders agreed upon some expansion in 1936, the Air Corps chief battled Secretary of War Harry H. Woodring, who was more interested in purchasing higher quantities of cheaper aircraft. The War Department pushed for procurement of smaller, less expensive, and less capable bombers to complete the expansion program, while Air Corps leaders desired the more expensive and more capable B-17, which the airmen viewed as the tool finally capable of meeting their recently developed strategic bombardment doctrine.[2] Resolution of these differences would not begin until after the 1938 Munich Conference.

In late 1938, Roosevelt vowed to increase aircraft production, especially the heavy bomber, which had become emblematic of American air power. FDR wanted to improve American military capabilities, but he also had an ulterior motive: increased aircraft production would arm friendly European countries, and American and European bombers might be a visible deterrent to Hitler. Chief of the Air Corps Oscar Westover and his assistant, Brigadier General Henry H. "Hap" Arnold, wanted a balanced air force with the proper ratio of bombers, pursuit, transport, and trainer aircraft, but one whose primary component would be the expensive strategic bomber. Even more adamant was the commander of the Army's air combat organization, Brigadier General Frank M. Andrews. With the commander in chief's true goals never clear to either his war secretaries or military leaders, civil-military conflict ensued, but primarily inside the administration, hidden from public view.

Fortunately, the Air Corps leaders returned to an attitude of moderation in public forums and within the Army after the retirement of Major General Benjamin D. Foulois. Bearing in mind the lessons from Billy Mitchell and Foulois, Air Corps leaders were determined to work out differences with the civilian leadership privately. Westover succeeded Foulois and, upon the former's death in an aircraft accident, Arnold, a reformed firebrand himself, took the reins of the Air Corps and guided it until after World War II. Rebellious elements still existed, led by General Andrews adamantly pushing for the procurement of the B-17 bombers, but most airmen wanted to work within the Army. Westover enforced this discipline from above, and many of the higher-ranking Air Corps officers had seen the failure of the confrontational tactics of Mitchell and Foulois. Arnold, from his previous experiences in the Capitol, knew the halls of Congress well and retained some contacts with air-friendly congressmen, but he used his political skills to avoid placing the service in the middle of public partisan battles. His primary disagreements would occur with cabinet members, with all parties trying to enhance their own departments and doing their best to carry out the often-concealed goals of the president.

THE FOULOIS FALLOUT, A NEW ATTITUDE, AND A NEW TEAM

The War Department reacted quickly to the major change recommended by the Baker Board by establishing the GHQ Air Force on 1 March 1935. The newly reinvigorated support for GHQ Air Force created an organization best suited to preparing for and carrying out strategic bombing, the mission that would dominate the Air Corps' preparations for war. All air combat units previously under the command of regional corps-area commanders now reported to GHQ Air Force. An airman, and not ground army leaders, commanded the geographically dispersed air units. Bulletin number one issued by GHQ Air Force stated: "The Air Corps has not been so organized in the past, as to permit the use in war of its tremendous striking power, to the best advantage. . . . The war mission of the GHQ Air Force will be to conduct offensive air operations against enemy air, ground and sea forces."[3] The commander of GHQ Air Force reported directly to the Army Chief of Staff—not the Chief of the Air Corps—during peacetime and to the theater commander during wartime. Veteran aviator Frank M. Andrews was appointed as the first GHQ commander and was promoted from his permanent rank of lieutenant colonel to the temporary rank of brigadier general.[4] Yet the new organization did not solve all of the Air Corps' problems.

The organizational structure created a two-headed organization. Supply and training functions for the air arm would remain with Chief of the Air Corps, but the GHQ commander controlled missions and execution. Billy Mitchell called the setup "nothing but a subterfuge [that] merely divides aviation into more parts,"[5] and the two generals commanding the separate sections would shortly disagree on Air Corps policies and on the procurement of aircraft. Another problem that the creation of GHQ Air Force did not relieve was the perceived domination of aviation needs by the General Staff and the War Department. The Baker Board did not recommend a separate budget for the Air Corps, which would have led to appropriations battles and procurement problems for years to come. Nor did the air officers receive separate promotion lists. Both Air Corps arms were still subordinate to the higher command authorities of the regular Army. The air leaders found that the organization for which they had labored did not cure all of their ills. The push for further improvements, however, soon took on an entirely new dynamic with multiple changes in military and civilian leadership.

Roosevelt delayed replacing MacArthur when his four-year term expired in mid-November 1934, and speculation arose about finding an able successor. One month later, Roosevelt issued orders allowing MacArthur to remain indefinitely, or, as the president stated, "until his successor has been appointed."[6] Secretary of War Dern, who remained close to MacArthur and relied on his judgment, actually encouraged FDR to keep MacArthur due to the unsettled world situation.[7] The War Department seemed pleased that MacArthur remained in office, and only a few congressmen voiced displeasure.[8] It took almost eleven months for Roosevelt and Dern to decide. The president meticulously gathered information before making the choice, insisting that the new Chief of Staff be able to serve the entire four-year term before reaching the mandatory retirement age.[9] Of all the generals on separate lists compiled by Dern, MacArthur, and Roosevelt, General Malin Craig, at age fifty-nine, was the youngest eligible. [10] Although not entirely pro-aviation, Craig was not as anti–Air Corps as other General Staff officers.

Soon after Craig's appointment, Roosevelt and Dern pondered Foulois' replacement. When Foulois began his terminal leave in September 1935, Westover became acting chief for the three months remaining on Foulois' term. Understandably, in light of the past year's events, FDR and Dern did not want a pro-independence firebrand. Dern prepared a chart listing all eligible officers, their past assignments and war records, and other pertinent information. Generals Westover and Henry C. Pratt ranked as the top two contenders for the position. Dern noted to the president Westover's "brilliant record" and his demonstrated ability to lead, emphasizing that "he has the ability to cooperate to a high degree."[11] Pratt's position as head of the Material Command during the procurement controversies and inquiries of the Foulois years undoubtedly hurt his chances with two civilian leaders. "[Pratt] has greater mental ability than others on this list but it is possible he is not as cooperative as Westover," noted Dern. [12] Obviously, Roosevelt and Dern wanted an air chief who would work smoothly with the administration and the War Department.

Dern and Roosevelt chose wisely, and Air Corps officers agreed with the civilians' assessment of Westover's style. Ira Eaker remembered that "General Westover would not tolerate any criticism of Army policy and leadership nor any snide or derogatory insinuations about other government departments, such as the Navy. His motto was 'everything constructive, nothing destructive or disruptive.'"[13] For the first time since Major General Charles Menoher in the early 1920s, the top-ranking air leader did not desire air independence at the soonest possible moment. In fact, Westover had opposed air independence since 1919, and he considered any officer who agitated for such a move insubordinate. [14]

Soon after taking office, and even before Foulois' formal retirement, Westover issued to the entire branch a statement outlining the new Air Corps attitude. He noted the need for the air arm to become a team player within the Army and to support the War Department and the General Staff. He specifically outlawed the tactics of confrontation and attacking people opposed to the aviators' views, and he theorized that many of the Air Corps' problems derived from "aggressive and enthusiastic efforts of some of its personnel."[15] Westover specifically called for cooperation with the president and Congress, support for the Drum and Baker Board recommendations, and the sending of any criticism or recommendation through proper channels. On the latter point, he emphasized that this restriction meant not going directly to Congress or the press. [16] The new chief believed that air independence would probably come eventually but that the tactics of confrontation only delayed the proper development of the Air Corps and hindered the service's ability to perfect the capabilities that would earn autonomy.[17]

Westover believed the Air Corps could better develop itself and its capabilities if it remained part of the Army. The Air Corps portion of the budget purchased aircraft and the infrastructure to conduct air operations, and it did not include items grouped with other Army elements (such as pay and medical care). Westover rightly understood how, by remaining part of the Army, the air service would benefit from being able to use the Army's service and support, thereby devoting the Air Corps' entire budget to the organization, training, and equipping of the air element. "We have to have places in which to live," Arnold recalled Westover saying, "all of which can be provided by the Army and the Navy. I believe that the time has not yet come when the Air Corps can demonstrate its fitness to sustain itself and operate independently of other units."[18] Therefore he urged his officers to bide their time and thus change the perception of the Air Corps. By becoming part of the Army team and supporting the president, Congress, the General Staff, and the War Department, those people and organizations would become more receptive to Air Corps desires.

The new air chief also realized the positive power of press coverage. Instead of using it to attack those who did not agree with the Air Corps, he wanted to work for coverage of positive Army aviation developments. He even urged his officers to become more involved in the local communities and take every opportunity to "enlighten the public" regarding the positive contributions the Air Corps made to the "Army team of national defense."[19]

Westover's leadership helped convert a former insurgent to accept cooperation as the way to improve the Air Corps and its situation. Westover asked for Arnold as assistant air chief. Dern approved and forwarded the recommendation to Roosevelt, even though Arnold's record contained "a number of derogatory remarks & reprimands," undoubtedly meaning his support of Mitchell and the infamous "circular" of 1925.[20] In recommending Arnold to Roosevelt, Dern admitted that another officer, Lieutenant Colonel Walter Kilner, had "the best record for his age and rank in the Corps,"[21] but he still recommended Arnold: "Although [his] record is spotted he is undoubtedly one of the outstanding officers of the Air Corps."[22] Arnold did have a strong background for the job and a demonstrated capability to command, but Dern's less-than-glowing endorsement allowed him to accept Westover's choice while still highlighting Arnold's past to Roosevelt.

Since leaving Washington in 1926, Arnold had learned firsthand that promotion of the service would serve the Air Corps better than griping about problems. He agreed with Westover's new approach and no longer supported immediate separation and independence. As Western Zone commander for the air mail in 1934, Arnold realized, with the help of Ira Eaker, the need to publicize the positive aspects. The mail fiasco convinced them of Mitchell's and Foulois' mistake of using negative rhetoric in an attempt to gain more funding. Instead, they believed that the Navy air arm had the right message: publicize the good things while noting the need for more and better equipment.[23]

Arnold's experiences also converted him to the view that neither immediate independence nor a radical reorganization of the national defense best served the Air Corps. In reply to a letter from Chairman McSwain, Arnold said he believed that those who sought "radical conclusions" for the air arm did so from looking only at the vastly increased performance in aircraft, not how the entire service fit into the national defense structure. The former insurgent told his pro-aviation congressman: "At one time I was for an immediate reorganization for a Department of National Defense with three equal branches—Army, Navy, and Air. Even today, I am still of the opinion that ultimately such an organization will come. The natural progress of the art alone insures it, but for the present it would be a step backward. . . . [Now] we must secure more equipment and concentrate on training personnel in command and operation duties before we can even successfully carry out the activities of a GHQ Air Force."[24] The new approach—use of positive press coverage and desire to remain as a branch of the Army for the foreseeable future while still advocating expansion and increased budgets—dovetailed perfectly with Westover's policies. Eaker, who came to Washington to work on Westover and Arnold's staff, recalled the two men as being "so different in personality and method but both dedicated to the same objective, a proper recognition, status and stature for U.S. military aviation."[25]

The activation of GHQ Air Force brought another powerful voice to the Air Corps (and very close to the capital area): the commanding general at Langley Field, Virginia. General Andrews organized the GHQ Air Force and became its first commander, initially working in Washington out of Davison's old office until offices in southeastern Virginia were ready.[26] With the dual command relationship and Andrews matching Westover in rank (and outranking Arnold), the GHQ Air Force commander immediately spoke out on aviation policy.

Unlike Arnold, Andrews did not always subordinate his desire for an independent air arm. Andrews believed air independence would come by procuring the heavy bomber. From the Tidewater of Virginia, he would advocate for the bombers, aided by another unreformed air power radical and his GHQ Air Force Chief of Staff, Colonel Hugh J. Knerr.[27] Those close to Andrews described him as a gentle, quiet, and extremely intelligent man whose soft-spoken nature and brilliance made him very persuasive. One officer on his staff believed Andrews withstood an extreme amount of pressure, placed on him by independence-minded officers (especially Knerr), to use his office and become a "recalcitrant and be a rebel and raise hell."[28] The time would soon come when Andrews would "raise hell" trying to get heavy bombers, but in the wake of Foulois' departure the Air Corps team appeared solid, with each general in a position to take advantage of his talents and training. One Air Corps colonel called this triumvirate the best possible case for the Air Corps: Westover's talents for details; Arnold's political abilities to "sell" ideas to the Budget Bureau, Congress, and the General Staff; and Andrews's expertise as a tactical commander heading up the fighting arm of the service.[29]

Dern himself did not live see the full impact of the changes in leadership and attitudes from the Air Corps and Army chiefs he helped pick. He battled health problems in the later stages of Roosevelt's first term and died prior to the 1936 elections. Roosevelt's delay in approving the elevation of Harry Woodring to the Secretary of War's office did not generate confidence and may even have undermined support for the Kansan. One month after Dern's death FDR designated Woodring as the temporary secretary, as required by law (barely beating the thirty-day requirement). "The President wants me to tell the Press [that the] appointment is temporary and necessitated by law," FDR's secretary confirmed.[30]

Roosevelt himself did little to reassure Woodring that he held the trust and confidence of the president. In a telegram, Roosevelt informed Woodring that Dern's position would not be permanently filled for at least two months, but the law required action within thirty days. Roosevelt emphasized to Woodring the temporary nature of the selection.[31] Administration insiders believed that Roosevelt preferred someone else as war secretary, but FDR wanted to secure the Kansas electoral votes in the upcoming election against Kansas governor Alfred M. Landon. The real choices were two other governors, Frank Murphy of Michigan and Paul McNutt of Indiana.[32] In fact, Roosevelt did not inform Woodring of his retention in the cabinet until after inauguration, and he did not send the nomination to the Senate until April 1937.[33] Roosevelt left the office of assistant secretary empty for almost another full year.[34]

The absence of a strong civilian team in the War Department did not adversely affect civil-military relations, as it could have had Air Corps insurgents still led the flying service. But air and congressional leaders could not have failed to note Roosevelt's lack of confidence in Woodring. This gap between the president and Woodring gave airmen and congressmen leverage when desires by aviation proponents would clash with those of the new secretary.

In addition to the new personalities in the War Department hierarchy, the battleground for budgets and Air Corps desires also changed. Instead of working for immediate independence, Westover and Arnold now moved to complete the overdue expansion program that had been recommended by the Baker Board. They wanted a balanced air service, but with enough striking power (bombers) to carry out the strategic bombardment doctrine being refined at the Air Corps Tactical School. On this issue they differed with Andrews, who wanted a more bomber-heavy force. Although Andrews often disagreed with Westover and Arnold, the latter's political connections, combined with his being closer in vision to the General Staff, limited Andrews's impact. Arnold and Westover would employ their new approach on an even more firmly entrenched and in-control Democratic Party.

For Arnold, returning to Washington after a ten-year exile, the capital was different. A Democrat sat in the Oval Office and Democrats firmly controlled Congress, gaining more power during the 1934 midterm elections. Republicans had lost fourteen seats in the House and eleven in the Senate, primarily due to Roosevelt's New Deal program and active leadership.[35] Democratic control of Congress and the White House prevented air leaders from using the partisan wedge as they had in the 1920s, and that in itself smoothed civil-military relations. In the Westover years only one serious legislative action appeared supporting an independent air force, and Westover himself helped quash the bill.

Democratic representative James M. Wilcox proposed making the Air Corps autonomous but still under the control of the Secretary of War. The Floridian drafted the bill, but he relied on the support and ideas of both Knerr and Andrews. Knerr traveled to the Sunshine state to help, and Wilcox often visited Andrews at Langley Field. The two airmen believed that a unified defense department would not pass the House, and Roosevelt would certainly not support the concept, so they and Wilcox tried a narrower approach. Predictably, Woodring and the General Staff reacted unfavorably to the bill, and Westover joined them, recommending to the Adjutant General against Wilcox's proposal. Chief of Staff Craig noted that only Wilcox and "a small group of dissatisfied Air Corps officers who were adherents of former Brigadier General William Mitchell" supported the measure.[36] Andrews stayed in the background, although the bill bore his fingerprints. In the new atmosphere, open support meant probable political and career suicide.

Once the House Military Affairs Committee scheduled hearings, Craig picked Westover to provide the War Department's official response, believing his opposition would kill the bill. Even though the chairman of the committee did not want to hold hearings on the proposal, Wilcox wanted Air Corps officers to testify publicly. Woodring asked the president to pressure Congress to drop the bill.[37] When Roosevelt called Alabama congressman Lister Hill and asked for the hearings' postponement, Wilcox's bill became doomed. The War Department did not know about Andrews's actions, but they did find out about Knerr's lobbying. When his next assignment came due, the War

Department sent him to San Antonio, to the same office and duties given to Mitchell more than a decade earlier.[38] Air Corps officers undoubtedly got the message. Westover's desire to work within the Army helped to stop quietly the only open attempt of the period to use Congress to support air independence. After fifteen years of controversy, all sides and the public were exhausted by the fight. Yet another public imbroglio over aviation in the wake of Foulois' recent actions would have been disastrous.

QUANTITY OR QUALITY: THE BATTLE FOR BOMBERS

One year after implementation of the GHQ Air Force, the Air Corps overall had changed little from its earlier years. It was a "force" in name only. The Air Corps Act of 1926 authorized the Secretary of War to equip and maintain 1,800 serviceable aircraft and authorized replacements for obsolete or unserviceable aircraft each year, not to exceed 400 aircraft annually.[39] However, the statute did not account for losses of aircraft by other means (crashes, for example). Therefore, the statute's language meant the Air Corps could never reach the 1,800-plane limit unless it received its full appropriations for aircraft procurement each year while also eliminating aircraft accidents.[40]

This anomaly created a debate over aircraft numbers. The Drum Board analyzed the 1,800-aircraft limit and, interpreting the Air Corps Act and the previous internal War Department debates, increased the number. Allowing for a 25 percent "war reserve" and a 12.5 percent adjustment for aircraft out of commission for repairs, the Drum Board set a new cap of 2,320 aircraft. The Baker Board authenticated that number but without specifying how it derived the bottom-line figure.[41] These boards presented the 2,320-aircraft program for the public as the authorized strength for the Air Corps, but Congress remained riveted on 1,800 as specified in the Air Corps Act. Thus the Air Corps needed increased appropriations and clarifying legislation to achieve the higher numbers.

Numbers seemed irrelevant when the key was money, but without a higher authorization no larger appropriations could be passed.[42] Chairman McSwain again assisted the Air Corps and offered legislation to increase the service's numbers on paper. Known officially as H.R. 11140, the McSwain bill proposed to amend the Air Corps Act to increase the authorized strength up to 4,000 aircraft and to inaugurate a new Five-Year Plan.[43] Because of this provision, the War Department withheld support for the amendment. Secretary Woodring agreed that a higher authorization for numbers of aircraft was needed, but he rejected an expansion plan that would have purchased so many aircraft in such a short time, saying such expenditures were "not in accordance with the financial program of the President."[44]

The administration did not want to increase the budget, especially to buy such a large number of expensive aircraft within a congressionally mandated period, despite Woodring's revelation that the Air Corps' strength as of December 1935 was 1,060, of which only 200 were considered modern types. Furthermore, Woodring projected an end-of-fiscal-year force of only 777 aircraft (by July 1936) due to estimated losses and lack of appropriations to augment the force.[45] Thus, McSwain's bill failed when Woodring, acting on orders from the president, asked to keep the number at 2,320 without a mandated expansion period.[46] The Conference Report reiterated the 2,320-aircraft limit as set by the Baker Board as the proper aircraft limit.[47] Again, money proved the main obstacle. Although Hitler continued to consolidate power, remilitarize the Rhineland (March 1936), and had cast aside the Versailles Treaty (1935) and begun rearming, he had yet to influence Roosevelt to alter his fiscal program. The president considered a radical increase in military budgets unwise, but he desired the possibility of increased production. To that effect, he saw the need for increased military appropriations and began a modest rearmament program, which would increase every year from 1936 onward.[48]

Still, until the Munich crisis in October 1938, aircraft numbers remained low. The testimony did not resolve a ten-year political quandary; Congress would support a robust Air Corps in theory but would not fully fund it. The debates also reasserted the priorities of the Roosevelt administration. The president did not openly comment on these struggles for increased Air Corps aircraft, but he made known, through his cabinet, his desires to spend money on his New Deal programs and not on expensive weapons (at least for the Army). From 1935 to 1938, while watching the world situation, Roosevelt kept his focus at home and inquired only sparingly about aircraft production.[49] Woodring toed the president's line and, although his biographer labeled him "a spokesman for airpower" and "a true friend of the Air Corps," during his tenure his actions diverged from the desires of the Air Corps leaders, especially in the type of aircraft the Air Corps wanted.[50]

THE FIGHT FOR THE LONG-RANGE BOMBER, 1936–1938

Air Corps budgets—and those of the War Department overall—continued to improve after the lows of the first Roosevelt administration. The 1936 War Department budget increased by 26 percent over that of the previous year, the highest since World War I. From that point until the attack on Pearl Harbor, departmental budgets continued to increase every year. The Air Corps' appropriation for augmentation, modernization, and replacement of equipment jumped from just over $13 million in 1935 to more than $30 million in 1936.[51] The increases immediately affected aircraft production, which increased from 459 aircraft in 1935 to 1,141 in 1936.[52] The air service's increases did not come about through congressional connections or improper lobbying, for few of the previous congressional friends of aviation remained on Capitol Hill and fewer still on the Military Affairs Committees. Historically, the Air Corps' funding problems came from the War Department, the Budget Bureau, and then the Appropriations Committees, each subsequently cutting more as the budget worked its way through the long process, though sometimes Congress would even add funding for military activities to the low Budget Bureau numbers.[53] After the Baker Board, which coincided with a slightly improved national economy, Congress and the president actually agreed on higher budgets for the Air Corps and the military overall. The Air Corps asked for more, trying to reach its aircraft limits and the supporting infrastructure. While Congress and the president would not grant the full amounts requested, budgets dramatically improved from 1936 onward.[54] The lack of any serious budget debate, combined with the absence of contentious bills advocating air independence or increased autonomy, dampened civil-military conflict involving the Air Corps and Congress. Instead, disagreement flared over the often-contentious issue of what types of aircraft to purchase.

Desiring to get the Air Corps to its full strength of 2,320, Woodring wanted cheaper airplanes. The Air Corps constantly insisted on a force built around the four-engine long-range bomber, specifically the Boeing B-17 Flying Fortress. Woodring wanted quantity, Andrews coveted quality, and Arnold, in the middle, worked for a balanced force with the B-17 at the center. Different Air Corps leaders displayed different commitments to all-out procurement of the Fortress, but they all wanted bombers even if their advocacy was carefully phrased due to the public's isolationist mood and the administration's matching policy.

The country's isolationist attitude forced military leaders always to frame their funding and equipment requests in terms of protecting and defending the continent. The elections demonstrated the national support for the Democrats and the party's platform, so air officers dared not openly oppose the administration. The public supported a more prepared military than in years past but wanted defensive weapons. The isolationist mood so permeated the nation that the Army altered its planning so as not to seem to be preparing for offensive operations.[55] The Air Corps therefore needed to cast the bomber, an inherently offensive weapon, in the role of defense while still developing the doctrine for strategic bombing.

The Air Corps publicly argued that bombers would protect the Western Hemisphere, including American overseas possessions. Still, the service continued its development of strategic bombing as its core mission, and the Air Corps Tactical School's doctrine began with the statement, "Air Forces must be employed offensively."[56] Understanding the public's mood, Air Corps leaders did not publicize their offensive doctrine. With the United States protected by vast oceans, the only way for enemies to attack would be by sea or by air. Bombers could provide coastal defense by striking ships long before they could reach the coast. The bombers' constantly increasing range of operation reignited the rivalry with the Navy, though all three Air Corps leaders worked with General Craig to keep the debate out of the press, unlike in the earlier years.[57]

For enemy bombers to reach the American coasts, they had to launch from air bases in the Western Hemisphere. American aircraft, the military leaders reasoned, could bomb these "nests" before the enemy could strike.[58] Internally, Army officers did not buy the Air Corps argument of defense, and a staff memo called the an experimental heavy bomber "distinctly a plane of aggression . . . [which] has no place in the armament of a nation which has a National Policy of good will and a Military Policy of protection, not aggression."[59] Still, the Air Corps stuck to the script. General Andrews, while arguing for this bombardment as defensive, actually upset Congress and the public by seemingly advocating preemptive bombing of neighboring countries.

In testimony before an executive session of the House Military Affairs Committee, Andrews warned Congress how an enemy could use areas close to the American continent, including Newfoundland and many Caribbean islands, as bases for bomber operations. Andrews recommended keeping such possible staging areas under surveillance and, if such bases appeared, bombing or occupying the area in order to defend the United States.[60] His faux pas of recommending preemptive attacks on other sovereign (and friendly) nations became even more evident when the committee released the secret proceedings. Andrews immediately clarified his remarks as being only in abstract terms in order to make a point about defensive use of aircraft and bombers' increasing capabilities.

In an unprecedented step, Roosevelt himself came to the general's defense and chided Congress for releasing the testimony. To amplify his point and to allay any implications of civil-military conflict or presidential pressure on officers' testimony, FDR notified the committee chairman that additional publications of secret testimony would prompt the president to limit future testimony and approve each and every military appearance before his committee. McSwain apologized and accepted responsibility.[61] A solid Democratic Congress supporting the president, coupled with a congressional realization of the obvious mistake, probably kept down any allusions of muzzling officers, as had occurred in previous years. Instead, FDR's forceful leadership and backing of his officers quickly muted the situation, but the Air Corps again realized the need to cast the bomber as a defensive weapon.

Soon after the minor flap over the Andrews testimony subsided, the Air Corps finally saw the aircraft of their dreams and doctrine. In just under one year after submitting its proposal, Boeing delivered an aircraft for testing that conformed to the minimum standards: a bomber that could deliver 2,000 pounds of ordnance, attain a minimum speed of 200 miles per hour (a desired speed of 250 miles per hour), and carry a crew of four to six out to a range of 2,200 miles. The Air Corps originally estimated total procurement at 220 bombers.[62] Glenn Martin, Douglas, and Boeing had submitted proposals for the bomber in August 1934.[63]

In May 1935, the Air Corps included 86 four-engine bombers in its estimates for the coming fiscal year prior even to testing the prototypes. To meet financial restrictions, the Air Corps reduced the quota to 60 Fortresses. Woodring further trimmed this number to 26 but increased the number of cheaper, two-engine bombers from 86 to 156.[64] Again Woodring concentrated on keeping the budget low while aiming to bring the service promptly up to its full quota of aircraft. Thus, from the outset of the debate over big versus small bombers the stage was set for conflict between the Air Corps and Woodring. The public rollout of the Boeing bomber and the performance it demonstrated even in its abbreviated testing only fueled the airmen's desire for the Fortress.

On 17 July 1935 Boeing rolled out the polished, silver-skinned B-17 for public view. Reporter Dick Williams of the *Seattle Times*, seeing the defensive armament of the five-gun turrets, labeled Boeing's Model 299 bomber a "Flying Fortress," and the name stuck. It also appealed to the Air Corps leaders' desire to sell the plane as a defensive weapon, a "flying fort" to defend American coasts.[65] Three companies had submitted their designs for the four-engine bomber contract, but the Boeing outperformed the Martin and Douglas entries in every category. Even before the competition ended, the Air Corps recommended purchasing 65 of the Boeing bombers in lieu of 138 other aircraft already authorized in the fiscal year 1936 budget.[66]

Bad luck kept the Air Corps from fully pressing for the object of its desire. On 30 October 1935, the only B-17 crashed and burned on takeoff, killing two test pilots. By not completing the competition, Boeing was legally disqualified from contract consideration. Douglas's B-18 won an order for 133 bombers. Having produced the bomber at an expense of $425,000, Boeing could have been ruined by the crash. Understanding the situation, the Air Corps astutely placed an order for thirteen of the now-designated YB-17 and one "static test article" to keep the company alive.[67] The bomber proponents of the Air Corps, notably Andrews, saw the future of air power and the aircraft that could fulfill the earlier prophecies of the air enthusiasts. "The B-17 was the focus of our air planning," Arnold remembered, and "the first positive answer to the need arising from the United States' modification of the Douhet theories," which the Air Corps Tactical School continued to teach.[68] Seeing two land at Langley Field, Virginia, led Arnold to think, "for the first time in history, [here was] Air Power that you could put your hands on."[69] Yet before the Air Corps could put its collective hands on the B-17 in large numbers, they still needed to convince the president, his Budget Bureau, the War Department, and Congress to approve the money.

Craig and Woodring originally agreed with the idea of ordering less-expensive bombers in larger numbers against the wishes of the air leaders. Craig believed that MacArthur had concentrated on theory at the expense of providing weapons for a viable force. In May 1936 the Air Corps included fifty B-17s in its estimates for fiscal year 1938. Five weeks later, a directive from the Adjutant General, by order of the Secretary of War, eliminated all B-17s from the 1938 program and substituted a standard two-engine model.[70]

A summer 1936 staff study clearly exposed the differences over bomber procurement between the Air Corps, the General Staff, and the Secretary of War. Westover wanted to purchase 61 long-range bombers, 11 Project A aircraft, an experimental large bomber (later became the XB-15), and 50 B-17s.[71] The General Staff response stated: "Until the international situation indicates the need for long range types of bombardment aviation as the Project A and the 4-Engine (Boeing) models, no more of that type should be procured except for experimental purposes."[72] The memorandum then recommended eliminating all Project A and B-17s from the 1938 program and substituting two-engine B-18s instead. The Assistant Chief of Staff agreed with the recommendations, but Westover dissented. Woodring overruled Westover and agreed with the General Staff's opinion.[73] Once again, Craig and Woodring opposed the recommendations of the Air Corps and purchased a medium bomber over long-range aviation.

The battle to obtain the Flying Fortress intensified in 1936. One month after Woodring's disapproval of heavy bombers for the 1938 projections, Arnold, as acting chief (Westover being away from Washington), fought to have twenty bombers retained in the 1937 fiscal program.[74] The War Department fiscal year 1938 plan included a provision for the twenty bombers never contracted for but authorized in the 1937 budget. However, a memo further stipulated that if Congress granted the appropriation, the purchase of the planes would still be subject to Woodring's approval.[75] In mid-December 1936 the War Department further reduced the Air Corps 1938 budget by 27 percent by substituting sixty-one two-engine bombers for the same number of requested four-engine bombers. "The War Department kept giving us the B-17s on paper," Arnold recalled, but "then they kept taking them away."[76]

The battle for the bombers intensified again the next year, but Arnold and Westover kept the confrontation private, pushing when they could, but avoiding antagonizing their opponents. Andrews, whose command would have to fight with the planes that Westover procured, was not as obliging as his Washington colleagues. Until his time as GHQ Air Force commander ended in March 1939, Andrews fought the War Department, Craig, and even Westover and Arnold for more large bombers, fervently arguing the superior effectiveness of the big over the medium bombers. He demonstrated that a B-17, with its larger bomb load, dropped more bombs on target for the dollar, even though it cost almost three times more than a B-18.

Andrews wrote prolifically during the latter half of 1937 stressing the bomber's defensive role and the Flying Fortress's better value. In his defensive pitch he especially concentrated on the Pacific. In Europe the United States would not face an immediate threat, but initially it would have to support its European friends against Nazi aggression. In the Pacific, on the other hand, Japan's rising militarism threatened American interests in the Philippines and the Hawaiian Islands. Protection of these possessions occupied a central portion of American foreign policy and military planning. Only the large bomber, Andrews pointed out, could reinforce Hawaii, and could do so with two engines out. The smaller bomber could not return to the American mainland with one engine out, even with a tailwind. As he wrote in one of numerous memoranda, even "reconnaissance and bombardment airplanes [should] be confined to the four-engine model. Mere numbers of airplanes, in my opinion, are of less importance than . . . airplanes that . . . [are the] safest, the most efficient, the most modern and the best basic air defense weapon that the industry can produce."[77]

Despite his passionate and dogged pursuit of the B-17, Andrews tried to adhere to civil-military norms; the majority of his letters and memoranda written to push for the B-17 circulated through the proper channels. Since GHQ Air Force operated under the General Staff, he directed his papers there. He did not bypass his superiors or appeal directly to the Secretary of War, nor is there evidence he collaborated with Westover or Arnold, who fell under the Secretary of War, although Andrews properly coordinated and informed the chiefs of GHQ actions.[78]

In June 1937 Andrews took his plea directly to Arnold (Westover being in Idaho), and there the disagreement came out into the open. Andrews talked with Arnold for about thirty minutes and passed along his belief that the War Department had made "a terrible mistake" in buying the Douglas bomber (B-18) rather than the Flying Fortress. "We should buy the planes based upon fire power and not based upon numbers," he asserted.[79] Andrews was reluctant to consider other arguments, dismissing a balanced procurement program, two-engine bomber tactical requirements, and "objection to the 4-engine bomber by the General Staff, Secretary of War and possibly the President . . . with a wave of the hand."[80] This meeting between the two air leaders epitomized their differences over tactics. Both wanted to buy the large bomber, but they also realized the General Staff's and the Secretary of War's desires to keep Air Corps funding low in order to augment other Army branches, while also bringing the Air Corps up to Baker Board numbers. Andrews more aggressively pursued his case and pushed for the large bomber at all costs, while Arnold acquiesced to a balanced force with the heavy bomber as an important part, though he still informed his civilian superiors of the need for the large and expensive aircraft.

Andrews's propriety was revealed in a 17 June 1937 memorandum to the General Staff. He pointed out that the War Department's contracts for two-engine bombers of different types (the older B-10s and B-12s and the newer B-18), added to those in service, made a total of 438, or more than enough to fulfill current requirements. Since the Air Corps did not need additional medium bombers to complete its units, Andrews urged purchasing a minimum of thirty-five four-engine bombers.[81] A few months later, he stepped up his efforts after Woodring hosted a conference and pronounced the procurement of ninety-one four-engine bombers for fiscal year 1939 as impractical. Andrews then asked Craig to bring to Woodring's attention Andrews's alternate program. Andrews used charts to demonstrate the "more bang for the buck" theory to support his case for the larger bombers, which could also save over half a million budget dollars. Additionally, the heavy bombers would benefit the service by requiring fewer officers to man the planes while delivering 60 percent more tonnage of bombs.[82]

At the November conference, Assistant Secretary of War Louis Johnson, whose job primarily involved procurement, invited Andrews to submit the GHQ Air Force's ideas for aircraft purchases, and he ordered Westover to revise the Air Corps' purchasing plan.[83] Andrews reviewed Westover's plan and sent Johnson a detailed letter filled with charts and comparisons outlining a program for purchasing all GHQ Air Force aircraft for the next five years (fiscal years 1940–45).[84] Under this scheme, the Air Corps would reach its Drum and Baker Board authorized strength by 1945, though it required an annual Air Corps budget equal to that of the entire War Department's appropriation. The Andrews plan also procured heavy bombers exclusively (as far as bomber purchases) for all years except 1943, when the obsolescence of those medium bombers in service would require replacement.

Overall, Andrews presented a blueprint for a balanced force of fighters, reconnaissance, transport, and bombers. His proposal pressed to acquire the large bombers he believed essential, but not to the total exclusion of the medium type.[85] This plan represented a moderate, almost Arnoldesque, approach and differed from his previous writings. Perhaps Andrews felt it necessary to temper his views. Or, he may have been using different tactics to respond directly to Assistant Secretary Johnson, who was not in Andrews's direct chain of command. Johnson normally worked closely with Arnold, and Andrews, wanting to make sure he demonstrated balance, undoubtedly realized that Arnold would submit a competing program. Andrews did, however, attach a study entitled "Types of Airplanes Required to Execute Air Force Missions," which more firmly advocated the need for the B-17. His study emphasized the defense of the United States and its Pacific possessions and also projected a possible future requirement to conduct transoceanic attacks. He mentioned the B-17 on occasion and never specifically attacked the medium bombers by name, but he constantly referred to the need for bombers with a tactical operating radius of at least a thousand miles.

Andrews's files reveal very little official contact with the Chief of the Air Corps concerning procurement and War Department debates and arguments. However, one letter demonstrated Andrews's frustration with the War Department, specifically Secretary Woodring, and reminded General Westover of the need for the heavy bomber. Westover asked Andrews to review an article the former had written, entitled "The Army Is Behind Its Air Corps," one of Westover's efforts to build a positive relationship with the ground leaders. Andrews questioned the thesis, as he remained unconvinced that the Army fully supported the Air Corps, and he thought that Westover should include more on the need for heavy bombers. "Mere numbers of airplanes without regard to types and models mean little," he asserted; "the backbone of our air defense should be the 1,000 mile coast defense bomber."[86] Having the best aircraft "regardless of expense" was the key, and the recently procured medium bombers were "ill suited" to Army missions. He bluntly stated that the latest purchase represented "the first time that an inferior airplane has been procured when a superior one was available."[87] Jabbing at the War Department, Andrews sneered, "If and when the War Department truly recognizes the need for adequate personnel and for suitable bombers as outlined above, and sincerely fights for their provision, then and then only, I think, can we say, without reservation, the 'the Army is behind the Air Corps.'"[88]

Although he did not proactively circumvent his chain of command or civilian leaders, Andrews rarely passed up an occasion to reassert his bomber beliefs to the decision makers. In the late fall of 1937 he wrote a reply to Roosevelt's military aide and trusted friend Colonel Edwin M. "Pa" Watson. Watson began his assignment in the White House in the first New Deal administration, and later he came to control whom the president saw.[89] Andrews expressed his dissatisfaction with bomber procurement and his dislike of the B-18's performance, especially when compared to the Flying Fortress. He also emphasized the B-17 as more economical per ton of bombs delivered and urged that the larger bombers be procured "as rapidly as practicable."[90] Andrews understood that to bend Pa Watson's ear could influence Roosevelt, but he wrote the letter only at Watson's request for information and thus acted within Army General Orders and the civil-military norms of the era.

The first eight months of 1938 replayed the previous year's conflict. Air Corps leaders continued to push for bombers—Andrews usually aggressively and Arnold within the context of a balanced force. Arnold, true to form, made his aggressive personal comments on the bomber in his confidential daily record, softening his stance in official memoranda. Andrews, on the other hand, continued to make assertive, overt statements. The year began with an analysis of the competing Air Corps procurement plans submitted by Westover and Andrews.

The Assistant Chief of Staff for Supply (G-4), Brigadier General George R. Spalding, analyzed the two plans for General Craig in late January 1938. Westover's program included personnel, maintenance, research and development, and aircraft—in character with Westover's and Arnold's "balanced force" and macro view of the Air Corps. Looking at Andrews's plan, Spalding said, *"This program is not a balanced program."*[91] In keeping with the GHQ Air Force's mission and organizational purpose, Andrews limited his plan to aircraft, combat strength, and required support. His plan called for a larger percentage of heavy bombers than Westover's proposal—over a quarter of the entire Air Corps' aircraft. Spalding commented: "The bulk of the bombers should be of the type that are cheap in cost, easy of replacement, and capable of close support to ground troops and readily responsive to their needs."[92]

This comment typified the view of the Army leadership and reflected Craig's view that the primary mission of the Air Corps' was to provide air support for ground combatant forces. Since Andrews's plan included more heavy bombers, his total cost and average costs per year were both slightly higher than Westover's projections. Spalding agreed that some bombers should be of sufficient type to make long flights over water (four-engine type), but he leaned toward the ground-support bombers. His overall recommendation sided with Westover's plan, which procured fewer of the larger aircraft.[93] One week after Spalding submitted his report, Andrews responded.

In addition to attacking other shortcomings noted by the G-4 report, Andrews provided more information to support the purchase, as soon as possible, of the heavy bomber. He argued that the Air Corps should equip itself with only two bombers: the attack bomber for ground support and the heavy, long-range bomber. Missing from his plan was any mention of the medium bomber preferred by the frugal-minded Woodring. Andrews believed both Spalding's and Westover's plans procured heavy bombers at too slow a rate and thus endangered national defense. Again he aggressively advocated heavy-bomber purchases to complete GHQ quotas by 1943, and Air Corps quotas by 1944, instead of the alternative plan delaying these allocations for more than a year. "It is believed that such a lapse of time in furnishing the tactical units with this essential vehicle for defense may prove unsound planning."[94]

In the summer of 1938, in his Confidential Record of Events, Arnold noted the possibility that the fate of the four-engine bomber could be decided by the Joint Board of Aeronautics, an offshoot of the Joint Army-Navy Munitions

Board that reviewed and coordinated industrial planning for the services.[95] Arnold was not optimistic about the board's upcoming review: "God help any further development in large airplanes by the Army Air Corps, because in my opinion we haven't a friend on the Joint Board."[96] The Joint Board confirmed Arnold's fears just one month later. It agreed with limited B-17 purchases but recommended against developing long-range aviation beyond current capabilities.[97] Around the same time, the War Council (consisting of Woodring, Johnson, Craig, and Craig's deputy, Brigadier General George C. Marshall) issued a mixed decision on bomber procurement. The group decided to cancel a circular proposal sent out to the aircraft manufacturers for a new two-engine bomber because none of them were building this type of plane. Instead, they substituted as many B-17s as could be procured—thirteen on a contract option from Boeing. After that bit of good news, the rest turned sour. The board recommended using the remaining savings from the canceled two-engine bomber proposal to purchase more of the older B-18s than even Arnold and Westover wanted. Westover had planned to purchase seventy-eight B-18s, but the War Department increased that number to eighty-eight, while canceling the Project A bombers. Arnold, in a memo to Westover, disagreed with purchasing more than seventy-eight B-18s and did not like canceling the four-engine experimental bomber project.[98] Arnold agreed that the B-18s offered minimal value as a combat aircraft, but he acquiesced with the War Department's purchase in limited numbers because B-18s could train bomber crews.

By this time, Johnson had been partially swayed by Andrews's arguments and advocated a bomber-heavy air force, though on an economy plan. When an officer asked about small bomber and pursuit aircraft purchases and the views of the Joint Board, Johnson replied that "it wouldn't do the Board a bit of good to come to any such findings."[99] Johnson did not want fighters or small bombers. He wanted to purchase the maximum number of bombers, and he mentioned, in words that would have caused Andrews and Douhet to smile broadly, "All we [the Army] needed was a mass of bombers."[100] Johnson's position, somewhere between those of Woodring, Arnold, and Andrews, did not totally agree with any of them. Like the Air Corps officers, Johnson wanted a large Air Corps, but, agreeing with his boss, he favored numbers over type. Disregarding Arnold's and Westover's quest for a balanced force, Johnson was willing to scrap purchasing anything but the cheaper bombers en masse.

In midsummer 1938, Andrews received word of the planned procurement of the seventy-eight B-18s and protested to the General Staff. Referring to seven previous recommendations he had written between August 1936 and November 1937, he argued vehemently against the B-18 from a performance standpoint. Using an analysis of "current wars," he indicated that between 60 and 70 percent of combat losses occurred from air combat, and a bomber's best defense was a combination of speed, altitude, and its own guns, with the first two being the most effective. The B-18 would be "at the mercy" of current pursuit aircraft. "To continue to equip our units with airplanes of low performance handicaps national defense, and is without justification; particularly in view of the fact that airplanes of greater performance are available."[101] The British would not purchase the B-18 because of its lack of speed; they wanted the Fortress. Andrews concluded by strongly recommending against "additional tactical aircraft of inferior performance."[102]

In the final month before Neville Chamberlain would hand part of Czechoslovakia to Hitler, the bomber debate entered its final stage. Woodring had stepped up his efforts to purchase B-18s as a cost-saving measure, and officers from his office urged the him to exercise options on the B-18.[103] Arnold countered that this would purchase five-year-old aircraft and invite criticism of the whole Air Corps and of Johnson, particularly from Andrews, "as the G.H.Q. is violently opposed to buying any more B-18's."[104] Just over a week later, Woodring's office ignored Arnold's advice and sent a memo outlining purchasing aircraft in 1939, including exercising options on more B-18s. Arnold requested an immediate conference on the matter. The G-4 then called to say that the purchase memo had been sent by mistake. Arnold still wanted his Air Corps plans office to provide a written response.[105] Finally, and with negotiations continuing in Munich, Arnold talked with Woodring and explained that only the most modern bombers should be bought because war was clearly coming.[106] When Woodring finally agreed to buy newer bombers, he still wanted to pursue a two-engine type because of insufficient funds to purchase larger aircraft.

Three days after he agreed to purchase the newer aircraft, Woodring notified Arnold that he would purchase no four-engine bombers in 1939 but instead ninety-one of the smaller bombers. Arnold responded that attack units would have too many aircraft and heavy-bomber units too few, a mistake because "bombardment aircraft are the backbone of all air force operations."[107] Arnold wanted to discuss the matter with Woodring further, even if it meant delaying the purchases until 1940, but Woodring wanted to have the full complement of 2,320 aircraft by 1940.[108] Once again, the Baker Board numbers—and the means to get to that figure as quickly and economically as possible—dominated Woodring's thinking. The General Staff also influenced Woodring, providing tactical reasons for the smaller bomber. The General Staff argued that ground forces remained the "Queen of Battle"—the basic combat element—and that therefore the Air Corps should concentrate on producing aircraft suitable for close air support of advancing troops.[109]

Thus by October 1938, the Air Corps, through the usual gentle prodding of Westover and Arnold and the aggressive statements of Andrews, remained in much the same situation as in 1935, when the Baker Board had reasserted the need to acquire aircraft to bring the Air Corps up to a 2,320-aircraft limit. Since that time, Woodring had pushed to obtain older and inadequate, but cheaper, medium bombers, while the Air Corps pressed for the long-range heavy bombers. Arnold and Westover had sought balance, while Andrews pushed for the heavies, using arguments ranging from economy to efficiency to needs for adequate defense and mission requirements. In those arguments, Woodring's view endured, and the air leaders grudgingly accepted his orders without reverting to the antics of Mitchell or Foulois. Their patience and adherence to civilian control of the military kept them in the administration's good graces, which would soon work in their favor. More significantly, the Munich agreement would focus Roosevelt's thoughts on air power and cause him to intervene in the bomber debate and change Air Corps procurement.

MUNICH AND THE AIR CORPS' "MAGNA CARTA"

After September 1938, the arguments over meager funds and types of aircraft changed. Roosevelt began overtly to favor the Air Corps over the other Army branches (and perhaps even the Navy) and signaled that the air arm would indeed become special when it came to appropriations. Also, another tragedy would alter the Air Corps leadership. On 21 September 1938, General Westover died when his plane spiraled into the ground on an attempted landing at the Lockheed plant in Burbank, California. With the plane still burning on the runway, a fellow officer called Arnold back in Washington.[110] Eight days later, after surviving rumors started by unknown rivals that he had a drinking problem, Arnold became Chief of the Air Corps.[111]

Probably more than anything else, Roosevelt's appointment of Arnold as chief validated the Air Corps' progress in efforts to work within the Army and with the civilian leadership. The former "Mitchellite"—whom Major General Mason Patrick had threatened with a court-martial for violating Army General Orders governing civil-military relations; who had helped operate a virtual intelligence gathering and propaganda at the Information Division offices of the air chief to further air power; who had been exiled to Kansas—now ascended to the chief's office largely because he had changed his tactics. Once an air insurgent who supported immediate independence from the Army and led efforts to circumvent civilian policy makers, Arnold now championed cooperation with civilian and military leaders and rejected immediate autonomy as not being in the air arm's best interest. Although he would temporarily fall out of Roosevelt's favor in 1939, for now the president trusted him to lead the service.

Perhaps Roosevelt selected Arnold because during the previous three years Arnold had demonstrated an attitude of cooperation as assistant chief. Perhaps FDR did not want to upset what had become a new tradition: since 1927, when Patrick gave way to Fechet, every assistant chief had taken over the top air job upon the chief's retirement. Since the establishment of the Air Service in 1920, Billy Mitchell was the only assistant chief who had not become chief. Or Roosevelt may have believed that Arnold's long and deep experience in every aspect of the air arm made him the most qualified. Arnold had learned to fly at the Wright brothers' school in 1911. He was skilled in dealing with the press and Congress on aviation issues dating back to his first Washington assignment in 1912, and he had significant command experience. Perhaps most importantly, Roosevelt knew that Arnold understood the intricacies of what was needed to build an air force, from research and development to training and organization to procurement and the aviation industry. By September 1938 Roosevelt probably knew he would have to expand the Air Corps, and he wanted an officer who he knew could do the job, even if that officer had "spots" on his record from a decade earlier. Arnold hardly had time to settle into his job when Neville Chamberlain, promising "peace in our time," signed the Munich agreement, giving parts of Czechoslovakia to Hitler without a shot being fired. The Munich Conference pulled Roosevelt to the forefront of the defense debate.

Within another month, and four days before the Munich Pact was officially signed, FDR expressed a desire to expand the Air Corps significantly, but gave no details. The conversation was with Assistant Secretary Johnson. Knowing Woodring's position on bombers, the president had sent Woodring on a political tour at that specific time so that Johnson might work with those in the War Department who would agree with a large Air Corps program based on heavy bombers.[112] Arnold, realizing that Woodring's plans would not satisfy the president, submitted a larger plan, but still not as large as Johnson envisioned.[113]

Not satisfied with the plans submitted to him by the War Department, and unhappy with the misunderstanding of what he actually wanted, Roosevelt called a meeting of key civilian and military leaders on 14 November 1938. In addition to Johnson, Craig, and Marshall, attendees included Secretary of the Navy Charles Edison, Secretary of the Treasury Henry Morgenthau Jr., and others involved in military affairs. Roosevelt, believing that the threat of the Luftwaffe had played a primary role in Chamberlain's appeasement, wanted an Air Corps that would influence Hitler. The president vented his unhappiness with the War Department's handling of expansion programs and emphasized that he wanted a large and powerful Air Corps heavy on big bombers.[114] Arnold remembered the president blaring, "A new regiment of field artillery, or new barracks at an Army post in Wyoming, or new machine tools in an ordnance arsenal . . . would not scare Hitler one blankety-blank-blank bit! What he wanted was airplanes! Airplanes were the war implements that *would* have an influence on Hitler's activities!"[115]

Arnold later called the president's statements the Air Corps' "Magna Carta." FDR called for actual production of 10,000 planes the first year and a capacity to produce 20,000 per year afterward. Arnold said this decree gave the Air Corps, for the first time in its history, "a definite goal of planes from the factories."[116] Years later, while writing his memoirs, he still could not contain his enthusiasm for the president's words that day: "A battle was won in the White House that day which took its place with—or at least led to—the victories in combat later."[117]

While FDR met with the other top leaders at the White House, Secretary Woodring sat at his residence, only ten minutes away, unaware of any conference. He had returned from his political trip, but Roosevelt did not invite him to this meeting because of his views for a balanced buildup including Army ground forces.[118] Woodring's opposition to a large air force based on heavy bombers lost out to the president's newfound focus. Roosevelt's words and actions also showed General Craig that he must change his views. Arnold, immediately after the meeting, drove Craig to the Air Corps offices to "give him a get-rich-quick course" on how to build an air force, and "He was a very apt pupil, and from then on until his tour was completed, fought for our program."[119]

THE WINDS OF CHANGE

For almost twenty years, from the early air power enthusiasts to 1935, the Army's air leaders fought with the ground leaders for independence, recognition, money, and organizational control. Even under the steady hands of Generals Patrick and Fechet, the air service hoped for independence at the soonest possible moment, which caused continued conflict with the General Staff. The new attitude and forceful leadership from Westover and Arnold limited the dreams of immediate independence in favor of cooperation within the Army and building up confidence in the service. The Air Corps used Arnold's and Eaker's lessons about good press coverage and expanded goodwill with air power demonstrations.[120]

The new strategy also kept civil-military relations inside the norms of the day. Even Andrews's passion remained limited to presenting his views in the proper forum. The Air Corps' moderation did not mean that the top officers became unquestioningly acquiescent to Woodring's desires, but it created a more appropriate working relationship. Civil-military conflict remained, but it occurred in the normal fashion of disagreement privately inside the executive branch, in proper channels and out of public view. That the Air Corps leaders were themselves divided undoubtedly helped to avoid confrontation and to enhance the civilians' authority. The conflicts were almost "inherent" in the circumstances: limited money and conflicting views of how best to spend it. Politics and economy closed with military requirements.

From 1935 to 1938 Harry Woodring stood in the eye of the civil-military storms. As Assistant Secretary, Acting Secretary, and Secretary of War, he never fully accepted the need for a quality air force and the Air Corps' reasons for pursuing the heavy bomber. Even when faced with statistics confirming that the large and expensive aircraft actually placed more bombs on target per dollar spent, he pushed for the less-expensive medium bombers. Roosevelt intervened only after it became apparent in late 1938 that the world situation required spending the money, and preparing his military and the civilian industry, for large-scale production. The president became convinced that only airplanes offered the hope of deterring German aggression.

However, these debates remained internal. The civilian leaders and military men did not fight their battles in the press or in investigations with partisan wrangling and congressional involvement. In fact, Congress played a smaller role in Air Corps affairs than it had in the previous decade and a half, and the need for Air Corps allies in Congress had abated. The Democrats fell in line behind presidential wishes, and only the failed Wilcox bill attempted to undermine Roosevelt's military desires. Without a galvanizing issue, the diminished and demoralized Republicans could not become a thorn on aviation to Roosevelt as the Democrats had been to the previous GOP presidents.[121] The approach of Westover and Arnold did not suddenly reverse the Air Corps' fortunes, but it inverted the previous patterns of civil-military relations. These politically astute flyers set the service on a more certain path toward viability, respect, and an appreciation, especially by the president, of air power's capabilities and promise.

CHAPTER SEVEN

The Politics of Air Corps Expansion, 1938–1940

"Airplanes—now—and lots of them!"[1] "Hap" Arnold vividly recalled President Roosevelt speaking these words at the "Magna Carta" meeting of 14 November 1938 in the White House. The then-unprecedented numbers of aircraft FDR wanted to roll off of American assembly lines reflected a dramatic shift at the top of the U.S. government regarding air power. The regular War Department budget for 1940, initially drafted two years earlier, had planned fewer than two hundred new aircraft for the Air Corps, yet the president now talked about purchases in the thousands.[2] The exact procurement numbers waxed and waned over the next two years as Congress and the administration sought to balance the Air Corp's expensive purchases with other military programs and American economic resources. The Army's expansion program also competed with Roosevelt's desire to sell aircraft to other countries and congressional partisan politics of isolationism and economic pump-priming.

In May 1940, German victories in Western Europe jolted the president to act again in favor of defense spending and aviation expansion programs. Roosevelt asked Congress to raise the Army and Navy air arms to 50,000 aircraft and increase annual production to that same number, and Congress responded to the German threat by giving the president $500 million more than he requested.[3]

The president's 1938 10,000-plane program favored the Air Corps over other branches of the Army, but aircraft did not begin rolling off the assembly lines and filling the service's ranks in 1939. With Republicans teaming with moderate Democrats and a strengthened isolationist bloc, a newly emboldened Congress moved to check Roosevelt's power. FDR's cabinet also differed on buying aircraft, on how many to sell abroad, and on how to utilize government funds for military preparations and expansion. Roosevelt's personal intervention reshaped the civil-military relationship regarding the building of the Air Corps, but the president's interest did not eliminate all of the problems. When Roosevelt came out strongly for the Air Corps, he minimized civil-military conflict, but when he pitted one subordinate against another, as he so frequently did, he fomented such conflict.[4]

The steadying influence and power of Arnold kept the Air Corps largely out of political trouble. Arnold maneuvered to stay clear of partisanship and convince military and civilian leaders of his plans for a balanced force and production capacity for expansion. Summoned repeatedly to congressional hearings, the air chief fought to better control the money appropriated to achieve the balanced force he—and the Army—supported. He also urged the president and his cabinet to equip the Army with newer and better aircraft before selling them to foreign countries, and in this arena he clashed with Treasury Secretary Henry Morgenthau Jr. Morgenthau's intervention in War Department affairs would cause Arnold grave difficulty with Congress and the president, and it almost cost the air chief his job.

The eighteen-month period from Roosevelt's startling "Magna Carta" to the proclamation of the 50,000-aircraft program in 1940 clearly demonstrated how the Air Corps had evolved from using insurgency tactics to becoming a steady, reliable, and trusted part of national defense—one that could, given the chance, prove that its doctrine and weapons might someday justify autonomy. Along the way, Arnold showed himself as being firmly in control of the air service, and he worked with, and even converted, the top Army leaders to understand and support the need for a strong and balanced force. The former air radical and Mitchellite now worked within the rules of the Army and the administration, but he did so without ever forsaking the doctrines and programs that would eventually win the air service independence.

THE FOX BECOMES THE HUNTED: THE 1938 POLITICAL CHANGES

The Democrats' dominance, begun in the 1932 elections and strengthened in 1936, began to slow within two years of the record-setting reelection victory. Roosevelt miscalculated trying to manipulate a change in a Supreme Court that had undercut many of his New Deal actions. The president's 1937 scheme to "pack" the Supreme Court wounded him politically, and Democrats who had chafed for five years under FDR's stern control moved to form a conservative coalition with the Republicans.[5] New Deal legislation slowed, and Congress no longer enthusiastically followed the president's lead.[6] The failed court fight and fissures in Democratic ranks led to Republican gains in the 1938 midterm elections. The GOP gained eighty seats in the House and seven in the Senate, though firm Democratic majorities remained in both chambers.[7] The election did not repudiate Roosevelt or the New Deal, but it resurrected the Republicans, who looked forward to the possibility of regaining the White House in 1940. By tradition, no U.S. president had stood for a third term. Thus Republicans acted more boldly, believing Roosevelt was losing his political power.[8]

As Europe edged closer to war, the isolationists in Congress increased pressure not only keep to the United States out of European entanglements but also not to arm either side. The isolationist coalition's strength rested in senators from the Midwest and Northwest and cut across party lines, though Republican William E. Borah of Idaho led the group. Other prominent isolationists included Republicans Gerald P. Nye of North Dakota and Arthur Vandenberg of Michigan, Progressive/Republican Robert La Follette Jr. of Wisconsin, and Democrats Key Pittman of Nevada and Burton K. Wheeler of Montana.[9] This group had helped pass the 1935 Neutrality Act, which shut off U.S. support for all belligerents in the event of war. Roosevelt wanted the discretion to impose the act selectively in order to deter aggressors, but Pittman warned him that such a provision would not pass the Senate, and Roosevelt gave the isolationists a victory he would later regret.[10] The president, and tangentially the Air Corps, would have to battle the isolationist elements for the remainder of the pre–World War II period.

The return of more influential partisanship, combined with Roosevelt's insistence that air power take the lion's share of defensive preparations, put Arnold in a difficult position. While the new air chief wanted a balanced program, Roosevelt clamored for bombers and the War Department worked to improve the non-aviation elements of the Army. Arnold needed to tread carefully—in the War Department, in public, and in all congressional appearances—lest the aviators forfeit the confidence of their supporters and superiors.

THE FIGHT FOR NUMBERS: EXPANSION PLANS AND POLITICAL CONSEQUENCES

From the late-1937 preparations for the expansion program to the outbreak of war in Europe in September 1939, the numbers of aircraft and what nation they would be sold to bounced around among Congress, the War Department, and the White House. Arnold wanted the aircraft to build up a still-understrength Air Corps. Congress supported U.S. defensive preparations, but isolationists and anti–New Dealers were wary of Roosevelt's true intentions. Isolationists worried that Roosevelt's interests meant producing aircraft for foreign sales, while the anti–New Dealers viewed the president's plan as another economic stimulus program costing massive amounts of tax dollars. Roosevelt never specifically stated his real intentions, but he apparently desired both: building a stronger military and supporting democratic European governments against Nazi aggression. Certainly he understood the positive economic benefits of government spending on a massive building program. However, by not clarifying his objectives to his staff, he triggered a conflict with the Air Corps in the middle.

Arnold later reflected that Roosevelt's aviation program signified an appreciation of air power and was not a spur-of-the-moment or haphazard statement.[11] In preparation for the 14 November 1938 meeting, requests began to trickle down to Arnold and his staff for planning recommendations. Upon his return to Washington on 13 October 1938, Roosevelt's ambassador to France and presidential confidant, William C. Bullitt, met late into the evening with the president and discussed the European situation. During a press conference the next day, Roosevelt admitted to an ongoing defense reevaluation, dating back to 1937, and said that further information would be available by 3 January 1939.[12] These statements led proponents of air power, seeing an opportunity due to the world situation, to lobby again for a larger air force.

Assistant Secretary of War Louis Johnson pushed both the president and Chief of Staff Malin Craig to increase aircraft production, noting that the previous year's production amounted to less than a thousand aircraft and that representatives from aircraft industries believed production could be doubled in one year. Johnson urged a thorough review and revamped planning for wartime production, calculating that the industry could not meet minimum initial war requirements until possibly two years after hostilities had begun. He wanted to stockpile production and war reserves.[13] Roosevelt agreed to the study and instructed Johnson to investigate ways to increase production, but the president looked toward arming Europe, not just the Army Air Corps. Within two weeks, Johnson presented a plan envisioning 31,000 planes by 1940 and an annual production of 20,000.[14] Although Arnold did not entirely agree with Johnson's plans and vision, the air chief averted any conflict by working within the War Department and through his chain of command to ensure that Craig and Secretary of War Harry Woodring stayed informed of Air Corps wishes.

Arnold responded to Johnson's request for ideas and numbers for a possible Air Corps expansion by urging balance within the Air Corps and in aviation's place in the overall defense structure. He informed Johnson that the air forces must remain coordinated with the other services, tied to national and international policies, and coupled with industrial resources. With a keen understanding of the political forces and public reactions, Arnold wrote that if the War Department planned on advocating any major changes in any of the services, both Congress and the American people would need to be convinced of the soundness of the program and the necessity to build over a period of years. He also urged, prior to any announcement, the appointment of a special council to advise the president on the expansion goals and numbers for national defense overall and the Air Corps in particular. In the absence of such a body, Arnold believed Roosevelt should "advocate for the Army Air Corps a round number objective of reasonable size, fully realizing as time goes on, the [final] number . . . must be changed."[15]

After briefly recapping the status of European air forces and the Air Corps' missions, Arnold put forth a tentative goal of 7,000 aircraft of all types and a supporting production capacity of 10,000 planes per year achieved within two years.[16] Arnold's plan, though far more conservative than Johnson's, still recommended a great leap over the production capabilities present in the late 1930s. While Arnold advocated balance, he also reiterated the need to build an adequate number of long-range bombers with "performance and numbers second to none," which meant the B-17s.[17]

The disparity between Arnold's and Johnson's numbers displayed obvious differences between the military and civilian planners of Air Corps expansion. It also highlighted the recent trend of improved working relations between the Air Corps officers and civilian officials. Johnson did not fully accept Arnold's balanced force plan. Although Johnson was an advocate for air power who earlier sided with the Air Corps in the big-bomber "quality versus quantity" debates, his change to even larger numbers probably came about due to his association with New Dealers who wanted to increase government spending in the private sector, not from a deeper understanding of air power.[18] However, Arnold's plans, and his dealings with Johnson, underscored the nonconfrontational approach that Arnold and the Air Corps had adopted after late 1935. Arnold would push for a solid, realistic program and adjust the numbers up or down as needed. He realized that Johnson's numbers, if accepted by Roosevelt, were not a balanced force. The exponential jump in numbers even of Arnold's smaller program represented more than a tenfold increase in annual production from the previous year within the space of two years, tripling the size of the current authorized strength of the Air Corps. The Air Corps staff required adjustment to larger numbers. Arnold had earlier asked the staff to use its imagination and estimate the essential number of aircraft needed by the Army, with an eye toward world events and American defense needs, including the Caribbean and the Philippines. The total from all the planners amounted to 1,500 aircraft.[19] If Arnold's staff could not foresee such a huge increase, the air chief knew that such numbers would be hard to sell to Congress and the public.

With the numbers bandied about the War Department and in the press, Arnold believed that of the participants in the November 1938 meeting, only he and Harry Hopkins, administrator of the Works Progress Administration, appeared unsurprised at the president's bold proposal for 10,000 aircraft. The announcement shocked General Craig. Craig and his deputy, Brigadier General George C. Marshall, clearly realized that the president's support for a bold Air Corps expansion program changed the defense equation and would lessen the intraservice rivalries of the previous decades, and thus began a more active integration of the Air Corps in War Department planning and an improved relationship between the Army Chiefs of Staff and the Air Corps.[20]

Arnold recognized Marshall as one of the three greatest supporters of air power in the War Department and the man who most helped the air chief in his job.[21] Marshall shared Arnold's views that aircraft alone did not constitute an air force, and he joined the air chief to obtain a balanced force with a proper support structure. Marshall supported the Air Corps' buildup, but he also warned against putting too much money in the air arm and not having the ground force to fight for and hold the ground for the planes to land upon.[22] Major General Frank Andrews, still the GHQ Air Force commander, had also forged a solid relationship with Marshall the previous summer. Andrews had escorted Marshall on a tour of GHQ Air Force and its operating units to teach him more about the Air Corps, its current situation and needs, and its capabilities. Marshall enjoyed the nine-day excursion and remarked how the knowledge he gained would serve him well in his Washington duties.[23] The relationships forged early between the Air Corps leaders, especially Arnold, and the Deputy and future Chief of Staff began to pay dividends immediately.[24]

In preparation for the president's January 1939 address to Congress, Marshall and Arnold worked closely to iron out the Air Corps numbers and budget requests, while Johnson worked a parallel path. Johnson's initial plan, sent to Roosevelt on 1 December, requested $1.3 billion for the air expansion program and another $543 million for Army ground force projects. Unhappy with Johnson's proposal, Roosevelt called a meeting of his military advisers and chided them for giving him a plan with everything but airplanes. He also affirmed his belief that Congress would not grant anything over $500 million. When some of the advisers stated that the aircraft might be obsolete before they could be used, Roosevelt countered that the British could use them if the United States could not. Again, Roosevelt showed only part of his hand. Andrews believed that the president wanted the bombers to build a more capable Army Air Corps, but Roosevelt also understood the tense situation in Europe. The Germans had conquered Poland quickly and had begun reinforcing their western borders. With nervous French sitting along the Maginot Line and Britain vowing to come to the aid of Belgium and France, Roosevelt knew he did not have long to reinforce Europe's democracies. Once hostilities began in Western Europe, the Neutrality Act would limit his ability to help Britain and France. Only FDR knew for certain why he wanted such massive and swift bomber production, but he clearly understood that the United States would have more time to arm.[25] As he kept his hand close to his vest, his military staff prepared plans for a massive U.S. air force.

In the aftermath, Marshall asked Arnold for the specific information to develop a complete plan. Marshall wanted to avoid Johnson's mistakes and give the president what he wanted while still retaining overall Army balance. Marshall's requests for information demonstrated his knowledge of the Air Corps and its requirements as well as a keen political awareness of Roosevelt's style and desires. Marshall asked more than twenty specific questions, ranging from per-unit costs and replacement estimates, to annual operations and support calculations, to costs of an average concrete runway.[26] Arnold's balanced views played well with the influential Marshall. The two men realized that any plan would have to be defended before a Congress that had begun to demonstrate a reluctance to approve any administration plan blindly.

The conservative alliance of Republicans and anti–New Deal Democrats formed a bloc in Congress that openly worried about increased defense spending and its budget-busting repercussions. The members also resented any New Deal measure to pump money into the economy at the expense of increased government deficits, which some congressmen labeled a "Federal spending-lending" program. The newly emboldened coalition led the charge.[27] Democratic congressmen who already openly opposed increased defense spending also leaned toward isolationism. Representative Louis L. Ludlow of Indiana sponsored an amendment requiring a referendum on any proposal to send U.S. troops abroad. The vote on this amendment awaited the next congressional session, as opposition by Roosevelt and Secretary of State Cordell Hull had forced the legislature to delay debates at the previous session. The amendment forewarned of a Congress probably not inclined to increased defense spending.[28] Senator Rush D. Holt of West Virginia, a first-term Democrat who had to wait for his thirty-fifth birthday before he could take his seat, insisted that before the legislature could approve increased defense spending, the executive branch must prove the "need for vast expenditures for armaments."[29] Senator William J. Bulow, a South Dakota Democrat, struck a similar tone and called the information coming from the White House "proposal propaganda."[30]

Republicans joined in and urged a serious need to look at new defense spending, which they said would sap government treasure at the expense of other domestic programs. Senator Borah, who served on the Committee on Foreign Relations, warned that massive expenditures on national defense would be a waste, as more pressing national matters loomed. Borah also advised against the United States' joining a "hysterical" world arms race and tacitly endorsed the Ludlow amendment. Senator John G. Townsend Jr. of Delaware reminded Americans that a reasonable program would receive the necessary attention, but he felt that the "scheme" increased governmental spending.[31]

The costs of the aircraft—the prime element in any new spending program —loomed large in the debate. The term "pump-priming" came into fashion, primarily due to Louis Johnson. He commented occasionally to the press and was a known collaborator with other New Dealers who wanted more government money flowing to the civilian industries. The Committee on Federal Finance of the U.S. Chamber of Commerce warned against using large military armaments for this purpose, since, by their nature, such programs took several years to complete, and controls to have them fill slack industrial work times proved extremely difficult. The U.S. Chamber of Commerce also warned Congress against deficit spending and asked lawmakers to cut the fiscal year 1940 budget and end deficits by 1941.[32] Though willing to fund certain military projects, some congressmen did not want defense spending as another New Deal project. Conservatives wanted to eliminate deficit spending. Thus, congressional lines became drawn a full month before the planned presidential address to Congress. Conservative Democrats would join Republicans and the isolationists to quell any military programs that exceeded the bare minimum necessary for defense of the United States.

During the second week in December, the *New York Times* began running editorials on the state and needs of national defense. The views struck a moderate line and reflected the popular mood on defense spending and the Air Corps. The *Times* called for a careful study of current capabilities and a balanced program of modernization. It endorsed the need for increased spending but not a "sudden sensational expansion." Three dangers existed: hysteria causing the nation to become "swept along on a high tide of oratory"; oversimplification of the problem leading to an unbalanced program (here, a specific warning was given on spending too much on one element of national defense, and the editor specifically noted aircraft as an example); and the "pump-priming" worry—unneeded armaments only to boost employment and industry output.[33]

On the fourth day of the defense editorial series, attention turned to the need for an "adequate air force." Calling the current air force the equal, if not superior, of any in the world, the editorial continued to advocate increased funding for research and development of aircraft and their instruments, supplemented by marginal spending on new aircraft. The paper said that although the Air Corps was under-strength, "no tremendous increment" was needed to ensure American defensive needs. In a statement that must have warmed Arnold's heart, the paper called for a balanced expansion, including not only aircraft but also personnel, training, equipment, and bases.[34]

A Gallup poll confirmed that the American public supported the view of the *New York Times,* providing even more support for Arnold and the Air Corps. Ninety percent of those surveyed wanted a larger air force. The survey demonstrated public approval for expansion in all military branches, but the air forces polled the highest: 4 percent above the Navy and 8 percent above the Army.[35] As it had since the end of the World War I, the American public supported the Air Corps, and now the leadership style of Arnold confirmed their confidence, and did not detract from the public's perceptions of the service, as Billy Mitchell had done.

During the final days before the president's address, Arnold and Marshall worked feverishly to complete the program. Johnson became the main impediment to achieving an amicable solution. Until the last hours before Roosevelt presented his plan to Congress, Marshall and Brigadier General Walter Kilner, acting in Arnold's place while the latter conducted a trip to Detroit, worked out final details and worried over Johnson's interference. The previous day, Johnson stated to Marshall his opposition to the "balanced force" proposal. Johnson wanted a ratio of 60 percent combat aircraft purchases to 40 percent support and training aircraft. He also wanted to add 4,000 combat aircraft to the plan, costing an additional $50 million. Marshall opposed Johnson and tried to convince him that the country needed a balanced force and that the $300 million request required the inclusion of expenditures for personnel, construction costs, bombs, and the remaining infrastructure needed to support the combat aircraft. Marshall prevailed and sent the balanced program to the president for his submission to Congress, and the War Department staffs prepared to defend it there.[36]

In his appeal to Congress for additional appropriations for military expenditures, Roosevelt reiterated the need to prepare before any war caught the nation off guard, as had occurred in 1917. The additional request, $525 million overall, allocated over $450 million for the Army alone, with $300 million of that solely for the Air Corps.[37] Calling the Baker Board's cap of 2,320 "completely out of date," the president appealed for a revision of aircraft needs. His proposed number, however, represented a one-third reduction from his "Magna Carta" announcement. He undoubtedly revised the figure downward to conform to the realities of American production capability, revealed by the Johnson and Arnold studies, as well as the public and congressional reactions over the previous month. He believed the money allotted to the Air Corps could purchase 3,000 additional planes, perhaps more if mass production reduced the per-unit costs. In addition to $300 million for the Air Corps, he asked for "educational orders" of $32 million for the Army to prime industries and prepare them for probable future large-scale contracts.[38] The debate now moved to Congress.

The Senate Military Affairs Committee began discussions on the expansion program and appropriations on 17 January 1939. In their testimony, Woodring, Craig, and Arnold presented a unified front and supported Roosevelt's program, which thoroughly encompassed Arnold's balanced force plan. Woodring and Craig even recommended raising the cap on aircraft numbers of the Air Corps Act.[39] Craig strongly advocated the balanced force plan and the need to build the infrastructure as well as the aircraft—again attesting to the influence of Arnold. Craig emphasized that training mainte-nance and flying crews took longer than building a plane and that a combat plane was useless without armament and bases.[40] While Craig and Woodring supported the air expansion program and demonstrated their conversion to the aviation cause, they left the specifics to Arnold.[41]

Arnold stressed the need to keep manufacturers employed and to use funds to purchase developmental aircraft not awarded contracts in order to keep the aircraft companies from going out of business. Here the lawmakers asked if the companies should be allowed to sell aircraft to foreign countries. This line of questioning led to a minor crisis that diverted the hearings for most of the session. During separate questioning, Arnold and Craig both agreed that manufacturers could, under certain circumstances, sell aircraft to foreign countries. Craig underscored the procedure: once the War Department de-termined that the model was no longer required by the U.S. military and that the aircraft's abilities would not outperform models being purchased by the Army, it then released the aircraft. After approval by the War Department, the State Department allowed the aircraft company to sell the plane to a foreign government.[42] Arnold agreed with this procedure and reminded the senators that this method provided a means to keep manufacturers employed and profitable without relying on American orders. Events soon provided an opportunity for opponents of foreign sales, squeezing Arnold and the Air Corps between the administration's desire to sell aircraft and the isolationist elements in Congress.

ARNOLD, MORGENTHAU, AND THE FRENCH

On 23 January 1939 one of the country's newest military aircraft, the Dou-glas attack bomber, crashed at Los Angeles Airport. The incident killed the Douglas civilian test pilot and ten civilians on the ground. However, one in-jured passenger, Paul Chemidlin, a representative of the French government, soon overshadowed the other casualties. Smacking of a cover-up, initial re-ports identified the Frenchman as "Smithson," a Douglas mechanic.[43] Dur-ing Arnold's third day before the committee, and the second straight day of being the prime witness, Senator Bennett Clark, a noted isolationist Demo-crat from Missouri, took the chance to grill Arnold over the incident.[44] The political controversy resulted in testimony by Woodring, Johnson, Arnold, and Morgenthau, along with several military officers, as to why a foreign official could observe a top-of-the-line piece of military equipment, who was responsible, and what, if any, military secrets were compromised. It took the next three weeks and monopolized the majority of the committee's session on the defense appropriations. The controversy also exposed a schism be-tween Arnold and Morgenthau and between senators for and against the Air Corps program and expanded defense spending. The resulting fracas ended with Roosevelt's having to openly defend a desire to sell aircraft to other countries and Arnold's "exile" from the White House for nine months for not "playing ball."[45]

Very early in the questioning, Senator Clark inquired if Arnold knew why a Frenchman flew in a sophisticated aircraft preparing for the Army's forthcoming contract-awarding competition. Arnold immediately replied that the foreigner participated "under the direction of the Treasury Department" for possible foreign purchase. Clark, alluding to the earlier comments by Arnold and Craig as to who controlled foreign purchasing, pointed out that the Treasury had no legal authority in the process. Arnold, indeed walking a thin line, remarked that until the United States purchased the aircraft, it remained property of the Douglas Company. In an exchange in which Clark repeatedly interrupted Arnold, Clark wanted to know who authorized the flight and what secrets may have been exposed.

Two days later, Morgenthau, along with four other Treasury officials, Woodring, and high War Department officials, appeared before the committee. Senator Clark again took the lead and asked Morgenthau why the Frenchman flew in the aircraft. Morgenthau reviewed events of the previous month in which the State Department, the proper authority for working foreign sales, accredited the mission. He also revealed that the president himself contacted the Treasury and the War Departments and informed them of his desire to assist the French. Morgenthau then read a memo from a War Department meeting that Arnold and Johnson had attended which reflected the president's wishes and agreement among those present to help the French delegation.

The memo revealed Arnold's concurrence to clear the French officer through his West Coast officers. According to Morgenthau, Arnold and Johnson also agreed that the aircraft's secret bombsight should be removed before any inspection. Morgenthau informed the senators that "the Treasury backyard is big enough for me, I am very, very careful not to go over to somebody's else's territory, and when I do I only do it on written instructions."[46] Morgenthau then produced the message from Arnold to the West Coast officer, Major K. B. Wolfe, and quoted: "They are authorized to inspect attack bomber secret accessories, fly in it, and negotiate for purchase."[47] This quote only inflamed Senator Clark's desire to find out what military secrets the French viewed. For the rest of Morgenthau's testimony, he underscored the Treasury Department's involvement only as facilitators and escorts, acting under the direction of the president.

Roosevelt, informed of the committee proceedings and undoubtedly worried about public reaction to his desire to arm the French, held a press conference the same morning. A Treasury official arrived on Capitol Hill soon afterward to announce Roosevelt's remarks to the committee. FDR openly admitted his desire to sell aircraft to foreign governments. He asserted that this plan would assist idle American manufacturers, and he reiterated Arnold and Morgenthau's assessment of the Douglas plane not being government property and of both departments' understanding and cooperation for the cash-only transaction. In private, however, the president displayed his anger toward all parties. He confronted Morgenthau, with whom he was very close, for intentionally aggravating Woodring, for exposing presidential desires for foreign sales, and for brandishing the Arnold memo.[48] That memo, and Morgenthau's insinuation of Arnold's culpability in the affair, represented one of the only times Arnold had to acknowledge publicly a difference with any member of the Roosevelt administration. Yet his defense focused more on preserving his reputation and not allowing Morgenthau to make Arnold take the fall for the administration's foreign-policy maneuvers and politics.

Morgenthau's actions and insinuations angered Arnold. The air chief resented Morgenthau's presentation of the order with Arnold's signature as proof of Arnold's complicity and of his being the lead actor in the scheme to allow the French access to the aircraft and secrets. Arnold especially fumed when he was asked if Morgenthau ran the Air Corps.[49] Another rub for Arnold must have been that the memo Morgenthau presented to the committee was neither an original nor a photocopy. After hours of testimony, the members solved the mystery of the memo and the "secrets." Arnold had indeed transmitted the message authorizing the Frenchman to see and fly the aircraft, but with the provision that they could see the plane "less secret accessories." Records confirmed that the message was correctly sent from Washington DC to San Francisco, and on to March Field. The final relay occurred when Major Wolfe received the instructions over the telephone and inadvertently omitted the word "less." However, Wolfe sent a message back to Arnold, after being informed of the argument, that although he, Wolfe, interpreted the message incorrectly, the aircraft contained no military secrets, and especially the classified bombsights. Another surviving passenger, Navy Captain Sidney M. Krause, confirmed this fact.[50] Craig returned to the committee room to support the air chief, telling the members that Arnold was only "one of the 'hands'" that the orders passed through and that he believed the orders came from the Treasury Department.[51] With this fact cleared in the committee's eyes, Arnold also reappeared to counter Morgenthau's insinuation that Arnold was giving the orders in the French affair.

From the very beginning, Arnold asserted, he disapproved of the proposed viewing and sale of the Douglas aircraft to the French. On 20 December 1938, when he first learned of the idea, he protested the matter to Woodring on the grounds that the contemplated actions did not conform to the legal release policy, and a demonstration flight might reveal performance characteristics of an aircraft being considered for purchase by the Air Corps. Arnold stated that he was not against selling bombers to the French per se, provided that Douglas delivered all Air Corps requests first.[52] With Arnold's clarification and Roosevelt's open admission of support for French purchases, the position of the isolationist senators weakened. It seemed that they had hoped to trap Roosevelt and his administration, but the president's admission defused their maneuvers. Five days after the Senate's pointed examination of the Army generals, the matter barely caused a stir in the House chambers. In fact, the events occasioned only a few partisan questions from Republican representatives Albert J. Engel and Chester C. Bolton. Engel tried to goad Arnold to implicate the president directly in ordering the access—probably to gain a further political advantage by painting Roosevelt as dragging the nation into a European war or finding fault with the legalities of the arrangements. However, the committee's focus quickly changed and Arnold was spared further questioning.[53]

The crash of a test aircraft caused a diversion in the Senate committee, yet its importance lasted beyond simply monopolizing a huge amount of time and effort by a committee supposedly conducting hearings only upon the measures outlined by the president's new defense requests. Undoubtedly, Morgenthau had extended his reach beyond his "own backyard" and had infuriated military officers by doing so and then denying it. His diaries also show that he and Woodring did not agree on the French purchase policy and did not get along personally, either.[54] The Treasury, when the deal became exposed, cited orders from the president to do so (correct, but not appreciated by Roosevelt) and tried to show Arnold as authorizing the mission, when the air chief was only following orders after registering protests.

Further incensed when the newspapers carried the story, Morgenthau also felt exposed and vowed not to allow Arnold to get away with placing any blame on the Treasury or its secretary. Morgenthau informed Woodring during a telephone call, "I'm not going to forget . . . Arnold has done this . . . and if the Army thinks that they can put me in a hole like this . . . they're just mistaken."[55] Mere minutes after Morgenthau ended his call with Woodring, Morgenthau talked to Senator Sheppard, chairman of the Senate Military Affairs Committee. With Arnold's testimony being conducted in executive session, Morgenthau wanted confirmation that the air chief had indeed implicated the Secretary of the Treasury as giving the order, and Sheppard confirmed this assertion, "in confidence," to Morgenthau. Morgenthau then arranged to clear his entire schedule if Sheppard could call him to testify at the soonest possible moment, which came less than two hours later.[56] Morgenthau spent the remainder of the morning in frantic phone calls lining up his testimony and evidence. He reveled in the chance to embarrass Arnold, at one point calling his coming testimony "good fun" and bantering with a Navy officer not to "miss a good show."[57]

Morgenthau even called Roosevelt to ask how far he could go in exposing the president's support for the French mission and sales. Roosevelt authorized Morgenthau to say that the president supported the sale but asked him to emphasize the economic benefits to the country. Knowing Morgenthau's plans, the president agreed he would tell the press the same story.[58] In the end, Morgenthau had wanted to direct the program and fulfill the president's wishes, but he did not want to suffer the political fallout when the crash occurred. Morgenthau felt betrayed by Arnold, and he used his close relationship with the president to get the Treasury Department off the hook—and perhaps let the Air Corps take the fall.

The entire incident, and how his War and Treasury staffs handled it, infuriated Roosevelt. Arnold believed his job was in jeopardy. In a meeting with Woodring, Morgenthau, and several Army and Navy officers, Roosevelt called for cooperation to sell aircraft, including the latest equipment, to foreign countries, and asked for his staff to be on guard when answering questions in front of congressional committees. FDR voiced his dissatisfaction with their latest performance, and Arnold remembered the president looking directly at him and saying that officers who did not "play ball" could find themselves in unwelcome locations, "such as Guam."[59]

After this meeting, Arnold became *persona non grata* at the White House. He was not invited to the White House even for meetings involving planning and procurement for the Air Corps. He remained abreast of inside events through the assistance of his friends Harry Hopkins and General Marshall. Understanding the need for Arnold's expertise and insight, Marshall knew that the air chief needed to remain in the decision loop, even if Roosevelt did not want to see the airman's face. Therefore, Marshall ordered a junior officer to attend all meetings and record the events—notes that were later shared with Arnold.[60] Nine months later, after Germany's invasion of Poland, Arnold was reinitiated into Roosevelt's graces over an old-fashioned, the general's first drink in twenty years.[61]

SELLING THE AIR CORPS PROGRAM
TO CONGRESS AND BOMBERS TO FRANCE

At the end of the Senate committee's deliberations, the president's program passed. Noteworthy for the Air Corps, the bill that emerged from the committee recommended adjusting the Air Corps Act of 1926 to allow an upper limit of 5,500 aircraft.[62] By February 1939 the House Military Affairs Committee introduced a bill to increase the Air Corps size up to 6,000, an odd number considering that Craig recently delineated a number just over 4,000 and the president requested some 3,000 aircraft in his program.[63] However, one historian credited the 6,000 number to the War Department through an extremely clever maneuver by Roosevelt.

The president realized that 10,000 might be too many aircraft for Congress to swallow, and he became increasingly aware of the program's costs. Also, pressured by Marshall and Woodring to increase funding for other areas of the Army, FDR scaled back the number to 3,000 *additional* aircraft, without specifically noting at the time of his speech if this number was a cap or an addition. The net strength of the Air Corps, minus soon-to-be-obsolete aircraft, was 1,446. Adding already-ordered aircraft to this number equaled 2,464. Roosevelt's request for 3,000 would bring the total to just less than 5,500.[64] Thus, from the "Magna Carta" meeting to mid-February, the president played the game of rumors and leaks to elevate expectations, then asked for a lower number, making it appear as a compromise and winning him political points.

A moderate approach also continued to benefit Arnold and the Air Corps. For two years Arnold enjoyed congressional and presidential support for building an air-force-in-being. Craig and Marshall supported Arnold's efforts, and the Army presented a unified front before the congressional committees. However, signs began to emerge of Roosevelt's intent to provide aircraft to other governments, particularly the French. The debate lasted well into the spring of 1940, just prior to the Germans' May invasion and the speedy French collapse, which then made the debate moot and presented a quandary for Arnold and the Air Corps. Every aircraft sold to the French would delay the Air Corps expansion program. On the other hand, foreign sales would increase industrial capacity and sell aircraft at premium prices. Thus the Air Corps could buy more aircraft later due to lower per-unit costs, while also receiving the latest production models. Arnold sought to balance the need to get aircraft to the flight line against obtaining improved models later, but Roosevelt and his staff, especially Morgenthau, became more adamant about sending as much as possible to France—to the detriment of the air chief's expansion plan.

Throughout the 1930s, the Air Corps approved of exporting military aircraft as a measure to support industrial capacity. In fact, as the decade wore on the Army liberalized its policies but still prohibited exportation of the latest and most secret models. Exporting American war implements ran counter to a national prejudice against becoming an arms supplier. The Air Corps agreed to sell to the British and French because they ordered models no longer desired by the U.S. Army, while keeping the latest versions rolling off the assembly lines for the Air Corps expansion.[65]

Arnold also touted the benefits to American industry, as the orders allowed plants to operate at capacity and hire workers. Using this line of reasoning, the air chief obviously tried to appeal to isolationist and anti-defense-spending congressmen, as it allowed representatives to bring work to their home districts with cash provided by a foreign power.[66] While congressman may have been concerned earlier about pump-priming, they might not object if their district received the benefit, and one that did not spend U.S. tax dollars. To Arnold, any action to get aircraft plants into operation would later benefit the Air Corps. He repeatedly espoused these beliefs at various congressional hearings. Although perhaps a simple philosophy easily followed while peace reigned, the outbreak of war in Europe further complicated matters and further intensified the debates.

The European war activated the Neutrality Acts. Passed against Roosevelt's wishes, the legislation penalized those nations trying to defend themselves against German aggression by limiting the type of aid the United States could provide. Roosevelt successfully lobbied Congress to lift the arms embargo, and after 4 November 1939 countries could purchase American munitions on a "cash and carry" basis. To better coordinate their purchases, the British and French formed the Anglo-French Purchasing Commission. During the last months of 1939 the two countries ordered 2,500 aircraft, and during the early months of 1940 they negotiated for 8,200 more.[67]

However, once the Air Corps began working on Roosevelt's expansion plans, the Europeans requested newer models, including the A-20A attack bomber (the Douglas "Havoc") and the Curtiss P-40 Warhawk.[68] On 8 January 1940, the president's military aide, Brigadier General "Pa" Watson, asked Arnold about the feasibility of selling the first P-40s to France. Arnold objected to this suggestion and asked if the French would be willing to accept a slightly delayed delivery so the Air Corps could get the first planes. Without having the answer, Arnold informed his staff the next day of his duties and obligations. He believed it was his job to call to his military and civilian superiors' attention the repercussions of any delays and restrictions arising from selling aircraft to foreign governments at the expense of their own program, but, if he was ordered to give the French the aircraft, he "would carry out the instructions 100%."[69] Arnold understood the president's keen desire to sell the aircraft, and he wanted to again "play ball" with Roosevelt while extracting benefits for the Air Corps. Four days later, Arnold agreed to a compromise that provided the French with aircraft and allowed the Air Corps to purchase more refined and improved models.

In early 1940, Roosevelt held several meetings in the White House to discuss foreign sales. At a mid-March conference, FDR spoke of wanting his staff's full cooperation and coordination on the sales. He again threatened those present with duty on Guam for not "playing ball" with him, meaning Arnold and Woodring. By 19 March 1940, only one week before Roosevelt approved a revised foreign sales policy, and in the midst of evening negotiations to iron out the policy in Woodring's office, the War Department staff received word of a presidential announcement concerning the sales. A radio broadcast quoted the president as supporting full release of all U.S. aircraft to foreign powers.

This policy caused immediate problems for the War Department, as it would negate the negotiated leverage with the industry to procure superior aircraft at low prices for the Air Corps. Officers and staffers began scrambling to find transcripts and ticker tape of the announcement and the press conference. As Woodring received these papers, the president called. Roosevelt confirmed that he wanted full release but that the government would handle each sale on an individual basis. Roosevelt again threatened his staff that any appearances before a congressional committee better mouth the party line and agree with the president on foreign release. Arnold repeated that, although he opposed it, once the commander in chief decided the course and his policies were clearly delineated, the War Department must salute and follow orders.[70] Roosevelt's open statement of his views and warning to his subordinates to support him clarified Arnold's and the Air Corps' duty. After months of playing with numbers and never providing clear instruction (and sometimes contradictory directives), the president revealed what had been his true intention all along: to build up U.S. forces but to provide as many planes as possible to those countries fighting Germany. During April and May 1940 the Anglo-French commission ordered approximately 6,000 aircraft. Soon, with the Germans slicing their way through Western Europe, Roosevelt would ask for an even bolder plan for American aviation expansion.

FROM 10,000 TO 50,000

Eighteen months and two days after the president shocked his staff, and made Arnold smile, with the ambitious 10,000-aircraft proposal of November 1938, Roosevelt quintupled that number in an announcement to Congress. Analyzing the state of affairs in Europe, he stressed the actions of the Luftwaffe and discounted the safety of oceans to protect the nation against the quick-striking lethality of modern combat aircraft. Roosevelt called on Congress to allow full delivery of aircraft to foreign nations, arguing that the failure to do so would "be extremely short-sighted" for America's self-defense. He then dropped the bombshell number of 50,000 aircraft for the Army and Navy, with the capacity to produce 50,000 per year.[71]

The president also continued to regard air power as a prime necessity for national defense, above all other War Department programs. Demonstrating his continued belief that ground forces would not scare Hitler, Roosevelt denied Marshall's request for $25 million for ground-force purchases, giving the Army $18 million instead. At the same time, the president approved Arnold's $186 million plan for 200 B-17 bombers and expanded pilot training.[72] Alarmed by the quick sweep of the German forces across Europe,

Congress agreed with Roosevelt's assessment and added $500 million over the president's request.[73]

In late 1938 Arnold had asked his staff to think big, and they arrived at a total of 1,500 aircraft. A scant eighteen months later, Roosevelt challenged Congress and the industry to provide 50,000. The massive change in so short a time undoubtedly resulted from the impetus provided by the seemingly unstoppable German war machine, but the foundation for the change had been developed through an American civilian and military bureaucracy during a stable period of civil-military relations. Arnold's only serious quarrel with civilian authority resulted from the Douglas bomber incident, and there he did not seek to hide from blame or implicate the administration, but only to defend himself against accusations by Morgenthau. Although Roosevelt often proved elusive about his true wishes, when he provided firm pronouncements he virtually eliminated any dissention among his civilian and military subordinates.

The intra-Army conflict over air power dissipated after the "Magna Carta" White House meeting in November 1938. Craig recognized Roosevelt's brash determination to build a U.S. air force. After that meeting, Craig and his successor, Marshall, worked closely with Arnold, but without abandoning attempts to gain additional monies and actions for the rest of the Army. These three officers also provided a unified front and mutual support during congressional hearings.

Not surprisingly, relations between Congress and the Army actually functioned smoothly during a period of drastic increases in defense spending. With money to meet most if not all of the competing defense needs, the conflicts over doctrine, equipment, and organization diminished. While several congressmen opposed the president's agenda, either due to party loyalty, isolationist sentiments, or opposition to the seeming fiscal irresponsibility of "pump-priming" New Deal schemes, their actions never seriously threatened the spending increases.

Although Morgenthau claimed he had plenty of work to do within the Treasury Department, his efforts to expand his influence into the War Department's procurement and foreign sales programs caused problems for Arnold. Morgenthau overstepped his authority during the Douglas affair, and he took Roosevelt's instructions to coordinate as permission to dabble further in Air Corps and War Department business. Worse still, when Morgenthau became embroiled in the congressional investigation, he washed his hands of involvement, presented misleading evidence, and passed the blame to Arnold. The air chief did not like the way Morgenthau handled criticism or suggestions from the Air Corps. If Arnold suggested an alternate plan to Morgenthau, the latter would often reply, "Then am I to tell the President that you would not comply with his directives?"[74] Arnold did not complain openly to his bosses, and it probably would have only worsened the situation. He knew Morgenthau remained close to the president, and that was not always the case for Arnold. More importantly, Arnold never tried insurgent methods, though ample opportunity existed. Instead, the air chief gave his honest opinions and defended his integrity and the orders he had been given. Although he briefly fell out of Roosevelt's good graces, Arnold did not adversely affect the Air Corps' continued growth and the ever-increasing respect for the service and its new attitude. He understood that Roosevelt supported a larger Air Corps, and he did not want to play the partisan politics of an earlier era.

Civil-Military Relations and Change in Military Organizations

The Air Corps attitude of cooperation with its military and civilian masters, initially instilled by Mason Patrick in 1922 but perfected and fully implemented under Oscar Westover and "Hap" Arnold, helped the air enthusiasts to achieve almost all of their dreams. The increasing size of the force and the importance of air power led to increased autonomy within the Army in 1941 when the Air Corps became the Army Air Forces. Arnold became the Commander, and later Commanding General, of the Army Air Forces. By March 1942 the service gained virtual autonomy within the War Department through the implementation of a streamlined command structure placing the Commanding General of the Army Air Forces on the Joint and Combined Chiefs of Staff. By 1941 the final elements of the team that led to aviation's success during the war were already in place.

George C. Marshall became Chief of Staff in September 1939, and Roosevelt replaced both Secretary of War Harry H. Woodring and Assistant Secretary of War Louis Johnson the next year. On 19 June 1940, Roosevelt asked for Woodring's resignation due to "a succession of recent events."[1] Woodring, in his reply to FDR the following day, wrote that he believed the final straw forcing his replacement was his stand against the president's desire to release B-17s to the British.[2] The next day the president recommended that the Senate approve his nomination of Henry L. Stimson to replace Woodring.[3] Johnson resigned on 25 July, upset that Roosevelt did not elevate him to the secretary's position.[4] In November 1940, Roosevelt filled the office of Assistant Secretary of War for Air, vacant since 1933, with Robert A. Lovett, whom Arnold later identified as being of "towering importance to our Air Force."[5] Arnold called Lovett a calming force whose strengths filled in perfectly in areas of the air chief's weaknesses. One air staff officer labeled the Arnold-Lovett relationship "perfectly wonderful," full of personal and professional respect and mutual admiration that allowed them to work "most harmoniously together."[6] For the first time in its history, the air service enjoyed the support and confidence of the president and the War

Department's civilian and military leadership, men who remained in their positions throughout the course of World War II.

The entry of the United States into World War II meant the almost unlimited flow of resources to the military. The Army Air Forces saw aircraft rolling off assembly lines in numbers that would have seemed impossible to comprehend only a few years earlier. The B-17, the B-24, and later the B-29 Superfortress provided the means to test the strategic bombing doctrine of the Air Corps Tactical School. By the end of World War II, air leaders believed that air power had significantly contributed to Allied victory, perhaps had even been the key margin of success, and the *United States Strategic Bombing Survey* claimed that Allied air power proved a "decisive" force.[7] Additionally, the development of the atomic bomb, the ultimate weapon to fulfill the strategic bombing doctrine, added potency to the service's belief that it now deserved independence.[8] On the basis of these arguments, and after many years of political controversy, the United States Air Force was officially created on 18 September 1947—with the full support of President Harry Truman and the Army staff.[9]

The twenty-three years of peace between the two world wars had been a tumultuous time for the U.S. military. The services struggled with modernization, challenges to doctrine, rapid technological advancement, and difficult economic conditions. For these and other reasons, air officers advocated aviation policies that often conflicted with those of the presidents and the majorities in Congress. Air leaders used publicity, lobbying groups, and congressional contacts to push their agenda.[10] In acting outside the norms of proper behavior expected in civil-military relationships in a democracy, they threatened to weaken civilian control.

No public pressure called for a large military, yet high-ranking military officers still had respect and great visibility. Due to their notoriety and heroic backgrounds, men like John J. Pershing, Douglas MacArthur, and Billy Mitchell often did swing key public and political support to programs they desired, but they could only go so far. In the typical American tradition following a war, presidents sought tax dollars for nonmilitary programs, especially after 1929. The lack of an internal or external (at least until the mid-1930s) threat to the nation caused factionalism to emerge in the Army as budgets shrank and the cost of modernization rose. The pro- and anti-air groups garnered political support for their cause and budgets.[11]

The odyssey of the air service from the end of the World War I to independence thirty years later verified that major changes in the structure of national defense must have presidential support. Neither the support of a public enthralled by the airplane nor coalitions of air-minded congressmen (and the partisan political alliances that joined them) persuaded any of the four chief executives that the United States should have an independent air force. The different Secretaries of War and Army Chiefs of Staff consistently supported a unified Army with an aviation support force. Congressional supporters of aviation during the 1920s included a mix of Democrats who sought to create political turmoil for GOP presidents, Republican aviation enthusiasts, and party rebels, yet this political concoction could not overcome presidential power and party loyalty. Meanwhile, the early air leaders became entangled in political intrigue in their quest for immediate independence.

In their attacks on the Army and the Navy, the air enthusiasts could not count on support from the nation's military leadership. Thus the early air leaders often turned to Congress and the press. Mitchell and young officers like Arnold and Spaatz used Republican congressmen in key positions in the Military Affairs Committees to aid the aviators' cause, and then Democrats joined in to thwart the Republican presidents' wishes. In most cases, the nation's newspapers supported the aviators and reported enthusiastically on air developments and advances. The airplane had enthralled the public for most of the twentieth century. Flamboyant airmen such as Eddie Rickenbacker and Charles Lindbergh joined Mitchell in pushing for aviation's advancement, and they used a friendly press to aid their cause.

Yet the press did not look kindly when the airmen spoke out against the commanders in chief. The early air insurgents indeed violated Army General Orders and acted outside the era's norms of civil-military relations. They argued publicly for a policy the administrations opposed, and they used public and political pressure to try and force their will on the civilians. Insurgent airmen submitted legislation covertly to congressmen and used personal relationships with legislators in an attempt to circumvent the policies of various administrations. But when the air leaders went too far in voicing opposing opinions or participating in partisan maneuverings within Congress, neither their congressional allies nor a fawning press could save them. Mitchell and Foulois both experienced such a downfall, and Arnold escaped a court-martial only because his actions did not receive wide press coverage. He was only a lower-ranking staff officer and not a major air leader at the time, plus he threatened to expose Patrick's involvement in controversial activities. The tactics of confrontation caused the press as well as the military and civilian leadership to look upon the air insurgents with suspicion and, in some cases, outright hatred.

Nevertheless, these rebellious air leaders succeeded in highlighting the importance of the new air weapon and in creating the political maneuvering room for the moderates to step in and gain concessions for advancing the air arm. Mitchell, Foulois, Arnold, and the others probably felt that they had a moral duty and a commitment to prepare the nation for future wars—and they believed that a nation without a strong air arm was doomed to defeat. The stirring of political controversies and heavy press coverage of Army aviation made sure that airplanes were never far from public consciousness and political life in the interwar years. Many of the aviators' methods crossed the lines of proper behavior and subordination, and those methods had, in some cases, pressured the government to act.

Fortunately, from the perspective of both civil-military relations and aviation history, the more moderate air power leaders came into power. Taking advantage of the controversies created by the insurgents, they helped piece together positive advances for the air arm. While the extremists at both ends of the debate fought for either total subordination of air power to the Army or, at the opposite end, complete independence, more temperate leaders accepted compromises. Patrick began the transition away from insurrectionist tactics, but he and his successor still voiced their opinion that air independence remained a goal to be obtained as soon as possible.

Not until the mid-1930s, after Foulois' tenure as Air Corps chief, did the air leadership suspend the crusade for immediate independence to concentrate on proving that their mission and capabilities deserved such a large advancement. Westover and Arnold focused on improving the service's attitude as well as its hardware, using the infrastructure provided by the Army to develop the doctrines and weapons that would eventually gain independence for the air arm. Not until the late 1930s did the service begin to receive the equipment capable of implementing the doctrine of strategic bombing.

From the postwar theorists to the battles for increased purchases of B-17s in the late 1930s, air service leaders understood that to gain independence they needed to obtain and perfect the means to bomb an enemy, even if the aviators had to sometimes couch that mission in defensive terms. Yet the forward-looking canons of long-range aviation placed the service in a quandary. The bomber remained an inherently offensive weapon costing large amounts of money. Air leaders argued for independence based on the defensive employment of an offensive weapons system, but the technology was lacking for most of the interwar period. War made the offensive application of air power palatable and gave the service an opportunity to try to prove that it could defeat an enemy nation through air power alone.

Throughout most of the interwar period, air leaders operated under the stigma of Billy Mitchell and the early insurgents. Those insurgents left a terrible legacy of civil-military relations. After 1925, Patrick and Fechet trod carefully so as not to upset a president, a Congress, and an American public angry about Mitchell's challenges to civilian supremacy over the military. Foulois reversed the goodwill built up during the first six years after the Air Corps Act and revived the image of the air arm as a radical and spoiled element that would undermine the system when it did not get its way. Not until Westover firmly established his authority and renounced immediate independence in favor of building a solid Army team (from which the means to obtain independence could be safely nurtured) did the air service stabilize its relationship with the president and Congress.

Still, some historians and modern aviation enthusiasts continue to support, or even confer martyr status on, the early air rebels. Until recently Mitchell retained a sacred status as one who gave up his career to advance early aviation. Studies of Arnold and Spaatz often overlooked their earlier rebellious actions, while the contributions of moderate leaders such as Patrick and Westover have been almost forgotten. [12] Many modern Air Force professional military education courses focus upon the fight for air independence and how history has vindicated the need for a strong aviation service. These disparate groups of historians and military professionals espouse the belief that were it not for the agitation of the early leaders, aviation may have been subjugated forever to the needs of the Army and never have achieved its rightful independent status. [13] Several writers contend that the pressure of the insurgents' political fights provided the basis for eventual autonomy, that without the confrontational attitude, air power may never have contributed to the victory of World War II or achieved independence. No study thus far has seriously challenged those beliefs or analyzed the impropriety of the insurgents' actions and the damage done to civil-military relations.

General George Marshall believed that when officers rebelled against their commander in chief, they harmed the underlying structure of a democratic society. [14] By Marshall's definition, during half of the last extended peacetime period in American history, the highest-ranking air officers tore at that societal underpinning. The American tradition of civil-military relations required officers to remain above the political fray and not attach themselves to a political party or display partisanship. [15] If officers disagreed with the administration, those differences should have been aired in private consultations and executive sessions. Thus, air officers could still voice their opinions without becoming involved in partisan politics and political infighting. Richard Brown believes that most twentieth-century American generals operated in this fashion and epitomized good citizenship by their obedience, respect for proper authority, and dedication to their duties. [16]

Although Patrick and Fechet, unlike Mitchell, did not overtly push for air independence and use their positions to breed unrest, they never hid their support for autonomy. Patrick still worked covertly with congressmen to aid air independence while disciplining the remaining Mitchell insurgents. Fechet suppressed his own desires, largely due to the recent fate of Mitchell and the lack of political and public support, and both Fechet and Patrick benefited enormously from having a civilian buffer and advocate who conducted the necessary political maneuvering. In fact, only one interwar air leader, Oscar Westover, truly represented the model of complete deference to civilian authority. He wanted an independent service, but he knew it did not have the necessary support. As a result, he worked through proper command relationships to strengthen Army aviation and set it upon the path to autonomy while still supporting the president's programs. Westover should be given credit for setting the air force upon the path to independence. Arnold rightfully adopted and continued Westover's methods.

Thus the actions of the majority of the interwar air leaders confirms one of the notions of Morris Janowitz, who postulates that military officers became more involved in twentieth-century internal military politics, defined as "the activities of the military establishment in influencing legislative and administrative decisions regarding national security policies and affairs."[17] The interwar years certainly typified Janowitz's belief that technological, social, and political changes altered the settings in which American society defined its civil-military relations. Due to that transformation, Janowitz called for a more politically involved officer corps, which has indeed developed during the last sixty years. Russell Weigley concludes that World War II and the Cold War suppressed rifts between civilian leaders and military chiefs because of the overriding threat to American security, and civilian governments, in turn, provided a steady flow of military resources.[18] In the first decade after the Cold War, civil-military tension and distrust reemerged.[19]

A recent study contends that although the principle of civilian control remains entrenched in the American military, many officers now believe in a more assertive stance in the policy process. They argue that on certain subjects they have the right to "insist rather than merely advise or advocate in private" when the president is considering using force.[20] The study concludes that the continuing civil-military relations "gap" gives cause for concern but not panic.[21] However, the events of 11 September 2001 may have again changed the dynamics of American civil-military relations. The war on terrorism has caused President George W. Bush to demonstrate that he is firmly in control of national foreign and military policy, with the support of the Secretary of Defense, the National Security Advisor, the Secretary of State, and the Joint Chiefs of Staff.

Both the interwar period and the current war on terror reinforce the belief that civil-military relations are, to a large degree, situational and personality dependent.[22] Billy Mitchell and Benny Foulois were both headstrong men who did not like the presidents they served. The ideas and partisan politics of the different Secretaries of War reflected those of the respective presidents—and disagreements with the president meant disagreement and conflict within the War Department. Conversely, Patrick, Fechet, and Westover all exhibited patience and understanding and showed great respect for the authority of the president. Even though these air leaders did not always agree with presidential policies regarding air power, they did not come into open conflict with their respective commanders in chief. On the eve of World War II, Roosevelt completed the "team" Arnold spoke of, and the top leaders worked very well together.[23] Although Arnold and Marshall sometimes disagreed with Roosevelt, they liked the president and understood their proper place. Even in the face of policy disagreements, the military leaders were willing to work with the president and carry out his orders without making their disagreements public record or matters of external controversy.

The struggles over integrating the Army's air power into the American military establishment provide insight into how military organizations transform themselves, especially during times of rapid change, and the implications for civil-military relations. Granted, the circumstances and the speed of change that occurred during the interwar years complicated the Army's ability to cope with converting its culture and doctrines from those of a post–Civil War frontier constabulary to those of a modern force.[24] Various scholars have highlighted how transformed technology, economic conditions, and national and international events exacerbated military leaders' difficulty in altering service organizational structure and doctrine.[25] But these changed circumstances did not give military chiefs the right to challenge the notion of civilian control.

When a small coterie of aviation leaders tried to force their wishes for a rapid organizational transformation upon the civilian and military hierarchies, the airmen were also threatening, by their behavior, the system of civilian control of the military in the United States. In a fundamental sense, sometimes directly and sometimes indirectly, the air leaders were insubordinate with their military superiors, attempting to overturn the policies of the civilian administrations they served, and trying to influence Congress to change the laws against the wishes of the civilian and military leadership in the executive branch. Whether convinced by the pressure of the airmen, their arguments, or the politics of the controversies, the national leadership supported additional autonomy for aviation in increments, and larger budgets for the purchase of more and more modern aircraft—but only when the airmen pursued these goals through the proper chains of command and began to prove that the continued maturation of the technology could indeed improve military capabilities.

The insubordination of Billy Mitchell and the early insurgents did not single-handedly produce the eventual independence of the air arm, but neither did the alternate extreme of cooperation and acquiescence. Instead, advances in aviation's status came about due to a complex, interactive process that involved public and political perceptions, the civilian government, and military leaders. Mitchell's antics and propaganda did help to focus national attention upon military aviation, but the public and political pressures he helped instigate were not the reason the lawmakers passed the reforms. Yet Patrick and the other moderates benefited from Mitchell's extremism. The politicians and the War Department wanted—and needed—to resolve the air issues, but they would not consent to independence for the air arm..Such a wide gulf existed between the positions of the War Department and Mitchell that Patrick needed only to work in the middle and push for interim measures to achieve some success for the airmen.[26] Patrick helped guide the passage of the major legislation of the period, the 1926 Air Corps Act, by using a moderate tone in his testimony and by making concessions (including dropping the bill he wrote at Congress's request) to Congress and the War Department.

Nine years after the Air Corps Act, the air arm took another large step toward its goals with the formation of General Headquarters Air Force. But it was not the Mitchell-like rants of Foulois that instigated the War Department to act. Foulois failed to perform his duty of informing Congress and the president of Air Corps limitations—particularly the likelihood of accidents and deaths—before agreeing to carry the mail. The subsequent investigation occurred not because of Foulois' reversion to the tactics of rebellion but because his leadership failures led to the death of Army aviators, which created a public outcry, and it was this outcry that helped to spur the changes. After Foulois' departure sealed the fate of insurgent tactics, Westover and a converted Arnold laid the foundation for the successes of World War II and the eventual independence of the United States Air Force. In the wake of investigations and uncertainty, these men accepted incremental changes and returned the air leadership to a more proper, nonconfrontational position.

For more than twenty years, the triangular relationship among politicians, the public, and the military was one of frequent conflict over such issues as doctrine, organization, technology, and budgets. The passionate beliefs of some air leaders caused them, at times, to act improperly toward their civilian superiors. A similar period of technological upheaval, economic turmoil, and public clamor, matched by strong-willed leadership in both the civilian and military ranks, would likely have a profound effect as well on the nature of American civil-military relations. While the interwar period of confrontation shows that internecine struggles between this country's political and military leaders are to be expected, it also demonstrates the enduring dominance of the principle of civilian control.

Notes

ABBREVIATIONS

AFHSO   Air Force History Support Office, Bolling Air Force Base/U.S. Navy Yard Anacostia Annex, Washington DC

FDRL   Franklin D. Roosevelt Presidential Library, Hyde Park NY

HHPL   Herbert Hoover Presidential Library, West Branch IA

LOC-MD   Library of Congress, Manuscript Division, Washington DC

NARA   National Archives and Records Administration, Washington DC

NARA II   National Archives and Records Administration II, College Park MD

RG   Record Group

USAFAL   United States Air Force Academy Library, Special Collections Branch, United States Air Force Academy, Colorado Springs

WD-DCS   War Department Decimal Classification System

INTRODUCTION

1. Allan R. Millett, "Patterns of Military Innovation in the Interwar Period," in *Military Innovation in the Interwar Period*, ed. Williamson Murray and Allan R. Millett (Cambridge, England: Cambridge University Press, 1996), 363.

2. For background on civil-military relations, see Richard H. Kohn, *Eagle and Sword: The Federalists and the Creation of the Military Establishment in America, 1783–1802* (New York: Free Press, 1975), 2–6. For a more in-depth analysis of the origins of American civil-military relations and the checks on military power purposely built into the Constitution, see Richard H. Kohn, "The Constitution and National Security: The Intent of the Framers," in *The United States Military under the Constitution of the United States, 1789–1989*, ed. Kohn (New York: New York University Press, 1991), 61–94. Political scientist Yehuda Ben-Meir wrote, "the United States has taken great pains to provide an adequate legal basis for civilian control." Ben-Meir, *Civil-Military Relations In Israel* (New York: Columbia University Press, 1995), 26.

3. Elias Huzar, *The Purse and the Sword: Control of the Army by Congress through Military Appropriations, 1933–1950* (Ithaca: Cornell University Press, 1950), 7–8.

4. Deborah D. Avant, *Political Institutions and Military Change: Lessons from Pe-*

ripheral Wars (Ithaca: Cornell University Press, 1994), 29. For information on the change of power between state and federal authorities, see Jerry Cooper, *The Rise of the National Guard: The Evolution of the American Militia, 1865–1920* (Lincoln: University of Nebraska Press, 1997), 128–79 passim. Cooper asserted that the War Department forced the Guard to come more and more under federal control in exchange for increased funding. For statutory changes in the organization of national defense structures and the role of state militias versus professional forces, see National Defense Act of 1916, *Statutes at Large*, 39:166–217, and National Defense Act of 1920, *Statutes at Large*, 41:759–837.

5. Kohn, *Eagle and Sword*, 81.

6. While the air arm represented the most significant "rebellious" element within the interwar Army, other branches also attempted to influence legislation and increase their importance and stature within the national defense system. Others included the Quartermaster Corps and Chemical Service. Michael L. Grumelli, "Trial of Faith: The Dissent and Court Martial of Billy Mitchell" (Ph.D. diss., Rutgers–New Brunswick, 1991), 16–17. For an overview of this tumultuous period for the Army and Navy, which also included arguments over other emerging technologies and doctrines, see Allan R. Millett and Peter Maslowski, *For the Common Defense: A Military History of the United States of America*, rev. ed. (New York: Free Press, 1994), 380–407.

7. David E. Johnson, "From Frontier Constabulary to Modern Army: The U.S. Army between the World Wars," in *The Challenge of Change: Military Institutions and New Realities, 1918–1941*, ed. Harold R. Winton and David R. Mets (Lincoln: University of Nebraska Press, 2000), 195.

8. R. J. Overy, *Air War, 1939–1945* (New York: Stein and Day, 1981), 11–12.

9. Wesley Frank Craven and James Lea Cate, eds., *The Army Air Forces in World War II*, vol. 1, *Plans and Early Operations, January 1939 to August 1940* (1948; reprint, Washington DC: Office of Air Force History, 1983), 76–77, 80–81.

10. Overy, *Air War*, 26–27. The USSR produced over 2,500 aircraft in 1932, with the closest competitor being Japan at 691. The United States stood third at 593. Seven years later the Soviets completed over 10,300 aircraft, over 2,000 more than the Germans and almost five times as many as the United States—then in sixth place in production with just under 2,200.

11. Overy, *Air War*, 12, 18.

12. Overy, *Air War*, 12; Craven and Cate, *Plans and Early Operations*, 87–88. James S. Corum's 1997 work dispels the notion that the Germans concentrated only on auxiliary air power and ignored independent (strategic) applications. See Corum, *The Luftwaffe: Creating the Operational Air War, 1918–1940* (Lawrence: University of Kansas Press, 1997), 5–7, 282–86.

13. Overy, *Air War*, 52; Craven and Cate, *Plans and Early Operations*, 82–85.

14. Craven and Cate, *Plans and Early Operations*, 91; Overy, *Air War*, 13.

15. Quoted in Overy, *Air War*, 13.

16. Overy, *Air War*, 13–14.

17. Overy, *Air War*, 13–14; Craven and Cate, *Plans and Early Operations*, 92–93.

18. President's Aircraft Board, *Aircraft: Hearings before the President's Aircraft Board*, 4 vols. (Washington DC: Government Printing Office, 1925), 3:1238, 1247.

19. Barry R. Posen, *The Sources of Military Doctrine: France, Britain, and Germany between the World Wars* (Ithaca: Cornell University Press, 1984), 55.

20. Posen, *Sources of Military Doctrine*, 224.

21. Harold R. Winton, "On Military Change," in Winton and Mets, *The Challenge of Change*, xiii.

22. Winton, "On Military Change," xiv.

23. William O. Odom, *After the Trenches: The Transformation of U.S. Army Doctrine, 1918–1939* (College Station: Texas A&M University Press, 1999), 4–5.

24. President's Aircraft Board, *Hearings*, 3:1241; Harry H. Ransom, "The Air Corps Act of 1926: A Study of the Legislative Process" (Ph.D. diss., Princeton University, 1953), 150–51.

25. Michael West discussed similar problems for the Navy during the same time frame in "Laying the Legislative Foundation: The House Naval Affairs Committee and the Construction of the Treaty Navy, 1926–1934" (Ph.D. diss., Ohio State University, 1980), 4. For an excellent summary of military spending during the interwar years, see Paul Kennedy, *The Rise and Fall of the Great Powers* (New York: Random House, 1987), 275–343.

26. Quoted in Ransom, "The Air Corps Act of 1926," 124. Also, the range of American bombers and their available airfields meant that the only "enemies" within range were Mexico and Canada. Millett and Maslowski, *For the Common Defense*, 402.

27. For a more thorough analysis of the American public's attitude toward aviation in the early twentieth century, see Joseph J. Corn, *The Winged Gospel: America's Romance with Aviation, 1900–1950* (New York: Oxford University Press, 1983), vii, 10–27.

28. For one example of this wavering support, see Jeffrey S. Underwood, *The Wings of Democracy: The Influence of Air Power on the Roosevelt Administration, 1933–1941* (College Station: Texas A&M University Press, 1991), 18.

29. For a short synopsis of rules related to restrictions on federal agencies lobbying Congress, see Stephen K. Scroggs, *Army Relations with Congress: Thick Armor, Dull Sword, Slow Horse* (Westport CT: Praeger, 2000), 3–13. Samuel E. Finer noted that military professionals sometimes feel that they alone are competent to judge organizational matters due to their expertise. Finer, *The Man on Horseback: The Role of the Military in Politics*, 2nd ed. (Boulder CO: Westview Press, 1988), 23. Peter Douglas Feaver argues that military men might not always be the experts, and even when they are civilians must decide the risks to the society and are

still "rightfully in charge." Feaver, *Armed Servants: Agency, Oversight, and Civil-Military Relations* (Cambridge: Harvard University Press, 2003), 6.

30. The division of responsibility in military policy and actions has been simplified for the purpose of brevity. For a more thorough coverage of different types of control in American national security policy and the implications for civil-military relations, see Peter Douglas Feaver, *Guarding the Guardians: Civilian Control of Nuclear Weapons in the United States* (Ithaca: Cornell University Press, 1992), 7–12; David C. Hendrickson, *Reforming Defense: The State of American Civil-Military Relations* (Baltimore: Johns Hopkins University Press, 1988), 11–28.

31. Louis Smith, *American Democracy and Military Power: A Study of Civilian Control of the Military Power in the United States* (Chicago: University of Chicago Press, 1951), 14–15. Smith's remaining two elements of civilian control: the heads of governments are civilian representatives removed only by the normal, legal process; and the courts can hold the military accountable for protecting the democratic rights of the nation's people.

32. Samuel P. Huntington, *The Soldier and the State: The Theory and Politics of Civil-Military Relations* (Cambridge: Belknap Press of Harvard University Press, 1957), 80–85. In a later work, Huntington noted that the domestic political side of military policy involves interest groups with "incompatible interests and goals" that battle for resources. *The Common Defense: Strategic Programs in National Politics* (New York: Columbia University Press, 1961), 1–23.

33. Kohn, *United States Military*, 7.

34. Richard H. Kohn, "The Erosion of Civilian Control of the Military in the United States Today," *Naval War College Review* 55, no. 3 (2002): 16. Kohn first used the term "situational" in his "Out of Control: The Crisis in Civil-Military Relations," *The National Interest*, no. 35 (Spring 1994): 16–17.

35. Michael C. Desch, *Civilian Control of the Military: The Changing Security Environment* (Baltimore: Johns Hopkins University Press, 1999), 3, 13–17.

36. Omar N. Bradley, *A Soldier's Story* (New York: Henry Holt and Company, 1951), 147.

37. Scroggs, *Army Relations with Congress*, 13.

38. Avant noted that budgetary changes and the rise of new technologies, notably air power, gave Congress the leverage it wanted to try and shift the balance of power in national security from the president to the lawmakers. Avant, *Political Institutions and Military Change*, 30–32. For additional information on this topic, see Edward A. Kolodziej, *The Uncommon Defense and Congress, 1945–1963* (Columbus: Ohio State University Press, 1966). Of course, soldiers had in the past opposed their presidents. See, for example, Robert Wooster, *Nelson A. Miles and the Twilight of the Frontier Army* (Lincoln: University of Nebraska Press, 1993), 232–37. General Miles had several confrontations with the War Department and the president, but the best-known was the beef scandal of the Spanish-American War.

39. Forrest C. Pogue, "George C. Marshall on Civil Military Relationships in the United States," in Kohn, *United States Military*, 194.

PROLOGUE

1. *New York Times*, 15 November 1925, IX, 3:1.

2. Alfred F. Hurley, *Billy Mitchell: Crusader for Air Power* (New York: Franklin Watts, 1964), 35.

3. See Hurley, *Billy Mitchell*, 34; Burke Davis, *The Billy Mitchell Affair* (New York: Random House, 1967), 36.

4. Until very recently, historical accounts of the flamboyant airman sometimes overlooked his insubordination and egotistical personality. Recent reexaminations have offered a new side. See Grumelli, "Trial of Faith"; James J. Cooke, *Billy Mitchell* (Boulder CO: Lynne Rienner, 2002).

5. R. Earl McClendon, *Autonomy of the Air Arm* (1954; reprint, Washington DC: Air Force History and Museums Program, 1996), 48.

6. James P. Tate, *The Army and Its Air Corps: Army Policy toward Aviation, 1919–1941* (Maxwell Air Force Base AL: Air University Press, 1998), 3 (this is a book version of Tate's 1976 doctoral dissertation from Indiana University. Changes made during the twenty-two-year interim period are unknown).

7. Technically, the Army did not demote Mitchell, as he retained his permanent rank of colonel. The rank of brigadier general came only to whoever held the office of Assistant Chief of the Air Service. However, everyone understood the reasons for his banishment from Washington and saw his no longer being in line to become the chief as a "demotion."

8. *New York Times*, 3 September 1925, 6:3.

9. *New York Times*, 3 September 1925, 6:3.

10. Tate, *The Army and Its Air Corps*, 39; Hurley, *Billy Mitchell*, 100–101. The *Shenandoah* incident came only three days after the publication of Mitchell's book *Winged Defense*, which compiled previously released statements and articles into his now familiar positions on the structure of national defense and the place of aviation.

11. "The Statement of William Mitchell Concerning the Recent Air Accidents," n.d., Box 38, Subject File and Court Martial, Brigadier General William Mitchell Papers, LOC-MD.

12. "The Statement of William Mitchell Concerning the Recent Air Accidents."

13. *New York Times*, 6 September 1925, 6:1. Secretary of War John W. Weeks suffered from medical problems and would tender his formal resignation to President Coolidge the following month, turning his four-year battle with Mitchell over to his able successor.

14. Mitchell issued, among others, another call for a unified air service; see *New York Times*, 9 September 1925, 14:1. Hines relieved Mitchell on Saturday, 20 September; see *New York Times*, 22 September 1925, 1:7.

15. *New York Times*, 10 September 1925, 1:3.

16. Commonly called the Morrow Board for its chairman, Dwight W. Morrow. The board's particulars and findings will be detailed in chapter 2.

17. Davis, *The Billy Mitchell Affair*, 227–28.

18. Copies of court-martial papers and the orders to remain in the District of Columbia area found in Box 38, Subject File and Court Martial, Mitchell Papers. On various occasions Mitchell requested leave from the area, when court was not in session, to attend certain functions. See also *New York Times*, 23 October 1925, 1:6, and 24 October 1925, 1:2.

19. Davis, *The Billy Mitchell Affair*, 235–36.

20. Of note, Mitchell was not part of the "West Point fraternity." Lists of the judges abound, but there are some inconsistencies due to the removals by defense challenges. The court originally included twelve general officers: six major generals and six brigadier generals. Defense challenges dismissed three of them, two of them major generals. Therefore the final panel consisted of four major generals, five brigadier generals, and a "Law Member," Colonel Blanton Winship. See Davis, *The Billy Mitchell Affair*, 240; Hurley, *Billy Mitchell*, 103–4; *New York Times*, 1:8; David E. Johnson, *Fast Tanks and Heavy Bombers: Innovation in the U.S. Army, 1917–1945* (Ithaca: Cornell University Press, 1998), 88; Cooke, *Billy Mitchell*, 187–88.

21. Quoted in Hurley, *Billy Mitchell*, 104. For additional examples of the familiarity of the officers, see Henry H. Arnold, *Global Mission* (New York: Harper and Brothers, 1949), 120.

22. *New York Times*, 29 October 1925, 1:8; Davis, *The Billy Mitchell Affair*, 242–43; Cooke, *Billy Mitchell*, 191.

23. *New York Times*, 29 October 1925, 1:8.

24. Years later, as a retired lieutenant general, Ira Eaker confided that the Chief of the Air Service, Major General Mason M. Patrick, told his officers to gauge their conduct during the trial carefully, as open support for Mitchell could ruin their careers. Eaker and the others decided "we'd rather stand with Mitchell for principle and for the future of airpower than to save our necks and skins." Eaker oral history interview, quoted in Johnson, *Fast Tanks and Heavy Bombers*, 89. Mitchell later wrote personal notes of thanks for the officers' help. Spaatz from Mitchell, 5 February 1925, Box 4: Diaries and Notebooks, General Carl A. Spaatz Papers, LOC-MD. Of note, the Spaatz collection contains many letters to and from Mitchell on a variety of topics, including personal communications. However, besides this note of thanks, the file contains no other information on the trial, despite Spaatz's prominent part in it and support for Mitchell. See also Marvin W. McFarland, *The General Spaatz Collection* (Washington DC: Government Printing Office, 1949), 28.

25. *New York Times*, 19 October 1925, 8:2.

26. *New York Times*, 17 December 1925, 8:2. The paper reprinted the entire

letter. For copy of original, see John W. Weeks, Secretary of War, to President Calvin Coolidge, 4 March 1925, Calvin Coolidge Papers, Series 1, Case File 25, Reel 34, microfilm, accessed at Walter Royal Davis Library, University of North Carolina at Chapel Hill.

27. Major A. W. Gullion, quoted in *Literary Digest*, 2 January 1926, 6.

28. Davis's account, although clearly slanted toward Mitchell and not fully documented, best captures the mood of the courtroom. According to his notes, he was the first to consult the formerly sealed records of the court-martial. One instance of Reid's humor came when two prosecutors rose to object; he objected to "tandem objections." On another occasion, the court reminded those present, "This ain't a vaudeville show." Davis, *The Billy Mitchell Affair*, 239–328.

29. Arnold, *Global Mission*, 158–59.

30. *New York Times*, 11 November 1925, 22:3; see also Hurley, *Billy Mitchell*, 105.

31. DeWitt S. Copp, *A Few Great Captains: The Men and Events That Shaped the Development of U.S. Air Power* (New York: Doubleday, 1980), 36–51 passim. The fifteen-page chapter 4 of Copp's book is entitled "Mitchell and His Boys." The disparaging remark came from Major General Benjamin D. Foulois, another proponent of air power, whom this study will later assess. Although they sought the same ends, and even used similar means, the two men had a long rivalry and a history of disagreements. In his memoirs, Foulois would repeatedly refer to those air officers who sided with Mitchell as "worshippers." For one example see Benjamin D. Foulois with Colonel C. V. Glines, *From the Wright Brothers to the Astronauts: The Memoirs of Major General Benjamin D. Foulois* (New York: McGraw-Hill, 1968), 111. On the animosity between Mitchell and Foulois see John F. Shiner, *Foulois and the U.S. Army Air Corps, 1931–1935* (Washington DC: Office of Air Force History, 1983), 10–11.

32. Arnold, *Global Mission*, 120. One courtroom observer believed that only General Patrick seemed surprised. See Cooke, *Billy Mitchell*, 217. On the expectations of the verdict, see also Mark Clodfelter, "Molding Airpower Convictions: Development and Legacy of William Mitchell's Strategic Thought," in *The Paths of Heaven: The Evolution of Airpower Theory*, ed. Phillip S. Meilinger, The School of Advanced Airpower Studies, with a foreword by General Ronald R. Fogleman, Chief of Staff, U.S. Air Force (Maxwell Air Force Base AL: Air University Press, 1997), 104.

33. Davis, *The Billy Mitchell Affair*, 327. A Mitchell supporter, Representative Fiorello H. LaGuardia, told of finding a discarded ballot in the room with "Not Guilty" scribbled in MacArthur's handwriting. MacArthur never fully acknowledged his part, but he did write, "I did what I could in his behalf and I helped save him from dismissal." Douglas MacArthur, *Reminiscences* (New York: McGraw-Hill, 1964), 85–86. See also D. Clayton James, *The Years of MacArthur*, vol. 1, *1880–1941* (Boston: Houghton Mifflin, 1970), 310–11. During an oral history in-

terview in 1977, retired Lieutenant General James H. Doolittle acknowledged that MacArthur "categorically" confirmed his lone acquittal vote. Doolittle interview by Murray Green, Los Angeles, 14 August 1970, MS 33, Box 63, Folder 4, p. 2, Murray Green Collection, USAFAL.

34. *Army and Navy Journal*, 26 December 1925, 395.

35. *New York Times*, 19 December 1925, 1:7; *Washington Post*, 19 December 1925, 1.

36. *Army and Navy Journal*, 26 December 1925, 395.

37. Quoted in *Army and Navy Journal*, 26 December 1925, 395. For biographical information on him, see, "Wainwright, Jonathan Mayhew, 1864–1945," *Biographical Directory of the United States Congress*, http://bioguide.congress.gov/scripts/bio display.pl?index=M000104, accessed 7 June 2001.

38. The *Army and Navy Journal* ran a weekly article, "As the Country Sees Our Service Problems," which quoted newspapers from around the country. The Hearst syndicate usually endorsed Mitchell and his positions. See also Grumelli, "Trial of Faith," 272; Cooke, *Billy Mitchell*, 194.

39. *New York Times*, 18 December 1925, 22:2.

40. *New York World*, 19 December 1925, 12.

41. *Milwaukee Journal*, 18 December 1925, 12. Grumelli covered these and additional press comments on the trial and verdict in "Trial of Faith," 272–74.

42. Coolidge statement, The White House, 25 January 1926, Coolidge Papers, Serial 1, Case File 249, Reel 109, microfilm. Reid blasted Coolidge's actions as the "most un-American sentence ever pronounced." Quoted in Grumelli, "Trial of Faith," 277.

43. Hurley, *Billy Mitchell*, 107. Hurley believed the administration wanted Mitchell to resign in order to reduce any martyr status he would obtain by a dismissal. Grumelli argued that Davis simply wanted him out of the Army in order to restore discipline and reduce Mitchell's stature. Grumelli, "Trial of Faith," 275. These goals were not mutually exclusive, and both elements probably swayed the decision on his punishment.

44. Johnson, *Fast Tanks and Heavy Bombers*, 89.

45. Grumelli, "Trial of Faith," 95.

46. Johnson, *Fast Tanks and Heavy Bombers*, 89.

47. Tate, *The Army and Its Air Corps*, 3.

48. Even contemporary Air Force officers are taught almost to worship him. The belief that Mitchell was a martyr is taught in the first-level officer education program (Squadron Officer School), in college Reserve Officer Training Corps classes, and at the United States Air Force Academy. The main dining hall at the Academy is named "Mitchell Hall" (with a large oil painting of the aviator adorning the second-level dais where the staff and distinguished guests speak), and the Academy's "bible" of new officer candidate learning, *Contrails*, reinforces how Mitchell sacrificed his career by laying the foundation for the future inde-

pendent Air Force. Freshmen cadets also must learn to recite quotes, and two of Mitchell's quotes are required learning. Mitchell's ideas and battles are recited on several pages, while those of Mason M. Patrick, who arguably did more for the service, garner only a paragraph. See Andrew M. Mueller, chief editor, *Contrails: Air Force Academy Cadet Handbook*, vol. 31 (Colorado Springs: United States Air Force Academy, 1985–86), 61–64, 183. Additionally, the vast majority of Air Force officers I surveyed never recalled hearing of the efforts of Patrick, except in relation to Patrick Air Force Base, Florida. However everyone in the service knew something about Mitchell, usually as a martyr for air power. For more insight on why the USAF elevated Mitchell and his actions, see Cooke, *Billy Mitchell*, 287.

1. BILLY MITCHELL AND THE POLITICS OF INSURGENCY

1. Arnold, *Global Mission*, 86.

2. This study will use the rank of the officer during the chronological time being discussed. For example, at the time of the Armistice, Arnold held the wartime rank of colonel, until he reverted to his permanent rank of major. Even during peacetime, officers' rank could rise and fall depending on their job. Where applicable, this study will use the following reference to determine the proper rank during the time under discussion: Robert P. Fogerty, *Air Force Historical Studies: No. 91: Biographical Study of* USAF *General Officers, 1917–1952* (1953; reprint, Manhattan KS: MA/AH Publishing, 1980). Pages are not consecutively numbered, and entries are arranged alphabetically.

3. Clodfelter, "Molding Airpower Convictions," in *Paths of Heaven*, 90–91.

4. Hurley, *Billy Mitchell*, 1.

5. "Mitchell, Alexander, 1817–1887," *Biographical Directory of the United States Congress*, http://bioguide.congress.gov/scripts/biodisplay.pl?index=M000802, accessed 12 January 2001. See also Davis, *The Billy Mitchell Affair*, 11–12; Hurley, *Billy Mitchell*, 2.

6. "Mitchell, John Lendrum, 1842–1904," *Biographical Directory of the United States Congress*, http://bioguide.congress.gov/scripts/biodisplay.pl?index=M000821, accessed 12 January 2001.

7. Hurley, *Billy Mitchell*, 2–3; Davis, *The Billy Mitchell Affair*, 12–13.

8. Quoted in Hurley, *Billy Mitchell*, 4.

9. Hurley, *Billy Mitchell*, 14–17; "Mitchell, William," in Fogerty, *Biographical Study;* "Hay, James, 1856–1931," *Biographical Directory of the United States Congress*, http://bioguide.congress.gov/scripts/biodisplay.pl?index=H000382, accessed 15 January 2001. For additional information on his finances and garnering influential friends, see Cooke, *Billy Mitchell*, 38–40.

10. Hurley, *Billy Mitchell*, 17; Clodfelter, "Molding Airpower Convictions," in *Paths of Heaven*, 83. The prosecution at Mitchell's court-martial would recall this opinion; Mitchell replied, "I never made a worse statement."

11. Mitchell, then thirty-four, had been married since 1903 to a woman from

a prominent New York family. Hurley, *Billy Mitchell*, 10; Cooke, *Billy Mitchell*, 37–38. Army aviation, being a new organization and full of youthful flyers, offered young officers access to influential contacts with Congress. When Congress required information from the established branches, the Army sent the higher-ranking officers who commanded that branch or their General Staff representative, who would rotate assignments. With the early mandates limiting the numbers and age of aviation officers, only certain individuals had the expertise desired. Thus, Congress and the Army actually bred, from an early point, a rapport between these young enthusiasts and the legislative branch that would continue for over thirty years.

12. Hurley, *Billy Mitchell*, 18–21; Clodfelter, "Molding Airpower Convictions," in *Paths of Heaven*, 83–84; McClendon, *Autonomy of the Air Arm*, 15–16.

13. Hurley, *Billy Mitchell*, 22–28; see also Cooke, *Billy Mitchell*, 54, 64.

14. Mitchell corresponded with Caproni about his bombers during the war, but it remains unclear if Caproni discussed Douhet's theories with Mitchell at that time. During mid-1917, Douhet remained in an Italian prison for criticizing his government, and he would not publish his great thesis, *Command of the Air*, until 1921. Hurley believes that Douhet's theories produced little effect on Mitchell, although they impressed him. Hurley is probably correct, for by the time Mitchell read the work he had already solidified and presented his ideas, which would change very little. See Hurley, *Billy Mitchell*, 31–32, 75. One historian claimed that Mitchell borrowed from Douhet to supplement the Air Corps Tactical School texts. See Phillip S. Meilinger, "Giulio Douhet and the Origins of Airpower Theory," in Meilinger, *The Paths of Heaven*, 33, 40 n. 74. The Italian's ideas did ring true with one American air officer, Benjamin Foulois, but this would not occur until later. See Giulio Douhet, *Command of the Air*, trans. Dino Ferrari, ed. Richard H. Kohn and Joseph P. Harahan (1942; reprint, USAF Warrior Studies, Washington DC: Office of Air Force History, 1983), ix.

15. Hurley, *Billy Mitchell*, 29, 33–34. Hurley believes the appointment was "probably" made in Washington, and Mitchell believed it was further evidence of politics interfering in the war from across the ocean. Foulois, also ever resentful of Mitchell, believed he came to France on a prearranged plan with Pershing prior to the latter's departure. Foulois remembers being asked to go to France, but he turned it down on the advice of Colonel George O. Squier in a meeting of the three in early 1917. Foulois remarked that he did not know that when he arrived in France he would also be fighting Mitchell. Foulois and Glines, *Wright Brothers to Astronauts*, 156–58; see also Cooke, *Billy Mitchell*, 56–66.

16. Foulois and Glines, *Wright Brothers to Astronauts*, 160–62; Hurley, *Billy Mitchell*, 33–34; Arnold, *Global Mission*, 80.

17. Arnold, *Global Mission*, 86–88. Arnold also wanted Mitchell in Washington, and Arnold asked him to take a job in the War Department. Mitchell, wanting to stay with "his boys" and drive onto German soil, turned down the request.

Before Arnold could depart Paris, Mitchell called and said he wanted to get back right away. Arnold arranged for the transfer. At the time, Arnold was assigned to the Office of the Director of Military Aeronautics in Washington DC.

18. Tate, *The Army and Its Air Corps*, 4–6. For resentment of Menoher not flying, see Arnold, *Global Mission*, 100.

19. Tate, *The Army and Its Air Corps*, 7; Eugene M. Emme, "The American Dimension," in *Air Power and Warfare*, ed. Alfred F. Hurley and Robert C. Ehrhart, Proceedings of the Eighth Military History Symposium (Washington DC: Office of Air Force History, 1979), 66.

20. Tate, *The Army and Its Air Corps*, 6–7. The board members included three other generals who would play an important role in the interwar air service arguments: Major Generals John L. Hines and William Lassiter and Brigadier General Hugh A. Drum.

21. Patrick to Pershing, 11 November 1919, Box 155, General Correspondence File, 1904–1948, General of the Armies John J. Pershing Papers, LOC-MD.

22. McClendon, *Autonomy of the Air Arm*, 36–38; Tate, *The Army and Its Air Corps*, 9; Craven and Cate, *Plans and Early Operations*, 24–25. Craven and Cate believe the Crowell report "seems to have been deliberately suppressed." Crowell submitted the report in July, but it did not reach Congress until December. However, it is more a case that Baker more widely circulated the other reports, which favored keeping aviation under the Army, than he would the one he disliked. He did send the Crowell report to Congress, along with a letter dissenting from its opinions, and Crowell appeared before Congress defending his findings.

23. U.S. Congress, House, Committee on Military Affairs, *United Air Service: Hearing before a Subcommittee of the Committee on Military Affairs*, 66th Cong., 2nd sess., December 1919, 31.

24. Craven and Cate, *Plans and Early Operations*, 23; Tate, *The Army and Its Air Corps*, 7–8.

25. Quoted in Isaac Don Levine, *Mitchell: Pioneer of Air Power* (New York: Duell, Sloan and Pearce, 1943), 317.

26. The power of the General Staff waxed and waned during its history, but it always served the purpose of advising the Secretary of War and of appearing before Congress to provide information and support for budgets and programs. Millett and Maslowski, *For the Common Defense*, 327–28.

27. Congress appeared even more zealous to cut the Army through the budget, and Wilson and Baker fought to the end of the administration to keep the budget low, but not totally at the expense of the Army. In fact, in Wilson's last hours in office he vetoed the Army spending bill as not providing for enough of a force. *New York Times*, 5 March 1921, 8:5.

28. Menoher noted that at the end of fiscal year (FY) 1921 the Air Service received only one-third of its requested appropriation and that the service included only 65 percent of its authorized enlisted strength. Menoher wanted to fill out

his service as authorized by the 1920 Army Reorganization Act. *Annual Report* [Air Service] *for the Fiscal Year Ending 30 June 1921*, 49, AFHSO. The annual report traditionally concluded with recommendations for actions, including legislative. Menoher's 1921 report did not contain any proposals sought by Mitchell and other proponents of air power. In fact, when officers commented upon the draft report, they noted the absence of a request for legislation for a separate Air Service promotion list. Menoher struck out this clause, and his executive wrote, "General Menoher desires to make no comment whatsoever regarding the single promotion list." Memorandum, Major Harmon from Major William F. Pearson, 1 September 1921, RG 18, General Correspondence, Box 5, WD-DCS 319.1, Annual Reports of the Air Service, 1921.

29. Tate, *The Army and Its Air Corps*, 9–12; McClendon, *Autonomy of the Air Arm*, 41–42. The Army trotted out the Menoher Board findings before Congress regularly to counter its considerations of air autonomy.

30. Daniel R. Beaver, *Newton D. Baker and the American War Effort,1917–1919* (Lincoln: University of Nebraska Press, 1966), 169.

31. *New York Times*, 31 January 1920, 10:8.

32. U.S. Congress, House, Subcommittee of the Committee on Military Affairs, *United Air Service: General Information: Hearings before the Subcommittee on Military Affairs*, parts 1–4, 66th Cong., 2nd sess., 4 December 1919 to 13 February 1920.

33. Other committee members included Representatives Hull and W. Frank James, both prominent pro-aviation supporters.

34. McClendon, *Autonomy of the Air Arm*, 40–45; Tate, *The Army and Its Air Corps*, 9–14. See also *Congressional Record*, 66th Cong., 1st sess., 28 July 1919, 3292, 31 July 1919, 3390, and 30 October 1919, 7738. The original Curry and New bills were numbered H.R. 7925 and S. 2693, respectively. When modified and reintroduced they were H.R. 9804 and S. 3348. Of note, Hull also introduced two bills for a unified Department of Aeronautics during the Sixty-sixth Congress; neither emerged from committee. Although he seemed an active supporter of aviation, he never received the attention of Curry and New. For a consolidated list of legislation introduced during the time, see RG 18, WD-DCS 321.9: Congressional Bills, Bureau of Aeronautics, 1919–1921, Box 8, NARA II.

35. Martha E. Layman, *Air Force Historical Studies: No. 39: Legislation Relating to the Air Corps Personnel and Training Programs, 1907–1939* (Washington DC: Army Air Force Historical Office, 1945), 17–20.

36. Memorandum of the Secretary of War, 9 April 1920, Box 8, General Correspondence, 1920, Mitchell Papers. The memorandum gives a brief background of Baker's policies in place at that time.

37. Memorandum of the Secretary of War, 9 April 1920.

38. During the hearings on the Army reorganization bill, Representative W. Frank James of Michigan, a Republican, asked Mitchell about the allusions by

Baker and March of Mitchell's writing the air sections of the bill. Mitchell answered that he was in Europe at the time. James, friendly to Mitchell, did not follow up the question. U.S. Congress, House, Committee on Military Affairs, *Army Reorganization: Hearings before the Committee on Military Affairs*, vol. 1, 66th Cong., 1st sess., 3 September 1919 to 12 November 1919, 907 [hereinafter cited as *Army Reorganization*].

39. For example, he congratulated the editor of the *New York Tribune* for an article showing the benefit of air power over sea power. Mitchell called the article "one of the best presentations" on the topic he had seen and lauded the paper on its stand. Mitchell to editor, *New York Tribune*, 29 November 1920, Box 8, General Correspondence, 1920, Mitchell Papers.

40. Bane to Mitchell, 13 September 1919, Box 7, General Correspondence, 1919–1933, Mitchell Papers.

41. Arnold to Mitchell, 3 January 1921, Box 9, General Correspondence, 1921–1922, Mitchell Papers. Arnold specifically asked Mitchell to influence the Aero Club of America to get behind the legislation. The general declined, holding out hope for national legislation instead of a "hodgepodge" of state control.

42. See various letters from Mitchell, 26 February 1921, Box 9, General Correspondence, 1921–1922, Mitchell Papers.

43. Various letters between Spaatz, Mitchell, and Patrick document the trips, their duration, and their purpose. 3 August 1922 and 6 March 1923, Box 2: Diaries and Notebooks, Spaatz Papers.

44. For example, in his official capacity in communicating with base and field commanders, the communications, if not routed through the corps-area commanders, would more properly also have informed them of goings-on. Spaatz's command at Selfridge, being in the Sixth Corps Area, would have gone through the air officer in that command, during this time Major H. S. Martin. Instead, Mitchell often communicated directly with his more active supporters.

45. Mitchell to Spaatz, 4 September 1922, Box 2: Diaries and Notebooks, Spaatz Papers. Spaatz also did favors for Mitchell, such as flying veterinarians from Camp Custer to tend to six horses Mitchell brought to northern Michigan. Spaatz would often obtain a car for Mitchell from Rickenbacker Motors. Various Spaatz-Mitchell correspondence, Box 2: Diaries and Notebooks, Spaatz Papers.

46. For examples of books sent, see letters of reply to Mitchell from Representative W. Frank James (R-Michigan), 10 February 1921, and Representative James R. Mann (R-Illinois), n.d., Box 9, General Correspondence, 1921–1922, Mitchell Papers. Mann served as minority leader in the Sixty-second through Sixty-fifth Congresses. "Mann, James Robert, 1856–1922," *Biographical Directory of the United States Congress*, http://bioguide.congress.gov/scripts/biodisplay.pl?index=Moo 0104, accessed 15 January 2001. Mitchell sent a letter to Major M. F. Harmon, Air Service officer in Panama, asking him to assist Representative Frederick C. Hicks, a New York Republican and chairman of the Naval Aviation Subcommit-

tee, on a tour of the Panama Canal area, and added, "Mr. Hicks is a great friend of ours here and has been a great help to us here in Washington." Mitchell to Harmon, 25 February 1921, Box 9, General Correspondence, 1921–1922, Mitchell Papers.

47. Arnold to Mitchell, 10 January 1921, Box 9, General Correspondence, 1921–1922, Mitchell Papers. Arnold referred to senator-elect Samuel M. Shortridge, a California Republican who took the seat formerly occupied by Democratic senator James D. Phelan.

48. Mitchell to McKellar, 13 May 1921, Box 9, General Correspondence, 1921–1922, Mitchell Papers.

49. Mitchell to Harris, 25 April 1921, Box 9, General Correspondence, 1921–1922, Mitchell Papers.

50. One possible reason was Foulois' resentment of Mitchell, and of reverting to a lower rank while his nemesis glorified in helping shape policy and continued to wear his general's star. Their background could also further explain the differences. Foulois' family background and education was working class compared to Mitchell's, and Foulois quit school at the age of sixteen. Finally, Mitchell had the all-important "fair foundation" of growing up around politicians. As Mitchell's frustrations increased over the next four to five years, his statements became more aggressive, but he was always more restrained in his language than Foulois when appearing before Congress. For Foulois' background see Shiner, *Foulois and the U.S. Army Air Corps*, 1–2; Foulois and Glines, *Wright Brothers to Astronauts*, 1–15. Foulois had shown his own brand of congressional negotiations earlier. In 1913 he used a poker game, drinks, and the help of friendly government employees to overcome his fear of approaching congressmen "via the back door." See Foulois and Glines, *Wright Brothers to Astronauts*, 105–7.

51. Army Reorganization, 927.

52. Army Reorganization, 961–69.

53. U.S. Congress, House, Committee on Military Affairs, *United Air Service: Hearing before a Subcommittee of the Committee on Military Affairs*, 66th Cong., 2nd sess., December 1919, 132.

54. Foulois and Glines, *Wright Brothers to Astronauts*, 188.

55. Foulois and Glines, *Wright Brothers to Astronauts*, 188. Although he had properly informed his chain of command and abided by all directives when providing congressional testimony, Foulois believed that, having alienated the General Staff, the Navy, and a powerful assistant secretary, he needed to leave Washington. He requested, and received, the air attaché position in Berlin.

56. The Quartermaster Corps represented one of the insurrectionary elements of the interwar period, though none of the others approached the level of the air officers. See Grumelli, "Trial of Faith," 16–17.

57. Unsent letter, Baker to Rogers, 23 March 1923, and Personal and Confidential letter, Baker to Rogers, 23 March 1923, both in Container 13 (microfilm

reel 10), Newton D. Baker Papers, LOC-MD. Baker assigned the Inspector General to investigate the problems and found at least two other officers sent out of the Quartermaster Corps for the purpose of influencing legislation. He demanded that Rogers provide a detailed report. Memorandum for Rogers from Baker, 30 March 1920, Container 13 (microfilm reel 10), Baker Papers.

58. Memorandum of the Secretary of War, 9 April 1920, Box 8, General Correspondence, 1920, Mitchell Papers.

59. General Orders No. 25, dated three weeks after the Baker memorandum, included verbatim quotes from Baker's instructions. See War Department, *General Orders and Bulletins, 1920* (Washington DC: Government Printing Office, 1921), General Orders No. 25, section V, 3–4.

60. Mitchell's accusation concerned spending on coastal defenses. Memorandum from the Aeronautical Board to the Secretary of War and the Secretary of the Navy, "Statement of Brigadier-General William Mitchel [*sic*] before the Committee on Military Affairs, U.S. Senate, May 4, 1920," 28 May 1920, Box 8, General Correspondence, 1920, Mitchell Papers.

61. Guy V. Henry, Adjutant General of the Army, to Director of the Air Service, 11 June 1920, Box 8, General Correspondence, 1920, Mitchell Papers.

62. 1st Ind., War Department Adjutant General to Director of the Air Service, 11 June 1920; Memorandum from the Director of the Air Service to All Officers on Duty with the Air Service, "Political Influence," 19 June 1920, with attached memorandum from Office of the Director of the Air Service, 16 June 1920; all in Box 8, General Correspondence, 1920, Mitchell Papers.

63. Memorandum, "Testimony of Officers before Committees of Congress," from Captain Oscar Westover, Executive Officer, Office of the Director of Air Service to the Adjutant General of the Army, 2 July 1920, Box 8, General Correspondence, 1920, Mitchell Papers.

64. Davis, *The Billy Mitchell Affair*, 85. Harding reportedly favored the unified department even after assuming the office, and he liked the idea, often espoused by Mitchell, of saving money by consolidating air planning and organization. *New York Times*, 16 July 1921, 3:2.

65. Mitchell to Major John F. Curry, 7 June 1921, Box 9, General Correspondence, 1921–1922, Mitchell Papers.

66. *Political Divisions of the House of Representatives (1789 to Present)*, http://clerk web.house.gov/histrecs/househis/lists/divisionh.htm, accessed 16 December 2001.

67. *Senate Statistics: Majority and Minority Parties (Party Division)*, http://www. senate.gov/learning/stat_13.html, accessed 16 December 2001.

68. *Political Divisions of the House of Representatives (1789 to Present)*, http://clerk web.house.gov/histrecs/househis/lists/divisionh.htm, accessed 16 December 2001. See also Robert K. Murray, *The Harding Era: Warren G. Harding and his Administration* (Minneapolis: University of Minnesota Press, 1969), 318–19.

The incoming House contained 225 Republicans, 207 Democrats, and 3 Independents.

69. *Senate Statistics: Majority and Minority Parties (Party Division)*, http://www.senate.gov/learning/stat_13.html, accessed 16 December 2001. See also Murray, *The Harding Era*, 318–19. Republicans lost six seats, with one being won by Farmer-Laborite Henrik Shipstead of Minnesota and the rest representing Democratic gains.

70. Robert K. Murray, *The Politics of Normalcy: Governmental Theory and Practice in the Harding-Coolidge Era* (New York: Norton, 1973), 41–47.

71. Murray, *The Politics of Normalcy*, 72–78.

72. Due to his support for Harding's election and the Republican Party, Weeks was assured a cabinet post. Most expected he would be the Secretary of the Treasury due to his service on the House and Senate banking committees and his Wall Street experience. He shunned the post of Secretary of the Navy, as he did not want to seem partisan in naval matters, especially participating in promoting officers he had served with while at Annapolis and during his subsequent naval service—his efforts got Edwin Denby that particular post. Thus, while familiar with the activities of the War Department, he was not ingrained in the culture. Benjamin A. Spence, "The National Career of John Wingate Weeks (1904–1925)" (Ph.D. diss., University of Wisconsin–Madison, 1971), 309–12.

73. *New York Times*, 11 June 1921, 3:3.

74. National Defense Act of 1920, *Statutes at Large*, 41:768.

75. *New York Times*, 11 June 1921, 3:3.

76. *New York Times*, 10 June 1921, 2:7, 11 June 1921, 3:3, and 18 June 1921, 2:4.

77. Hurley, *Billy Mitchell*, 66–67.

78. *New York Times*, 10 June 1921, 2:7. The quote represents the paper's paraphrase of Weeks, not an exact quote from Weeks.

79. *New York Times*, 12 June 1921, 3:5. The paper mentioned the rumblings of unnamed senators. One must also keep in mind that Menoher would have had influential friends, if not in Congress, at least among the "old Army" and General Staff, not to mention his West Point classmate, General Pershing. Mitchell did receive a personal letter of support from Representative Hubert F. Fisher, a Tennessee Democrat, who stated that Weeks would have to stand behind "the one flying general of the army who has done so much for aviation." Representative Hubert F. Fisher to General William Mitchell, 11 June 1921, Box 9, General Correspondence, 1921–1922, Mitchell Papers.

80. *New York Times*, 11 June 1921, 12:2.

81. Curry to Mitchell, 22 July 1921, Box 9, General Correspondence, 1921–1922, Mitchell Papers.

82. Mason M. Patrick, *The United States in the Air* (Garden City NY: Doubleday, Doran, 1928), 80.

83. *New York Times*, 22 July 1921, 1:8, 2. *Washington Post*, 22 July 1921, 1, 3. For expanded coverage of the various bombing tests and specific results, see Mauer Mauer, *Aviation in the U.S. Army, 1919–1939* (Washington DC: Office of Air Force History, 1987), 113–20.

84. *New York Times*, 20 August 1922, 1:8 (report), and 14 September 1922, 1:2 (Mitchell reply).

85. Hurley, *Billy Mitchell*, 65–69.

86. Hurley, *Billy Mitchell*, 69.

87. *New York Times*, 17 September 1921, 3:5.

88. Robert Paul White, "Air Power Engineer: Major General Mason Patrick and the United States Air Service, 1917–1927" (Ph.D. diss., Ohio State University, 1999), 141–42. White believes Pershing engineered this "masterstroke" with Weeks and once again called his classmate to solve their "intractable problem." Patrick was not sure he wanted to take the job and "straighten out a tangled mess" in the Air Service again, and this time under more limited peacetime conditions. Patrick, *The United States in the Air*, 83.

89. Millett and Maslowski, *For the Common Defense*, 389.

90. Patrick, *The United States in the Air*, 86.

91. Patrick, *The United States in the Air*, 85–86.

92. Mitchell to the Adjutant General of the Army, 17 September 1921, Box 9, General Correspondence, 1921–1922, Mitchell Papers. The Army's Deputy Chief of Staff immediately informed Mitchell that Weeks wanted him to stay on through the next series of bombing tests beginning 20 September but would take actions he deemed appropriate afterward. Memorandum for General Mitchell, Major General J. C. Harbord, 17 September 1921, Box 9, General Correspondence, 1921–1922, Mitchell Papers. For Patrick's stand, see Patrick, *The United States in the Air*, 86–88. For press reaction to Patrick's appointment, see *New York Times*, 22 September 1921, 4:3. After Patrick presented his new rules, Mitchell again threatened to resign. Patrick escorted him to Harbord's office (10 October), where the Adjutant General agreed to accept the resignation this time. Mitchell backed down. White, "Air Power Engineer," 149; Patrick, *The United States in the Air*, 86–88.

93. "Memorandum for General Patrick," from Mitchell, 8 October 1921, section 10, 4–5, RG 18, WD-DCS 321.9, Reorganization of Office, Chief of the Air Service, NARA II.

94. Arnold, *Global Mission*, 105–6.

95. A quick analysis of the *New York Times* proves Patrick and Weeks's newfound control over Mitchell. After dominating defense news in 1921 and being in the paper on a regular basis since the war, Mitchell appeared only once in 1922, commenting on tests at Hampton Roads "proving" airplanes could defend the coasts and replace coastal artillery. *New York Times*, 8 November 1922, 10:2.

96. Murray, *The Politics of Normalcy*, 130–32. Coolidge even repeated Harding's

veto of a second, and still controversial, Bonus Bill sent up by Congress. Congress would later override the veto with the vast majority of Republicans siding with Democrats against Coolidge.

97. Arnold, *Global Mission*, 111.

98. Murray, *The Politics of Normalcy*, 132–33.

99. *New York Times*, 7 December 1923, 1:4.

100. Weeks was forced to defend the band expenditure, and he outlined to the president the average amount spent per recruit ($46.36) and demonstrated how the Army had reduced these costs. Weeks closed by asserting, "The War Department is not spending any government money improvidently as I think I can demonstrate without difficulty." John W. Weeks to President Calvin Coolidge, 20 November 1924. Coolidge sent the note back with a handwritten reply, "I am glad to tell you that I found out your recruiting bill was reasonable." Coolidge Papers, Series 1, Case File 25, Reel 34, microfilm.

101. Fritz Morstein Marx, "The Bureau of the Budget: Its Evolution and Present Role, I," *American Political Science Review* 39, no. 4 (1945): 653–84. Marx gives the history behind the bureau as well as its functions and limitations. Specifically, it compiled, correlated, and revised the estimates of all departments to present a comprehensive presidential program (668).

102. Marx, "The Bureau of the Budget," 669. Additionally, each department identified a budget officer to simplify and coordinate the efforts of the department with the bureau. Michael West points out that the bureau acted as another filter to control the military departments, shifted appropriation responsibility back toward Congress, and forced departments to submit different legislation for appropriations and authorizations. The unhitching forced closer and independent scrutiny of new programs. West, "Laying the Legislative Foundation," 484.

103. Memoranda from the Adjutant General of the Army to the Chiefs of All War Department Branches and Bureaus, "Procedure to be followed in regard to legislation affecting the War Department," 7 January 1922, RG 18, WD-DCS 032, Box 52, Acts of Congress, November 1919 to October 1926 and Air Corps Act 1926 (2 July) and Proposed Amendments, 1929–1941, NARA II. The Legislative Branch worked for the Deputy Chief of Staff.

104. Memoranda from the Adjutant General of the Army to the Chiefs of All War Department Branches and Bureaus, "Procedure to be followed in regard to legislation affecting the War Department," 7 January 1922.

105. Quoted in Arnold, *Global Mission*, 111.

106. Patrick's limited papers show that his most recurrent contacts with congressmen occurred between 1923 and 1925, when he contacted no less than sixteen members trying to get his son an appointment to West Point. The air chief noted his willingness to make any arrangements for his son to have an address in their district. Bream Patrick entered the Military Academy in July 1926 with an appointment given by Senator Guy D. Goff, from his father's home state of

West Virginia. See various letters to and from congressmen in the Major General Mason M. Patrick Papers, RG 18, Boxes 2–8, Correspondence File, NARA II.

107. Layman, *Air Corps Personnel and Training Programs*, 21–24. Appropriation acts reduced both the enlisted and commissioned strength. The 30 June 1922 bill reduced the Army's branches to 70 percent of their 1920 strength. The War Department lightened the burden on the air arm, but since Congress had established the overall number, the Army made additional cuts in other branches. Patrick counted a total of 8,500 enlisted men as his authorized strength in his 1922 annual report, but he remarked on how the reduction represented an amount roughly equal to the Army's overall cut. *Annual Report* [Air Service] *for the Fiscal Year Ending 30 June 1922*, 6, AFHSO.

108. Patrick believed that the truth "lies somewhere between those two views." Robert Frank Futrell, *Ideas, Concepts, Doctrine: Basic Thinking in the United States Air Force, 1907–1960*, vol. 1 (Maxwell Air Force Base AL: Air University Press, 1989), 40.

109. White, "Air Power Engineer," 153–54.

110. *Annual Report* [Air Service] *for the Fiscal Year Ending 30 June 1922*, 10–11 and 41, AFHSO.

111. *Annual Report* [Air Service] *for the Fiscal Year Ending 30 June 1922*, 41–45, and *Annual Report* [Air Service] *for the Fiscal Year Ending 30 June 1923*, 78, both at AFHSO.

112. This study has already highlighted several instances of how youth dominated in the Air Service as it could not in others. One has to look only at the youth of officers, in rank and age, testifying before congressional committees as compared to the traditional branches. The commissioned strength numbers from the annual reports shows how lieutenants made up the bulk of the service. By 1925, Patrick noted how the situation caused junior officers to command units normally assigned to field grade officers. *Annual Report* [Air Service] *for the Fiscal Year Ending 30 June 1925*, 38, AFHSO.

113. Patrick to Major General Francis J. Kernan, 28 September 1921, quoted in White, "Air Power Engineer," 155.

114. White, "Air Power Engineer," 154–60. However, Patrick did recognize the power of publicity, and he would later support and promote in the press the Around-the-World Flight, the first in-flight refueling, and the first dawn-to-dusk flight across the continent, and he would boast about new altitude records (273–75).

115. White, "Air Power Engineer," 160–69. White credits Patrick with formulating the foundations of air doctrine later expanded by the Air Corps Tactical School during the important period from 1926 to 1931. He credits Patrick with defining the roles of a supporting "air service" and a more independent-operating "air force" conducting offensive operations. White believes Patrick helped the service "find itself" during the "defining moment in the history of American air power."

116. Patrick, *The United States in the Air*, 99–100; White, "Air Power Engineer," 175–76.

117. See memorandum from Major General Patrick to Chief, Information Division, 22 September 1924, and reply and speech, 27 September 1924, RG 18, Patrick Papers, Articles and Speeches, 1922–1927, Box 1. This same box also had copies of speeches by Davis delivered in St. Louis and prepared by air officers on 24 September 1923 and 1 October 1923.

118. *Annual Report* [Air Service] *for the Fiscal Year Ending 30 June 1922*, 8–9, AFHSO. Of the 1,318 aircraft on hand at the end of the year, trainers and nontactical aircraft totaled 884, while the service flew only 163 pursuit aircraft and 21 bombers (or 14 percent of the force) (*Annual Report*, 21).

119. Futrell, *Ideas, Concepts, Doctrine*, 41–43; McClendon, *Autonomy of the Air Arm*, 50–52. White called the Lassiter Board "an extremely important incremental success" with Patrick realizing that complete victory was "impossible at this point." White, "Air Power Engineer," 207.

120. Until the Lampert Committee in 1924, Mitchell remained as quiet with the press as he had been since Patrick took office. For all of 1923 and 1924 he published only 5 articles (out of a total of 3 books and 108 articles for his career). White, "Air Power Engineer," 282. Although he never initiated legislation on the Lassiter recommendations, Weeks became irritated that Mitchell did not recognize the secretary's efforts to gain funding for implementing improvements. In truth, Weeks never forcefully pushed either for the Lassiter program or for additional funding. He continued to reject a separate air force and relied on the advice of those army leaders opposed to such a proposition. As one biographer pointed out, "a more flexible Secretary of War would have given greater attention to Mitchell's basic contention rather than over-relying on the opinions of those who would naturally oppose an independent and unified air service." Spence, "The National Career of John Wingate Weeks," 378–79.

121. White, "Air Power Engineer," 145.

2. THE POLITICS OF INVESTIGATIONS

1. Layman, *Air Corps Personnel and Training Programs*, 25–26. Still, the War Department began to use the Lassiter findings as its basis for wartime organization of air assets. For more information on disagreements between Weeks and his naval counterparts, Edwin Denby and his successor Curtis D. Wilbur, see Tate, *The Army and Its Air Corps*, 30–31.

2. *New York Times*, 28 January 1925, 5:5.

3. Tate, *The Army and Its Air Corps*, 34–35. Martin believed that the Air Service colluded with manufacturers in order to pick certain aircraft, and then negotiated the bids to keep out competition. Patrick believed that Martin suffered from "a persecutory mania."

4. *Congressional Record*, 68th Cong., 1st sess., 29 January 1924, 1625–33.

5. *Congressional Record*, 68th Cong., 1st sess., 29 January 1924, 1633.

6. H.R. 192 and 243, 68th Cong., 1st sess. Of note, Nelson's resolution (H.R. 163) to begin the inquiry did not pass. H.R. 192, offered by Representative Bertrand H. Snell, passed 160–0 by the House with only one amendment, which expanded the membership from seven to nine members. See *Congressional Record*, 68th Cong., 1st sess., 29 January 1924, 3126, 3293, 4815–17.

7. "General Topics of Investigation Which Seem to Cover the Reasonable Scope of Inquiry by the Select Committee of Inquiry into the Operations of the United States Air Services," Report of Alexander M. Fisher, Chief Investigator and Statistician, Lampert Committee, 7 December 1924, Records of Select Committees, *Of Inquiry into Operation of the U.S. Air Services*, RG 233, Box 331, NARA. Inquiry began with the purpose of investigating seven specific areas, all dealing with patents, engineering matters, and bids and contracts.

8. Republicans: Florian Lampert (Wisconsin), Albert H. Vestal (Indiana), Randolph Perkins (New Jersey), Charles L. Faust (Missouri), and Frank R. Reid (Illinois, and who would later serve as Mitchell's court-martial defense counsel). Democrats: Clarence F. Lea (California), Anning S. Prall (New York), Patrick B. O'Sullivan (Connecticut), and William N. Rodgers (New Hampshire).

9. For information on all members, see respective entries in the *Biographical Directory of the United States Congress*, http://bioguide.congress.gov, accessed January 2001. Lampert served in Congress during the war, but for only six days. On the same day he was elected to finish the term of James H. Davidson (who died while in office), he also won the seat to the Sixty-sixth Congress. He took his seat in the Sixty-fifth Congress on 5 November 1918. Vestal wielded the majority whip during the Sixty-eighth Congress (beginning March 1923) and four succeeding Congresses.

10. Reid, Prall, O'Sullivan, and Rodgers entered Congress in March 1923, and the latter two served only one term. Rodgers returned to Congress after the 1932 elections and served for five years, leaving after a failed Senate bid. Perkins and Faust came to Capitol Hill in 1921.

11. One Army study later proposed that the Lampert Committee existed solely to corroborate Mitchell's beliefs and implement his desires. Chase C. Mooney and Martha E. Layman. *Air Force Historical Studies: No. 25: Organization of Military Aeronautics, 1907–1935* (Washington DC: Historical Division, Assistant Chief of Air Staff, Intelligence, 1944), 64.

12. At this time, no government agency supervised commercial or civil aviation. As mentioned in chapter 1, Arnold drafted such a bill for oversight in California, but Mitchell wanted a national agency. The 20 May 1926 Air Commerce Act created a Bureau of Civil Aviation under the Commerce Department. See Air Commerce Act, *Statutes at Large*, 44:568–76. For more on the early history of

civil aviation and efforts to regulate the industry, see Nick A. Komons, *Bonfires to Beacons: Federal Civil Aviation Policy under the Air Commerce Act, 1926–1938* (Washington DC: U.S. Department of Transportation, 1978).

13. U.S. Congress, House, Select Committee of Inquiry, *Inquiry into Operations of the United States Air Services*, 68th Cong., 1925, 2110. The hearings were published in six parts and sequentially numbered, but they are hard to find in bound form. They are available in microfiche as "U.S. Cong. Hearings, Senate Library, 1925, vols. 379–381." Therefore, only the title and page number will henceforth identify these hearings. See also Futrell, *Ideas, Concepts, Doctrine*, 44–46.

14. Spaatz had probably developed a relationship with Stout during the airman's time as commander of Selfridge Field, which is just over twenty miles northeast of Detroit. Traditionally, base commanders worked closely with the business and community leaders of their areas in order to foster good relations and a community supportive of the local military unit. Spaatz to Stout, 20 January 1925, Box 3: Diaries and Notebooks, Spaatz Papers. No evidence existed in Spaatz's files of any reply by Stout, nor if Stout or other businessmen directly contacted congressmen.

15. *Inquiry into Operations of the United States Air Services*, 521.

16. White, "Air Power Engineer," 292–93.

17. *Inquiry into Operations of the United States Air Services*, 521.

18. When considering Reid's questioning and the sometimes confrontational attitude he took with Patrick, one must remember the tensions that existed between Mitchell and Patrick going back to World War I. At certain points it seemed as if Reid intended to grill Patrick for his "controlling" of Mitchell and the fact that Patrick, and not Mitchell, led the Air Service—wrongly in both the congressman's and the controversial airman's eyes.

19. Reid brought up Wilbur's statement about the next war being fought in the air as "an absurdity, partaking of the Jules Verne type of literature." *Inquiry into Operations of the United States Air Services*, 535.

20. *Inquiry into Operations of the United States Air Services*, 535.

21. Mitchell and Patrick presented different figures on the numbers of available aircraft. Mitchell wanted to present the number as being very low in order to create the image of a national defense emergency. Patrick wanted to show the numbers as accurate but not adequate. Reid also took Patrick to task on purchasing and manufacturing of aircraft.

22. *Inquiry into Operations of the United States Air Services*, 544. Reid's confrontational style with Patrick extends from pages 535 to 555.

23. *Inquiry into Operations of the United States Air Services*, 555. After Patrick declined to amend any information, Reid noted, "I want to be fair, you know, General."

24. *Inquiry into Operations of the United States Air Services*, 2246.

25. *Inquiry into Operations of the United States Air Services*, 2246, 2248. Major

Raycroft Walsh also testified along the same lines and with eloquent testimony supporting the Patrick plan as the best step toward the believed ultimate solution (1709–10). Lieutenant Charles B. Austin did not mention the Patrick plan, but he stated that he knew 75 percent of the fliers, and he and they wanted independence (2241).

26. Arnold left his assignment in California as the commander of Rockwell Field on 15 August 1924 and was sent as one of only two air officers to the second class of the Army Industrial College (thirteen total graduated). The class graduated in January 1925, and Arnold became Patrick's Chief of Information in the Air Service headquarters. See "Arnold, Henry H.," in Fogerty, *Biographical Study;* Dik Alan Daso, *Hap Arnold and the Evolution of American Air Power* (Washington DC: Smithsonian Institution Press, 2000), 109–10. Daso also offered me the other stated possible reasons for Arnold's not testifying. Daso to author, e-mail, 19 April 2001.

27. From 1924 to 1925 Foulois attended the Command and General Staff School at Fort Leavenworth. See Foulois and Glines, *Wright Brothers to Astronauts*, 198–200. However, the committee had brought in officers from different locations to testify, and it even traveled to gain more information. Foulois, ever bitter over Mitchell's taking the limelight and exaggerating facts to highlight his argument, undoubtedly bristled at Mitchell's comment about being with Army aviation since it began. See *Inquiry into Operations of the United States Air Services*, 1673. As Foulois often pointed out, he was one of the original Army fliers, and Mitchell arrived later in the air arm's history. See Foulois and Glines, *Wright Brothers to Astronauts*, 46, 124–25, 139–40. Patrick's biographer also noted how Foulois struggled with the academics at Leavenworth, compounded by problems with his eyes. White also called James Fechet (who would follow Patrick as chief) a "company man" who would not disrupt the chief's plans. However, one cannot discount Foulois' desire to become the air leader in the future as a factor in his low profile during the politically charged investigation. As for some of the others, Eaker, new to the capital, had not yet established connections. Frank M. Andrews, who also wanted independence and a force heavy with strategic bombers, was known for his patience and cool demeanor. He remained in Texas from 1923 to 1927, but he and Mitchell conversed often. See "Andrews, Frank M.," in Fogerty, *Biographical Study;* Copp, *A Few Great Captains*, 124. Oscar Westover, a future Chief of the Air Corps, was stationed just down the Virginia coast at Langley Field, but he always insisted on rigid obedience to orders, strict discipline, and a moderate position on air power. Tate, *The Army and Its Air Corps*, 150.

28. Mitchell testified on five different days between 17 December 1924 and 19 February 1925, and on some occasions he appeared before the committee more than once during the same day (due to breaks and being recalled to clarify certain aspects). He also submitted a post-testimony letter to the committee as further evidence. For the beginning points of his testimony, see *Inquiry into Operations*

of the United States Air Services, 291, 331, 1669, 1886, 1899, 2110, 2149, 2757, 2815, and letter on 3064.

29. *Inquiry into Operations of the United States Air Services,* 339–41. For information on H.R. 10147 see U.S. Congress, House, Committee on Military Affairs, *Air Service Unification: Hearing before the Committee on Military Affairs,* 68th Cong., 2nd sess., 8 January to 17 February 1925.

30. This second day of testimony came on Saturday, 31 January 1925, six weeks after his first visit. So far the committee had heard from the heavyweights in the War and Navy Departments, including Weeks, Davis, Patrick, Wilbur, and Moffett. Of interest, they did not call the Chief of Staff, General Drum, until very late, and after the current fire had been set ablaze by Mitchell and his congressional supporters.

31. *Inquiry into Operations of the United States Air Services,* 1673–74.

32. *Inquiry into Operations of the United States Air Services,* 1674.

33. *Inquiry into Operations of the United States Air Services,* 1675–76.

34. The statements and line of questioning here do not assert that Mitchell's statements were truthful. As pointed out earlier, he often exaggerated his own experience and embellished statements about the Navy and aircraft versus battleships. One must also recall that his court-martial was properly targeted toward the charges of insubordination, not because of his questionable congressional testimony.

35. *Inquiry into Operations of the United States Air Services,* 1682–83.

36. *Inquiry into Operations of the United States Air Services,* 1683.

37. War Department, *General Orders and Bulletins, 1922* (Washington DC: Government Printing Office, 1923), General Orders No. 20, section VI, 3–4.

38. War Department, *General Orders and Bulletins, 1922,* General Orders No. 20, section VI, 3–4.

39. *Inquiry into Operations of the United States Air Services,* 1683. It remains unclear to whose permission Mitchell referred. The Secretary of War did not require permission to issue orders to his officers and department. Perhaps, following the line of "free speech," Mitchell meant to imply these violated legal rights, but neither he nor the committee followed up on that remark.

40. *Inquiry into Operations of the United States Air Services,* 1683.

41. *Inquiry into Operations of the United States Air Services,* 1683.

42. *Inquiry into Operations of the United States Air Services,* 1804–16.

43. *Inquiry into Operations of the United States Air Services,* 1806. Drum replied to Reid with sarcastic appreciation, saying, "I appreciate your compliment."

44. *Inquiry into Operations of the United States Air Services,* 1848.

45. *Inquiry into Operations of the United States Air Services,* 1883–84. Drum's answer ran counter to General Order No. 25. Obviously, he knew the order and agreed with it and the department's policy.

46. *Inquiry into Operations of the United States Air Services,* 1884.

47. The decision to recall Weeks originated on 19 February during Mitchell's fifth visit to the witness stand. O'Sullivan offered the motion to recall, and it carried unanimously. *Inquiry into Operations of the United States Air Services*, 2774.

48. The committee sent the memorandum to Weeks on 26 February 1925. Weeks replied the next day and appeared on the second day after the request. Randolph Perkins, Examiner for the Committee, to Weeks, 26 February 1925, and Weeks to Perkins, 27 February 1925, both in Records of Select Committees, *Of Inquiry into Operation of the U.S. Air Services*, RG 233, Box 339, NARA.

49. Weeks to Perkins, 27 February 1925, RG 233, Box 339, NARA.

50. Weeks to Perkins, 27 February 1925, RG 233, Box 339, NARA.

51. The need for promotion reform, increased appropriations, and some manner of reorganization remained a staple among Patrick's annual complaints. See also *Annual Report of the Chief of the Air Service* from 1922 to 1925, AFHSO.

52. *Inquiry into Operations of the United States Air Services*, 3020.

53. *Inquiry into Operations of the United States Air Services*, 3020.

54. *Inquiry into Operations of the United States Air Services*, 3020. For a copy of Coolidge's original reply to Mitchell, see memorandum regarding Mitchell, 12 November 1924, Coolidge Papers, Series 1, Case File 25, Reel 34, microfilm.

55. *Inquiry into Operations of the United States Air Services*, 3020. Mitchell rebutted the charge that he disobeyed orders of the president or the War Department by sending a letter to the committee (2 March 1925), telling his side of the publication story. He asserted that he took all appropriate actions, and Patrick gave him approval after the president diverted the authority to Mitchell's supervisor (3064–65).

56. *Inquiry into Operations of the United States Air Services*, 3023.

57. *Inquiry into Operations of the United States Air Services*, 3048.

58. *Inquiry into Operations of the United States Air Services*, 3056–59.

59. *Inquiry into Operations of the United States Air Services*, 3056.

60. Senator Bingham, a huge air supporter and former air officer (more will follow later in chapter on Bingham), testified, as did former airman and Mitchell subordinate Fiorello LaGuardia. Both supported an independent air service, and LaGuardia supported the muzzling claim, stating that Army and Navy men could not testify in support of the Curry bill. See *Inquiry into Operations of the United States Air Services*, 1667, 2382, 2752–54.

61. *Inquiry into Operations of the United States Air Services*, 2774.

62. See previous chapter. Mitchell withdrew his threat to resign, agreed to cease his publicity campaigns, and agreed to follow the orders of his new boss, Patrick.

63. The House Military Affairs Committee conducted hearings on Curry's latest bill from 8 January to 17 February 1925. Mitchell testified before the committee on four separate days and made his usual recommendations. Of note, besides Mitchell, only Rear Admiral Hilary P. Jones, Commander of the Battle Fleet,

appeared more than once (Jones testified twice). General Drum represented the War Department at the hearings as the highest-ranking official, military or civilian, to appear. Weeks did send in a letter, which carried his familiar arguments of unity of command and cited the Dickman, Menoher, and Lassiter Boards and letters (from 1919 and 1920) from Menoher and Pershing. Curry, unable to attend the hearings due to illness, astutely responded to Week's letter, writing, "[it] throws no new light on this important subject." U.S. Congress, House, Committee on Military Affairs, *Air Service Unification*, 68th Cong., 2nd sess., 8 January to 17 February 1925, 378. For Mitchell's testimony see 10–55, 70–90, 376–413.

64. Weeks to Coolidge, 4 March 1925, Coolidge Papers, Series 1, Case File 249, Reel 109, 2, microfilm.

65. Weeks to Coolidge, 4 March 1925, Coolidge Papers, Series 1, Case File 249, Reel 109, 4, microfilm.

66. *New York World*, 14 February 1925, 12. For other papers' comments, see *Dayton Daily News*, 13 February 1925, 39; *Washington Post*, 16 February 1925, 46. Grumelli also discussed the press's reaction to Mitchell's tactics; see Grumelli, "Trial of Faith," 95, 116 n. 49.

67. *New York Times*, 20 February 1925, 20:1.

68. Futrell, *Ideas, Concepts, Doctrine*, 45–46; "Fechet, James E.," in Fogerty, *Biographical Study*. Fechet entered the flying service on 5 October 1917 at Scott Field in Illinois.

69. White, "Air Power Engineer," 188.

70. *New York Times*, 15 March 1925, IX, 12:5. The paper even quoted officers as asking what he looked like so that they would know him when they saw him. Some officers may not have known him, but the paper undoubtedly exaggerated this point. Fechet served in an influential position as chief of the Training and War Plans Division in the Air Service from 1920 to 1924—the same position Mitchell had earlier held under Menoher. "Fechet, James E.," in Fogerty, *Biographical Study*.

71. *New York Times*, 15 March 1925, IX, 12:5.

72. Perkins to Patrick, 10 March 1925, RG 18, WD-DCS 333.5, Investigations, Box 529, NARA II. Of note, Perkins's letter replied to a 5 March 1925 letter from Patrick, who offered any further assistance and information as the committee compiled its report (same file and location). It seems Patrick was desperately trying to contact Perkins before he left town, as the congressman's letter also expressed sorrow for missing Patrick's many calls to Perkins's home office (New Jersey). Patrick, it seems, attempted to further sway the committee even though the time for testimony had formally ended.

73. See *Political Divisions of the House of Representatives (1789 to Present)*, http:// clerkweb.house.gov/histrecs/househis/lists/divisionh.htm; and *Senate Statistics: Majority and Minority Parties (Party Divisions)*, http://www.senate.gov/learning/ stat_13.html, both accessed 23 November 2001.

74. *New York Times*, 13 September 1925, 1:3.

75. The board members and their positions from Coolidge's original announcement included retired Major General James G. Harbord, then president of the Radio Corporation of America; retired Admiral Frank F. Fletcher; Dwight W. Morrow, lawyer and banker (with the powerful J. P. Morgan and Company); Howard E. Coffin, consulting engineer and expert in aeronautics; Hiram Bingham, formerly in the Air Service (a lieutenant colonel) and member of the Senate Committee on Military Affairs; Carl Vinson, member of the House Committee on Naval Affairs; James S. Parker, chairman of the House Committee on Interstate and Foreign Commerce; Arthur C. Denison, judge of the Sixth Circuit Court of Appeals; and William F. Durand of Stanford University, president of the American Society of Mechanical Engineers and member of the National Advisory Committee for Aeronautics. President Calvin Coolidge to Secretary of the Navy Curtis D. Wilbur and Acting Secretary of War Dwight F. Davis, 12 September 1925, contained in President's Aircraft Board, *Hearings*, 1:1–2.

76. President's Aircraft Board, *Hearings*, 1:1.

77. Gabriel A. Almond, "The Political Attitudes of Wealth," *Journal of Politics* 7, no. 3 (1945): 229–31. Morrow's service to Coolidge would pay off, as he was appointed ambassador to Mexico in 1927, a position he held until his election to the Senate as a Republican from New Jersey in 1930, where he served until his death in 1931. "Morrow, Dwight Whitney, 1873–1931," *Biographical Directory of the United States Congress*, http://bioguide.congress.gov/scripts/biodisplay.pl?index=Moo 1002, accessed 7 March 2001. Morrow and Coolidge were also classmates at Amherst College, class of 1895. See Dumas Malone, ed. *Dictionary of American Biography*, (New York: Scribner, 1934), 7:234–35. Morrow's daughter, Anne, would marry Charles Lindbergh in 1929.

78. *New York Times*, 13 September 1925, 28 (continuation of story that began on 1:3).

79. *New York Times*, 13 September 1925, 28.

80. "Bingham, Hiram, 1875–1956," *Biographical Directory of the United States Congress*, http://bioguide.congress.gov/scripts/biodisplay.pl?index=B000470, accessed 7 March 1925.

81. Bingham sent a complimentary copy of this book to Patrick soon after its publication and wished the new chief the best in improving the Air Service. At that time he was still teaching at Yale. Bingham to Patrick, 17 October 1921, and Patrick to Bingham, 20 October 1921, both in RG 18, Patrick Papers, General Correspondence File, 1922–1927, Box 2. In the final section of the book, "The Future of Aviation," Bingham does not lay out a plan for military aeronautics, rather believing the services would work these out themselves. At the time he was more concerned with civil and commercial aviation. See Hiram Bingham, *An Explorer in the Air Service* (New Haven: Yale University Press, 1920), 242–44.

82. The depth of the article and the inclusion of pictures suggest that it was prepared prior to the previous day's announcement and was probably a reaction

to the lost Hawaiian flight, the *Shenandoah* crash, and Mitchell's remarks the previous week. *New York Times*, 13 September 1925, IX, 3.

83. *New York Times*, 13 September, IX, 3.

84. Bingham to Patrick, 13 January 1925, RG 18, Patrick Papers, General Correspondence File, 1922–1927, Box 2. Bingham was elected to the Senate in December 1924 to fill the vacancy created by the death of Senator Frank F. Brandegee. He won reelection to the seat in November 1926 and served until March 1933. Bingham and Patrick corresponded over the "good work" being done by Arnold in promoting the accomplishments of the Air Service. See Bingham to Patrick, 20 April 1925, and Patrick to Bingham, 22 April 1925, RG 18, Patrick Papers, General Correspondence File, 1922–1927, Box 2.

85. Patrick to Bingham, 14 September 1925, RG 18, Patrick Papers, General Correspondence File, 1922–1927, Box 2.

86. Patrick to Bingham, 14 September 1925.

87. Harold Nicolson, *Dwight Morrow* (New York: Harcourt, Brace, 1935), 281.

88. President's Aircraft Board, *Hearings*, 1:4.

89. Clark G. Reynolds, "John H. Towers, the Morrow Board, and the Reform of the Navy's Aviation," *Military Affairs* 52, no. 2 (1988): 79.

90. President's Aircraft Board, *Hearings*, 1:9–10.

91. President's Aircraft Board, *Hearings*, 1:10–60 passim.

92. President's Aircraft Board, *Hearings*, 1:17.

93. Patrick requested a total for the period of $94 million, and Congress appropriated slightly over $54 million. The overall percent reflects the typical annual reduction. For instance, he requested $26 million in 1923 and received just under $13 million. During FY 1924 the service received its highest percentage, with over $12 million allotted out of an $18 million request. President's Aircraft Board, *Hearings*, 1:64–65; see also *Annual Report of the Chief of the Air Service* from 1922 to 1925, AFHSO.

94. President's Aircraft Board, *Hearings*, 1:69.

95. For example, he noted how the Cavalry, once the Army's premier reconnaissance force, could not reconnoiter as far in advance of the troops as quickly and efficiently as aircraft, and how the Air Service had minimized the need for coastal defenses. President's Aircraft Board, *Hearings*, 1:69.

96. President's Aircraft Board, *Hearings*, 1:69–70.

97. President's Aircraft Board, *Hearings*, 1:72.

98. President's Aircraft Board, *Hearings*, 1:72. Patrick also voiced his displeasure with any plan placing military and civilian air assets under one cabinet position in a department of aeronautics, thus disapproving of Mitchell's plan without naming his former assistant.

99. President's Aircraft Board, *Hearings*, 1:79–81.

100. Most members of the committee questioned Patrick to one degree or another. Bingham's questions seemed well planned and easy to answer, and Patrick

was well prepared for them. General Harbord followed Bingham, and his line of questioning, while civil, demonstrated his bias against air power and put him in line with the old guard of Hines and Drum. Vinson asked only a few questions, the majority meant to clarify that Patrick still saw a place for the Navy in national defense (to counter Mitchell's oft-repeated statement that air power would make the Navy obsolete). President's Aircraft Board, *Hearings*, 1:75–93 passim.

101. President's Aircraft Board, *Hearings*, 2:495–587.

102. Arnold, *Global Mission*, 119–20.

103. Nicolson, *Dwight Morrow*, 283–84.

104. President's Aircraft Board, *Hearings*, 2:569–70.

105. President's Aircraft Board, *Hearings*, 2:593–618.

106. President's Aircraft Board, *Hearings*, 1:4.

107. Additionally, at the beginning of the second week of testimony (Monday, 28 September), which was reserved to hear from "the actual flying men," Chairman Morrow announced that the officers would be free to give their personal opinions, and he read letters from the Navy and War Secretaries inviting their officers to testify "fully and freely their individual views." President's Aircraft Board, *Hearings*, 2:365–66.

108. President's Aircraft Board, *Hearings*, 1:397. Other officers countering the notion of coercion and suppression included Major Leslie MacDill (2:635), Major J. H. Pirie (1:431), Lieutenant H. L. George (1:434), Major B. Q. Jones (1:462), Major W. G. Kilner (1:367), and Major Ralph Royce (1:382).

109. Bernard C. Nalty, ed., *Winged Shield, Winged Sword: A History of the United States Air Force*, vol. 1, *1907–1950* (Washington DC: Air Force History and Museums Program, 1997), 103.

110. Patrick advised Mitchell's followers not to become too closely associated with Mitchell during the coming court-martial, as it could jeopardize their careers. Daso, *Hap Arnold*, 112.

111. President's Aircraft Board, *Hearings*, 4:1623. Arnold was brought back before the board to clarify certain points. He also agreed with the Patrick concept at his earlier questioning (2:634).

112. President's Aircraft Board, *Hearings*, 2:634, 655. Other air officers supporting Patrick included Major W. G. Kilner (1:366), Major Harvey B. S. Burwell (1:412), Major T. G. Lamphier (1:424), and Lieutenant H. L. George (1:440–41).

113. President's Aircraft Board, *Hearings*, 1:398–99.

114. President's Aircraft Board, *Hearings*, 1:398–99. Here he did show some deference to Patrick's plan of intermediate steps, but in further questioning he reasserted his full support for an independent service within a department of national defense (see 1:401–2, 406).

115. *Army and Navy Journal*, 10 October 1925, 121.

116. *Army and Navy Journal*, 10 October 1925, 121.

117. Quoted in *Army and Navy Journal*, 17 October 1925, 146.

118. Quotes from both papers from *Army and Navy Journal*, 17 October 1925, 146.

119. Quoted in *Army and Navy Journal*, 17 October 1925, 146.

120. Patrick prepared the report during the interim between the Lampert Committee and Morrow Board hearings, but, in line with normal procedure, the War Department did not release it until December. At the time of the writing, Patrick undoubtedly felt emboldened by the congressional inquiry and knew the basics of their forthcoming findings and recommendations. Therefore he allowed himself to take a stronger line in his report than usual, though he repeated the same general shortfalls of previous reports.

121. *Annual Report of the Chief of the Air Service for the Fiscal Year Ending 30 June 1925* (Washington DC: Office of the Chief of the Air Service, 1925), 203.

122. *Annual Report of the Chief of the Air Service for the Fiscal Year Ending 30 June 1925*, 7, 101. The service numbered 786 aircraft, or 950 short of planned requirements and 2,180 under the Lassiter recommendations. Patrick wanted the majority of aircraft to be "air force," or offensive planes, instead of "air service" support aircraft. At the end of FY 1925, almost 80 percent of the service, in men and planes, worked the support side of aviation.

123. *Annual Report of the Chief of the Air Service for the Fiscal Year Ending 30 June 1925*, 2.

124. The legislative section remained active for FY 1926, but future annual reports deleted the Legislative Section. See *Annual Report of the Chief of the Air Service [Air Corps as of 1927] 1925–40*, AFHSO.

125. Daso, *Hap Arnold*, 111–13. Arnold admitted to making "many friends" in the press and in Congress. Arnold, *Global Mission*, 122.

126. *New York Times*, 3 December 1925, 1:8.

127. *New York Times*, 3 December 1925, 12:1. The *Times* printed the full text of the report on pages 12–13, including the appendices with aircraft numbers and comparisons.

128. For a summary of these and other papers' reactions, see *Army and Navy Journal*, 12 December 1925, 346.

129. *Army and Navy Journal*, 12 December 1925, 346.

130. *New York Times*, 9 December 1925, 1:5. The committee did defeat two of Reid's proposals. One would have created an air academy on similar lines as West Point and Annapolis, and the other an aviation committee in Congress to rival the Military Affairs and Naval Committees.

131. U.S. Congress, House, *Report of the Select Committee of Inquiry into Operations of the United States Air Services*, 68th Cong., 2nd sess., 14 December 1925, 6–7.

132. *Report of the Select Committee of Inquiry into Operations of the United States Air Services*, 24.

133. *Report of the Select Committee of Inquiry into Operations of the United States*

Air Services, 25–41. The writing and details of Reid's suggestions sounded as if they had came directly from Mitchell, who undoubtedly took an active part in drafting the addendum.

134. Fechet to Mr. Mentor Entyre, Kansas City MO, 1 October 1925, RG 18, General Correspondence File of General Fechet, Box 1, NARA II.

3. LAST ACTS OF THE REBELS

1. W. Frank James, "Handling Military Legislation in the House of Representatives," lecture given before the Army War College, Washington DC, 16 June 1927, file number 340A-4, Army War College Archives, Carlisle Barracks PA, 15.

2. *Literary Digest*, 12 December 1925, 10.

3. Quote and other information in *Army and Navy Journal*, 23 January 1926, 489. Although not officially published by the services, the *Journal* served the purpose of informing military officers and represented their views. A similar publication also lamented that "nothing radical" would occur with this Congress. George S. Carll Jr., "Congress Struggling with the Air Problem," *U.S. Air Services* 11, no. 3 (1926): 45.

4. Historians often refer to 1912 as the "floodtide" or "high tide" of the Progressives. See Robert H. Wiebe, *The Search for Order, 1877–1920* (1967; reprint, Westport CT: Greenwood Press, 1980), 208.

5. Those prominent in the swing bloc of the Sixty-eighth Congress included La Follette; Smith W. Brookhart (formerly on the Progressive ticket, but switched to the Republican Party in March 1925), Iowa; Lynn J. Frazier, North Dakota; Edwin F Ladd, North Dakota; Robert B. Howell, Nebraska; and William E. Borah, Idaho. Carroll H. Wooddy, "Is the Senate Unrepresentative?" *Political Science Quarterly* 41, no. 2 (1926): 235 n. 1. Information on the senators obtained from *Biographical Directory of the United States Congress*, http://bioguide.congress.gov, accessed 12 June 2001.

6. "La Follette, Robert Marion Jr., 1895–1953," *Biographical Directory of the United States Congress*, http://bioguide.congress.gov/scripts/biodisplay.pl?index =L000005, accessed 12 June 2001.

7. The Republicans held a 54–41 majority in the Senate, but at least 11 of them often acted against party wishes. The House majority was larger, 247–183, but 20 GOP members were considered "irregular." The GOP sought to reconcile especially with new members from the farm states. In addition to La Follette Jr. the party targeted Senator Gerald P. Nye, appointed to fill the seat vacated by the death of Senator Ladd. Senators Brokhart and Frazier "remain in outer darkness." *Literary Digest*, 19 December 1925, 9. See also *Senate Statistics: Majority and Minority Parties (Party Division)*, http://www.senate.gov/learning/stat_13.html, accessed 12 June 2001.

8. *Literary Digest*, 19 December 1925, 9. For additional information on the Republican "irregulars" and the party's attempt to quell the schism in the ranks,

see also Clarence A. Berdahl, "American Government and Politics: Some Notes on Party Membership in Congress, II," *American Political Science Review* 43, no. 3 (1949): 492–503.

9. President's Annual Message to Congress, *Congressional Record*, 69th Cong., 1st sess., 7 December 1925, 459.

10. *Army and Navy Journal*, 30 January 1926, 513, 6 February 1926, 539.

11. Notable Democrats included John M. McSwain of South Carolina and Hubert F. Fisher of Tennessee. McSwain would later rise to prominence within the committee, but for now he represented the minority party and was still a friend of aviation.

12. Morin's 1919 bill was H.R. 11206, introduced in the second session of the Sixty-sixth Congress. Ransom, "The Air Corps Act of 1926," 284–85. The bills introduced in the Sixty-ninth Congress will be detailed below.

13. Both are printed in U.S. Congress, House, Committee on Military Affairs, *Department of Defense and Unification of Air Service: Hearing before the Committee on Military Affairs*, 69th Cong., 1st sess., 19 January to 9 March 1926. For the Hill bill see H.R. 46, 1327–29, and for the Curry bill see H.R. 447, 1329–47.

14. *Department of Defense and Unification of Air Service*, H.R. 4084, 1348–67.

15. Morin introduced H.R. 7916 for the War Department on 18 January; Wainwright brought forth H.R. 8533, drafted by General Patrick, on 28 January; and James sponsored two bills within five days in February (H.R. 8819 on the 3 February, H.R. 9044 on 8 February). *Department of Defense and Unification of Air Service*, 1367–88.

16. W. F. James, "Handling Military Legislation in the House of Representatives," 4.

17. *Department of Defense and Unification of Air Service*, 416–17.

18. "Air Service Report on S 2614," Subject File, Box 26, Major General Benjamin D. Foulois Papers, loc-md. The War Department's bill was submitted to both chambers of Congress. The House Morin bill was identical to S. 2614, also known as the Wadsworth bill.

19. Department of Defense and Unification of Air Service, 258–62.

20. *Department of Defense and Unification of Air Service*, 321. Only two days after Patrick's testimony, Wainwright introduced the legislation, H.R. 8533, for Patrick. See also Robert Paul White, *Mason Patrick and the Fight for Air Service Independence* (Washington dc: Smithsonian Institution Press, 2001), 123–24.

21. *Department of Defense and Unification of Air Service*, 134–36. Davis did not answer affirmatively, but said he would have to seek legal counsel. If testimony by Mitchell was legal, Davis emphasized, he would allow him to appear.

22. *Department of Defense and Unification of Air Service*, 168.

23. *Department of Defense and Unification of Air Service*, 169–71.

24. *Department of Defense and Unification of Air Service*, 256–301 passim.

25. *Department of Defense and Unification of Air Service*, 396–435.

26. *Department of Defense and Unification of Air Service*, 429.

27. *Department of Defense and Unification of Air Service*, 416.

28. Arnold, *Global Mission*, 122.

29. The entire contents of the circular were reprinted in *Army and Navy Journal*, 13 February 1926, 562, and in Exhibit A of "Report of the joint investigation . . . concerning the alleged secret publication in, and distribution from, the office of the Chief of the Air Service of a document intended to influence military legislation which had not been approved by the War Department," 13 February 1926, and associated evidence and testimony, all contained in MS 17, Box 7, Series 4, Folder 4, Benjamin Foulois Collection, USAFAL [materials in this folder hereafter cited as Arnold-Dargue Investigation].

30. "Report of the joint investigation . . . ," 13 February 1926, Arnold-Dargue Investigation, 1. The circulars contained two pieces: the aforementioned letter and the six-page "Comments on General Patrick's Proposal for an Air Corps" [hereafter referred to as the circular and the "Comments," respectively].

31. Copp, *A Few Great Captains*, 48–49; White, *Mason Patrick*, 126.

32. Lieutenant Colonel Thorne Strayer represented the Inspector General's office. See "Report of the joint investigation . . . ," 13 February 1926, Arnold-Dargue Investigation.

33. "Report of the joint investigation . . . ," 13 February 1926, Arnold-Dargue Investigation, 1.

34. Fechet conducted the investigation in three phases. The first phase involved interviewing the officers in charge of the Information and War Plans Division in the Office of the Chief of the Air Service—Arnold and Dargue, respectively. Fechet called this phase unsuccessful in his report and moved on to examine personnel, machines, and records of the Information Division. This phase, annotated as "partially successful," led him to the final phase of formal examination of officers under oath. "Report of the joint investigation . . . ," 13 February 1926, Arnold-Dargue Investigation, 3.

35. Arnold, *Global Mission*, 122. Copp wrote how Arnold considered himself the "leader in a very one-sided battle that pitted a handful of 'undisciplined flying officers of junior rank' against the massive bulk and power of the War Department and the Coolidge Administration." Copp, *A Few Great Captains*, 48.

36. Arnold, *Global Mission*, 122. Arnold did not mention these events in his memoirs. He only mentioned the circular once in passing as "'irregular' correspondence."

37. "Report of the joint investigation . . . ," 13 February 1926, Arnold-Dargue Investigation, 4. After Arnold, Fechet interviewed Dargue, who also denied his involvement, but Fechet noted a lack of any signs of mental or physical stress.

38. Fechet found that Dargue prepared the "Comments" letter the day after Patrick's 27 January testimony in Congress, ostensibly for the inevitable re-

quests for information. "Report of the joint investigation . . . ," 13 February 1926, Arnold-Dargue Investigation, 1.

39. "Report of the joint investigation . . . ," 13 February 1926, Arnold-Dargue Investigation, 5.

40. "Report of the joint investigation . . . ," 13 February 1926, Arnold-Dargue Investigation, 5–6.

41. "Report of the joint investigation . . . ," 13 February 1926, Arnold-Dargue Investigation, 9–10. Arnold denied any responsibility for the circular's preparation during the first two phases of the investigation and during his first interview under oath during the third investigative phase. During his testimony on 6 February he acknowledged receiving the circular and then copying it for distribution, but he denied any role in drafting the paper. Arnold requested another chance to testify and clear up inconsistencies, and he testified again on 10 February. At this time, although continuing to hide the totality of his deception, Arnold admitted to drafting the letter. He attributed his passive attitude earlier to not wanting to interrupt the investigation or not wanting to outwardly confront witnesses while the two main investigators worked. Strayer and Fechet hammered away at Arnold for not coming clean initially—especially when he remained silent during the questioning of Rowe. Arnold explained how an officer naturally makes mistakes in judgment when faced with the stress and shock of confrontation by the Inspector General. Testimony of Major Henry H. Arnold, 6 and 10 February 1926, Arnold-Dargue Investigation, 1–10.

42. "Report of the joint investigation . . . ," 13 February 1926, Arnold-Dargue Investigation, 9–10.

43. Arnold, *Global Mission*, 122; *New York Times*, 18 February 1926, 25:7.

44. Daso, *Hap Arnold*, 113–14. See also White, "Air Power Engineer," 320. Arnold's wife believed that her husband had done nothing outside of normal practice and that, in the past, officers had been encouraged to lobby for local support. Daso correctly pointed out that while such lobbying may have been accepted on the West Coast, it would not be accepted in Washington. What Daso did not reiterate was how this activity had occurred within two months of Mitchell's trial for similar activities and how Patrick had warned the young officers under Mitchell's sway to sever their ties with the controversial figure. Daso, *Hap Arnold*, 112–13. Copp cleansed Arnold of guilt, believing he had only done what Patrick had earlier encouraged him to do, namely, "use public relations as a means to sell the Air Service." Copp, *A Few Great Captains*, 48.

45. White, "Air Power Engineer," 320. Arnold's wife believed Patrick remained jealous of the younger airman for his congressional contacts, and thus Patrick felt "a little bit outside of it." Quoted in Daso, *Hap Arnold*, 268–69 n. 65. Arnold wrote very little about the incident in his memoirs, but he did mention the first night at his new post. The commander, General Ewing E. Booth, who had served on the Mitchell court-martial, cordially greeted Arnold and in front of other officers

told the airman, "I know why you're here, my boy. And as long as you are here you can write and say any damned thing you want. All I ask is that you let me see it first!" Arnold, *Global Mission*, 123. Dargue's participation seemed not to have soured Patrick's attitude toward him too much, as Patrick and Fechet selected him to lead the highly promoted Pan-American Goodwill flight, which began some nine months later, on 26 December 1926: a South American tour by five Air Corps planes visiting all the countries of South America (except Bolivia, due to the altitude) and several Caribbean islands. The tour received major press coverage, and the eight surviving aviators (two died in a collision over Buenos Aires) received the Distinguished Flying Cross from Coolidge. James Parton, *"Air Force Spoken Here:" General Ira Eaker and the Command of the Air* (1986; reprint, Maxwell Air Force Base AL: Air University Press, 2000), 51–61.

46. Testimony of Major Herbert A. Dargue, 6 February 1926, Arnold-Dargue Investigation, 4.

47. Testimony of First Lieutenant Burnie R. Dallas, 6 February 1926, Arnold-Dargue Investigation, 4.

48. Testimony of First Lieutenant Burnie R. Dallas, 3–4. Some people did receive the letter in plain envelopes and with stamps. Captain Montgomery testified that he sent out approximately forty-five and paid for the postage himself. He initially refused to detail whom he sent them to beyond "friends" and newspapermen (he later named only four other reserve officers, including Eddie Rickenbacker) but said he did not use any list given to him by any other person. Testimony of Captain John K. Montgomery, Air Service Reserve, 8 February 1926, Arnold-Dargue Investigation, 2–4.

49. As noted in the previous chapter, Patrick had written the bill at the request of Congress and had properly tendered the legislation through Davis. Still, Davis remained incensed that a subordinate had submitted a competing plan to Congress.

50. *Army and Navy Journal*, 13 February 1926, 561.

51. *New York Times*, 9 February 1926, 27:1.

52. Quoted in *Army and Navy Journal*, 13 February 1926, 562.

53. *New York Times*, 9 February 1926, 27:1. Mitchell added, "He has taken my place, and now they are going after him."

54. *New York Times*, 10 February 1926, 22:4. Other papers supporting Coolidge and his offensive against new Air Service propaganda included the *Des Moines Register* and the *Morning Oregonian*. See *Army and Navy Journal*, 13 February 1926, 562.

55. *New York Times*, 10 February 1926, 22:4.

56. The *Army and Navy Journal* published Davis's letter in its entirety. See Dwight F. Davis, Secretary of War, to Honorable John M. Morin, Chairman, Committee on Military Affairs, House of Representatives, 10 February 1926, in *Army and Navy Journal*, 13 February 1926, 564. The letter also entered the official

record of the hearings. See also *Department of Defense and Unification of Air Service*, 496–99. Davis also accused Patrick of flip-flopping on his approach, saying that he pushed for the changes for tactical reasons before the Morrow Board but now advocated them for administrative reasons. Davis charged that the real reason for the proposals was the desire for increased funding. Admiral Moffett also disparaged Patrick's bill, calling it "at variance" with the two departments and the president. *Department of Defense and Unification of Air Service*, 687.

57. *New York Times*, 11 February 1926, 6:1. LaGuardia's earlier testimony before the committee had blasted the General Staff as "either hopelessly stupid or unpardonably guilty" for their continued and dogged rejection of any realignment of national defense or change in the status of Army aviation. *Department of Defense and Unification of Air Service*, 383.

58. *New York Times*, 11 February 1926, 6:1. Some members still believed Davis wanted to, at least in part, implicate Patrick.

59. *New York Times*, 18 February 1926, 25:7. The quote is from the paper and is not a direct quote from Davis.

60. "Statement given to Press Relations by Gen. Patrick," February 1926, RG 18, Patrick Papers, Articles and Speeches File, Box 1.

61. For an example of Patrick's style and clarifications of his opinions, see U.S. Congress, Senate, Committee on Military Affairs, *Reorganization of the Army Air Service: Hearing before the Committee on Military Affairs on S. 2614*, 69th Cong., 1st sess., 5 February 1926, 5.

62. *New York Times*, 28 February 1926, 21:1, 4 March 1926, 23:1.

63. *New York Times*, 4 March 1926, 23:1.

64. *New York Times*, 4 March 1926, 23:1. These maneuvers, although undocumented by either the War Department or Patrick, seemed in harmony with Patrick's character and modus operandi. He supported his position to a point but would not press his case if it met with the disapproval of Davis and Coolidge. Patrick knew that the congressmen understood his true desires, and it is likely that he discussed the proposal's prospects of passage with air-friendly congressmen.

65. *New York Times*, 4 March 1926, 23:1; W. F. James, "Handling Military Legislation in the House of Representatives," 4. Although this represented the closest such a measure came to passing until the 1940s, it would have survived neither the Senate nor the president's veto. Ransom unofficially worked out the voting on the bill (for specifics and names see Ransom, "The Air Corps Act of 1926," 318). Only four Republicans voted for the measure establishing a unified defense structure, joining six Democrats. Two Democrats voted with the remaining nine Republicans. No sources reveal who switched his vote overnight. Most likely, one of the two Democrats split from his party. Still, all understood that Coolidge, while not overtly threatening a veto, would not hesitate to block any such bill. *Army and Navy Journal*, 13 February 1926, 561.

66. *Army and Navy Journal*, 13 February 1926, 561.

67. W. F. James, "Handling Military Legislation in the House of Representatives," 4.

68. W. F. James, "Handling Military Legislation in the House of Representatives," 4–6. James's lecture provided the best insight into the wrangling over issues and some of the political compromises that need no further elaboration here. For the bill being submitted to the House, see *Congressional Record*, 69th Cong., 1st sess., 6544.

69. Patrick to James, 4 May 1926, *Congressional Record*, 69th Cong., 1st sess., 8751.

70. W. F. James, "Handling Military Legislation in the House of Representatives," 6–7.

71. *New York Times*, 24 February 1926, 1:1.

72. *New York Times*, 6 March 1926, 5:5.

73. James told Senator James W. Wadsworth Jr., chairman of the Senate Committee on Military Affairs, that he recognized nothing about the bill except the name and number. W. F. James, "Handling Military Legislation in the House of Representatives," 7, and 8–9 for the differences between the House and Senate versions; for the latter see also Layman, *Air Corps Personnel and Training Programs*, 30–31. The entire Senate version is contained in U.S. Congress, Senate, Committee on Military Affairs, *To Increase the Efficiency of the Army Air Service*, Report No. 830, to accompany H.R. 10827, 69th Cong., 1st sess., 1–2. The Senate committee substituted H.R. 10827 for another bill it had previously reported.

74. For more insight on the debates, which revolved around wording and amendments, see Ransom, "The Air Corps Act of 1926," 328–29; *Congressional Record*, 69th Cong., 1st sess., 10403–13. Among the differences of opinion between the two, Robinson wanted to protect the new Air Corps from having too many non-flying officers and to ensure that the officer selected as chief would always, in peace and during war, be a flying officer. Bingham wanted to stick closer to the Morrow recommendations and protect the leeway of the president in selecting general officers, especially during wartime, when available personnel could require selecting an officer from another branch.

75. *Congressional Record*, 69th Cong., 1st sess., 10498.

76. *Congressional Record*, 69th Cong., 1st sess., 10498. No known records exist among the prominent air officers of pushing for the legislation or visiting senators, but the lack of such is not surprising given the recent actions against Arnold and a desire to leave no paper trail behind. One cannot discount the possibility that air officers stationed in the District (at the War Department and at Bolling Field) may have taken Arnold's circular to heart and approached congressmen.

77. For example, an amendment on whether the president must appoint an aviator as the Air Corps chief carried 33–23 in favor of allowing the president to choose without making aviator status mandatory. Only Morris Sheppard of Texas defected from the Democrats to vote with the Republicans, and only Progressive

Republican La Follette joined the Democrats. Ransom, "The Air Corps Act of 1926," 331–32.

78. *An Act to provide more effectively for the national defense by increasing the efficiency of the Air Corps of the Army of the United States, and for other purposes* [hereafter referred to as the Air Corps Act], Statutes at Large, 44:780–90.

79. *New York Times*, 20 February 1926, 4:5.

80. Ransom, "The Air Corps Act of 1926," 303–7. Mitchell drew large crowds only when newspapers bought out the lectures and then gave the tickets away.

81. W. F. James, "Handling Military Legislation in the House of Representatives," 15.

4. THE IMPACT OF AN "AIR-MINDED" CIVILIAN

1. Ransom and McClendon agreed that the act represented a victory for Coolidge and the War Department. Ransom, "The Air Corps Act of 1926," 338; McClendon, *Autonomy of the Air Arm*, 59–60. Tate viewed the long-awaited legislation as a true compromise with no winners, but he agreed that it more closely followed the Morrow recommendations. Tate, *The Army and Its Air Corps*, 47–48. White believed the Air Corps gained much from the law, as they finally had obtained legal guarantees and a step toward more respect and autonomy, but he gives a slight political advantage to the War Department. White, *Mason Patrick*, 129.

2. One of the two assistant secretaries would take charge of the War Department when the Secretary of War left the District. The Assistant Secretary of War (whose primary duty involved overseeing procurement matters for the department), and the Assistant Secretary of War for Air shared the duty.

3. W. Frank James, "A Five-Year Development Program for the Air Corps at Last," *U.S. Air Services* 11, no. 6 (1926): 45–46.

4. Biography of F. Trubee Davison, RG 18, General Correspondence, 1926–1933, Assistant Secretary of War (Air), Box 7, NARA II.

5. Copp, *A Few Great Captains*, 64. For more information on Davison's family background, see Thomas W. Lamont, *Henry P. Davison: The Record of a Useful Life* (New York: Harper, 1933).

6. Assistant Secretary of War (Air) and Assistant Secretary of War to Secretary of War, 20 September 1926, RG 18, General Correspondence, 1926–1933, Assistant Secretary of War (Air), Box 7, NARA II. By existing law, the assistant secretary oversaw all procurement activities for the War Department. Davison asked to be included in the decisions regarding Army aviation, and the joint letter demonstrated that Assistant Secretary Hanford MacNider agreed. In fact, MacNider was undoubtedly relieved to accept a reduction in his workload.

7. Two memoranda in the files of the air secretary provided information on Davison's specific duties. The first, though undated and unsigned, was located

nearby the previous request by Davison and MacNider. It seems probable that either they submitted this description to Davis for his approval or Davis assigned Davison these specifics in a reply. Another memorandum, given to Davis on 16 March 1929, specified much the same information but noted that an executive memorandum codified these duties. It outlined eight specific areas of responsibility, with the first being "The carrying out of the Five Year Program" and the third ordering him to represent the Secretary of War on all Air Corps budgetary matters. Unsigned and undated memoranda and "Duties of the Assistant Secretary of War (Air)," both in RG 18, General Correspondence, 1926–1933, Assistant Secretary of War (Air), Box 7, NARA II.

8. Those present from the Military Affairs Committee included John Morin, Frank James, Jonathan Wainwright, and Allen Furlow. See invitations, responses, and other letters associated with the preparation in RG 18, General Correspondence, 1926–1933, Assistant Secretary of War (Air), Box 8, NARA II.

9. U.S. Congress, Senate, Subcommittee of the Committee on Appropriations, *War Department Appropriation Bill for 1931: Hearing before the Subcommittee of the Committee on Appropriations*, 71st Cong., 2nd sess., 1930, 7. For other examples of Davison's taking the lead in the hearings, see U.S. Congress, Senate, Subcommittee of the Committee on Appropriations, *War Department Appropriation Bill, 1929: Hearing before the Subcommittee of the Committee on Appropriations*, 70th Cong., 1st sess., 1928, 81–92; U.S. Congress, House, Subcommittee of the House Committee on Appropriations, *War Department Appropriation Bill for 1930: Hearing before the Subcommittee of the House Committee on Appropriations*, 70th Cong., 2nd sess., 1928, 421–584 passim.

10. Davison seemed especially close to Bingham and provided him with information, with the proper push for the Army Air Corps. See Bingham to Davison, 6 May 1929, RG 18, General Correspondence, 1926–1933, Assistant Secretary of War (Air), Box 7, NARA II.

11. U.S. Congress, House, Committee on Military Affairs, *Air Corps: Progress under Five-Year Program: Hearing before the Committee on Military Affairs*, 69th Cong., 2nd sess., 19 January 1927, 23.

12. Fechet took Davison hunting to Ft. Bragg, North Carolina, on at least one occasion. See Fechet to Colonel H. W. Butner, 9, 21, and 25 January 1929, RG 18, Correspondence of General James E. Fechet, 1925–1930, Box 1, NARA II.

13. Davison's file includes many speeches, requests for and replies after speeches, and notes to officers on speeches. Most requests came from men's associations and academic societies. For two examples see requests from National Aeronautic Meeting of the Society of Automotive Engineers in St. Louis for their meeting on 18–20 February 1930 and a request from the Overbrook Men's Association, Overbrook Presbyterian Church (Philadelphia), for their meeting on 8 November 1926. See also address given by Lieutenant Colonel H. E. Hartney, "Aviation as a Factor in the Restoration of Prosperity," given at

the American Institute Luncheon, Astor Hotel, n.p., 10 January 1931. All items contained in RG 18, General Correspondence, 1926–1933, Assistant Secretary of War (Air), Box 45, NARA II.

14. *New York Times*, 4 October 1928, 4:4, 14 October 1928, 27:5, 16 October 1928, 1:2, 29 October 1928, 8:3.

15. *New York Times*, 12 January 1930, 3:6.

16. *New York Times*, 1 February 1928, 2:6.

17. *New York Times*, 14 May 1926, 4:6.

18. Air Corps Act, *Statutes at Large*, 44:783–84. The act increased the number of officers by 403 and of enlisted by 6,240.

19. Air Corps Act, *Statutes at Large*, 44:784. Emphasis added. The original wording of the bill, H.R. 10827, mandated that the president "shall" submit a supplemental spending bill, but it was struck out for the more liberal verb. U.S. Congress, Senate, Committee on Military Affairs, *The Army Air Service: Hearing before the Committee on Military Affairs on H.R. 10827*, 69th Cong., 1st sess., 10 May 1926, 4. In not submitting the additional appropriation, Coolidge reasoned that providing the necessary money into the remainder of the fiscal year would "crowd" fiscal estimates. Edwin H. Rutkowski, *The Politics of Military Aviation Procurement, 1926–1934: A Study in the Political Assertion of Consensual Values* (Columbus: Ohio State University Press, 1966), 29. The Air Corps requested a total of $25,794,000 for FY 1927 and received only $15,900,000 in appropriations. *Annual Report of the Chief of the Air Corps for the Fiscal Year Ending 30 June 1927*, 62–63, AFHSO.

20. Conner made similar remarks during congressional debate on the air bills, but these sentiments appeared again in the *New York Times*, 9 August 1926, 5:4. See also Senate Committee on Military Affairs, *The Army Air Service*, 43.

21. The first signs that Coolidge would not accept large expenditures came when he submitted his FY 1928 budget on 11 December 1926. Budget shortfalls the previous year had forced the Army to cut its enlisted strength to just under 111,000, not even close to the 118,750 authorized. The new budget did not restore funding for the authorized strength but provided only enough for a force of 115,000. Congress later restored the money to bring the Army up to the 118,750, but these numbers still affected the Air Corps' expansion plan. The act's vague language did not specify if the additions to the Air Corps would come at the expense of the Army or if it would fund additional men—Coolidge interpreted the former, much to the Army's disdain. Rutkowski, *Politics of Military Aviation Procurement*, 23–25; Senate Committee on Military Affairs, *The Army Air Service*, 40–41.

22. Tate, *The Army and Its Air Corps*, 61, 83.

23. For examples, see statements of Fechet and Hinds, *New York Times*, 13 May 1927, 3:1, 16 May 1927, 5:4.

24. Since Congress and the presidents would not increase the total number

of men in the Army, the Army transferred men to the Air Corps in order to meet the mandated increases in air personnel strength. In 1929 the Army deactivated five battalions of infantry in order to allow the air increases. The number of men transferred or inactivated from the non-flying Army for the first years of the program totaled, by year, 1,248 (1927), 536 (1928), 1,960 (1929), and 1,248 (1930). *New York Times*, 21 August 1929, 4:5, 16 November 1930, II, 2:3. Secretary of War Patrick Hurley warned that unless the Army obtained overall increases, continued transfers would result in the Army's being unable to perform its missions, and the entire Army and the National Defense Act would require reorganization and modification. *Report of the Secretary of War to the President* (Washington DC: Government Printing Office, 1930), 3. Accessed at AFHSO.

25. During the budget battles for FY 1930, Congress tried to use this tactic. The War Department disagreed with this numbers-juggling act. Congress did attempt to make up for program deficiencies and tried its best to complete the expansion program on time. Irving Brinton Holley Jr., *Buying Aircraft: Matériel Procurement for the Army Air Forces*, United States Army in World War II, Special Studies, ed. Stetson Conn (Washington DC: Office of the Chief of Military History, Department of the Army, 1964), 64–65. See also U.S. Congress, *War Department Appropriations Bill, 1929*, 70th Cong., 1st sess., H. Rept. 497, serial 8835, 13.

26. Senate, Subcommittee of the Committee on Appropriations, *War Department Appropriation Bill, 1929*, 85.

27. Davison reported in 1928 that the vast majority of wartime equipment, except for 3,000 dreadfully outdated Liberty engines (of the 15,000 produced during the war), had been removed from service. By adding newer models of aircraft and eliminating the obsolete and some of the obsolescent models, the Air Corps reaped the benefits of safety, doubling the number of flying hours per fatal accident. "Annual Report of the Assistant Secretary of War (Air)," contained in *Report of the Secretary of War to the President*, fiscal year ending 1930, 69, 72–73, accessed at AFHSO. For information on the Liberty engine and its production, see Irving Brinton Holley Jr., *Ideas and Weapons* (Washington DC: Air Force History and Museums Program, 1997), 124. In order to save money, the House wanted the Air Corps to continue to use the Liberty engines, instead of spending $7,000 apiece on newer engines, in its observation aircraft. U.S. Congress, *War Department Appropriations Bill, 1929*, 13.

28. Approved congressional appropriations by year: $24.891 million (FY 1928); $27.150 million (FY 1929); $32.441 million (FY 1930); and $35.823 million (FY 1931). The service actually added more than 400 aircraft during FY 1930. From *Annual Report of the Chief of the Air Corps* and *Report of the Secretary of War to the President* for fiscal years ending June 1928 to June 1931. All reports accessed at AFHSO.

29. "Report of the Assistant Secretary of War (Air)," contained in *Report of the Secretary of War to the President* (Washington DC: Government Printing Office,

1932), 39. One must also realize, though, that the Air Corps budget represented the costs for material, training, buildings, and so on, but the pay for the men (always the single largest budget item) and other personnel costs (medical care, etc.) remained part of the overall Army budget.

30. U.S. Congress, *War Department Appropriations Bill, 1931*, 71st Cong., 2nd sess., H. Rept. 97, serial 9190, 13–14. Previously the view tended toward the program's being legally mandated. This statement, however, gave Congress an excuse for its budget shortfalls to complete the program—though many in the service and in Congress continued to view the Five-Year Plan as a requirement.

31. War Department, "A Study Presenting the Details of the Promotion System, U.S. Army," 1926, 1, located in 69A-F30.6, RG 233, NARA. The army brought in a large number of officers in 1916 and 1920 but then cut the army by 66 percent in 1922. The reductions did not occur evenly throughout the ranks, thus creating a situation where a large number of officers occupied the same general area on the promotion lists, with the majority in the "hump" separated by less than two years. War Department, "A Study Presenting the Details of the Promotion System, U.S. Army," 43–44.

32. The act directed the secretary of war to "investigate and study the alleged injustices which exist in the promotions list" and recommend changes by the second Monday in December 1926. See section 4, Air Corps Act, *Statutes at Large*, 44:782.

33. Annual Report of the Chief of the Air Corps for FY Ending June 30, 1930, 3–4, and "Annual Report of the Assistant Secretary of War (Air)," contained in Report of the Secretary of War to the President, fiscal year ending 1930, 58, both accessed at AFHSO.

34. "Annual Report of the Assistant Secretary of War (Air)," contained in *Report of the Secretary of War to the President*, fiscal year ending 1930, 58, accessed at AFHSO.

35. The Air Corps chiefs and Davison both stressed the need for promotion legislation and urged a separate list for the air arm in their annual reports throughout the Five-Year Plan. See *Annual Report of the Chief of the Air Corps* and "Annual Report of the Assistant Secretary of War (Air)," contained in *Report of the Secretary of War to the President*, from fiscal years ending 1927 to 1932, accessed at AFHSO. The 1926 War Department report made no mention of a separate promotion list for the Air Corps.

36. "Annual Report of the Assistant Secretary of War (Air)," contained in *Report of the Secretary of War to the President*, for fiscal year ending 30 June 1929, 87, accessed at AFHSO.

37. *Congressional Record*, 70th Cong., 1st sess., 9576, reprinted in *Air Corps News Letter*, 5 June 1928, 197. See also "Secretary Davison's Statement RE Air Corps Promotion Bill," *Air Corps News Letter*, 20 June 1928, 219, and "Memorandum for Assistant Secretary of War, Mr. Davison," from Major Delos C. Emmons (Office

of the Chief of the Air Corps), 21 May 1928, in RG 18, General Correspondence, 1926–1933, Assistant Secretary of War (Air), Box 14, NARA II. Newspaper editorials also supported the Furlow bill and its potential to help the Air Corps. See *Washington Post*, 27 January 1929, II, 1.

38. This study will not discuss the myriad of promotion legislations and their histories. For information on the different bills and their outcomes, see Layman, *Air Corps Personnel and Training Programs*, 91–117. Furlow's first bill died in committee, but he reintroduced similar legislation soon thereafter, and it passed the House with unanimous consent (106). Representative Wainwright introduced his own measure, which contained a promotion overhaul for the remainder of the Army (106 n. 33).

39. Layman, *Air Corps Personnel and Training Programs*, 105–8. Robinson, an Arkansas Democrat, had earlier championed air causes and even supported promotion legislation for inclusion in the Air Corps Act. The filibuster occurred over a controversial bill on the Boulder Dam. *Air Corps News Letter*, 5 June 1928, 193.

40. Quoted in David M. Kennedy, *Freedom from Fear: The American People in Depression and War, 1929–1945* (New York: Oxford University Press, 1999), 46.

41. D. M. Kennedy, *Freedom from Fear*, 11. Kennedy noted how Hoover brought a more progressive spirit into the White House, especially regarding labor policy.

42. Herbert Hoover, *The Memoirs of Herbert Hoover*, vol. 2, *The Cabinet and the Presidency, 1920–1933* (New York: Macmillan, 1952), 133–34.

43. See press releases dated 3 October and 3 November 1928 in Campaign and Transition File, Box 193, Presidential Papers, HHPL.

44. For examples of Hoover touting his aviation policies, see *New York Times*, 12 September 1928, 29:4, 16 September 1928, 25:1, and 23 September 1928, II, 6:1. *Aero Digest* published a similar article, with Hoover again summarizing civilian aviation expansion during his tenure as commerce secretary. Herbert Hoover, "Civil Aviation's Rapid Progress," *Aero Digest*, April 1928, 509, 696–97. The previous year, after Charles Lindbergh's feat and the frenzied news coverage, Hoover lunched with the flier upon his return from Paris. The newspapers covered this meeting, and Hoover stood out among those having prominent and lengthy contact with the hero. *New York Times*, 24 June 1927, 3:1. Hoover rarely mentioned military aviation, but in 1927 he worked with the paper and published a feature article, "Hoover Foresees a Greater Air Service," in which he mentioned the military benefits derived from a strong aviation manufacturing base and civilian operations and airports. *New York Times*, 26 June 1927, IX, 1.

45. Bingham to Hoover, 13 August 1930, note to Hoover recording Bingham's call and invitation, 29 July 1932, and Lawrence Richey, Secretary to the President, to Bingham, 29 July 1932, all in Personal File, Box 1: Aeronautics, 1930–1932, Presidential Papers, HHPL.

46. Proposed letter, n.d., with annotations in Personal File, Box 1: Aeronautics, 1930–1932, Presidential Papers, HHPL.

47. Hoover to Wright Memorial Committee, 18 November 1932, Personal File, Box 1: Aeronautics, 1930–1932, Presidential Papers, HHPL.

48. See Frederick B. Rentschler to Davison, telegram, 3 June 1929, Davison to George Akerson (Secretary to the President), 4 June 1929, and Akerson to Davison, 4 June 1929, all in Subject File, Box 56: Aeronautics Correspondence, 1929–1932, Presidential Papers, HHPL. By the time of the 1932 campaign, Hoover's speeches barely mentioned national defense, much less aviation's place. Even when Hurley campaigned for him, he did not tout Hoover's military policies and accomplishments. See various speeches by Hurley in the E. French Strother Papers, Box 9, HHPL.

49. Secretary of State Henry L. Stimson made the official announcement of Good's death and recounted his background. Official announcement dated 19 November 1929, Cabinet Offices File, War [Department] Correspondence, Box 54G, Presidential Papers, HHPL. On Good's qualifications, see also the *Rockford* IL *Daily Republic*, 19 November 1929, 6.

50. Tate, *The Army and Its Air Corps*, 72–73. See also "Extract from *Who's Who in America*," Cabinet Offices File, War [Department] Correspondence, Box 54F, Presidential Papers, HHPL. Tate argued that Good took an easygoing approach, especially regarding the primary problem of his tenure, that of coastal defense concerns with the Navy. Good also gave his Chief of Staff, General Charles P. Summerall, more of a free hand in the department than Hurley did. A confrontation with Senator William H. King (D-Utah) provided a good example of Hurley's style and attitude. King questioned Hurley's integrity, and Hurley replied that senators often "browbeat" witnesses. On the differences between witnesses' testimony and senators' statements, he sarcastically announced, "I realize when a witness appears before this committee it is a stump speech, but if one of the members of the committee speaks, it is the height of statesmanship," after which he stormed out of the room, excusing himself from the other senators. King apologized and removed his attack on Hurley from the record. Individuals File, Box 147 (Hurley folder), Presidential Papers, HHPL.

51. Eaker to Major Walter Kilner, 9 December 1929, Box 3: Personal File, General Ira C. Eaker Papers, LOC-MD.

52. For Davison's resignation letter see Davison to Hoover, 4 March 1929, Cabinet Offices File, War [Department] Correspondence, Box 54F, Presidential Papers, HHPL. For information about the "Little Cabinet" and the retention of the three men see *New York Times*, 27 February 1929, 2:1.

53. For example, see the following speeches (audience and date): Advertising League of Omaha, Omaha, Nebraska, 26 November 1928; Community Forum, Boundbrook, New Jersey, 15 January 1928; and Men's Club, Garden City, Long

Island, 6 December 1927. All are in Box 28: Speech and Article File, General Charles P. Summerall Papers, LOC-MD.

54. Box 28: Speech and Article File, Summerall Papers.

55. *New York Times*, 7 August 1930, 20:4, 10 August 1930, II, 6:1. Actually, Hoover's administration botched the announcement and caused a brief embarrassment. Hoover first proclaimed that MacArthur was the only officer who could serve the entire four-year term before reaching the mandatory retirement age. In fact, eleven major generals had sufficient time remaining, but MacArthur still had seniority on those men (including Malin Craig, who would follow MacArthur, and Fox Conner, who had fought budget battles against the Air Corps).

56. Hoover, *The Cabinet and the Presidency*, 339.

57. John R. M. Wilson, "The Quaker and the Sword: Herbert Hoover's Relations with the Military," *Military Affairs* 38, no. 2 (1974): 45–46.

58. *New York Times*, 16 June 1927, 11:1.

59. See *Annual Report of the Chief of the Air Corps* for fiscal years ending June 1928–31, AFHSO.

60. *Annual Report of the Chief of the Air Corps for Fiscal Year Ending 30 June 1930*, 1, AFHSO.

61. Copp, *A Few Great Captains*, 73.

62. Various letters, November to December 1929, Cabinet Offices File, War [Department] Correspondence, Box 54G, Presidential Papers, HHPL.

63. Hoover, *The Cabinet and the Presidency*, 338. The president agreed with disarmament, but he did not like the isolationists' attitudes that caused the county to neglect "our proper responsibilities" in world affairs (330–31).

64. Wilson, "The Quaker and the Sword," 41.

65. "Notes on the Survey," Summerall to Hoover, 26 October 1929, 1, in Cabinet Offices File, War [Department] Correspondence, Box 54F, Presidential Papers, HHPL.

66. John W. Killigrew, "The Impact of the Great Depression on the Army, 1929–1936" (Ph.D. diss., Indiana University, 1960), 20–29.

67. Summerall, "Notes on the Survey," 21.

68. Summerall, "Notes on the Survey," 22, 32.

69. Summerall, "Notes on the Survey," 21–22.

70. Tate, *The Army and Its Air Corps*, 105 n. 5.

71. Tate, *The Army and Its Air Corps*, 61–70. Hurley and Secretary of the Navy Charles F. Adams even argued over who started the controversy (76).

72. See notes in Presidential File "War, Secretary of, filed February 21, 1930," in Subject File, Box 57, Presidential Papers, HHPL. The notes pertained to a full letter from Hurley to the president three days earlier.

73. Hurley to Hoover, 29 May 1930, Cabinet Offices File, War [Department] Correspondence, Box 54G, Presidential Papers, HHPL.

74. Tate, *The Army and Its Air Corps*, 71–79. For the official press release, see "Army and Navy Agree on Spheres of Activities of their Air Forces," 9 January 1931, contained in RG 165, Records of the War Department General and Special Staffs, War Plans Division General Correspondence, 1920–1942, Box 54, NARA II.

75. Hearings before the Joint Committee on Aerial Coast Defense, 7 May 1929, in RG 165, Records of the War Department General and Special Staffs, War Plans Division General Correspondence, 1920–1942, Box 54, NARA II, 75–119.

76. As noted earlier, Davison also passed Bingham information to ensure that the senator clearly understood the Army's view, and Bingham gladly accepted the help. See Bingham to Davison, 6 May 1929, RG 18, General Correspondence, 1926–1933, Assistant Secretary of War (Air), Box 7, NARA II.

77. See http://www.senate.gov/learning/stat_13.html and http://clerk web.house.gov/histrecs/househis/lists/divisionh.htm, both accessed 18 August 2001.

78. Herbert Hoover, *The Memoirs of Herbert Hoover*, vol. 3, *The Great Depression, 1929–1941* (New York: Macmillan, 1952), 101.

79. D. M. Kennedy, *Freedom from Fear*, 60–61; Killigrew, "Impact of the Great Depression," 92–93.

80. Hurley joined MacArthur, the Deputy Chief of Staff (Major General George Van Horn Moseley), the Chief of Engineers, and the Quartermaster General, among others. "Notes Taken during the Conference at the President's Rapidan Camp, May 9, 1931," in Cabinet Offices File, War [Department] Correspondence, Box 54K, Presidential Papers, HHPL. For information on the Rapidan Camp see Hoover, *The Cabinet and the Presidency*, 322–23.

81. "Notes Taken during the Conference at the President's Rapidan Camp, May 9, 1931," 1–2.

82. "Notes Taken during the Conference at the President's Rapidan Camp, May 9, 1931," 2. MacArthur advised: "Money spent on Army improvements [buildings, post improvements, etc.] creates permanent value."

83. "Notes Taken during the Conference at the President's Rapidan Camp, May 9, 1931," 3. Hoover's questions did not clearly explain if his "air fleets" included both the Army and the Navy or only the latter (as was last being discussed). His advisers recommended that a joint board look at the possibility of aircraft reductions. Also of note, Hoover held another Rapidan Conference with various (and additional) Army and Navy officials on 6–7 June 1931. This conference held more general discussions on how to limit the military budget and did not produce the drastic changes of the more limited conference one month earlier. See "Naval Conference at Rapidan" in Cabinet Offices File, War [Department] Correspondence, Box 54K, Presidential Papers, HHPL.

84. "Notes Taken during the Conference at the President's Rapidan Camp, May 9, 1931," 28.

85. "Notes Taken during the Conference at the President's Rapidan Camp, May 9, 1931," 1. Top corner of front page noted, "Only three copies made." One for the president, one for MacArthur, and one for Lieutenant Colonel C. B. Hodges, likely the note taker.

86. "The Next War Will Be Won in the Future," in Douglas MacArthur, *A Soldier Speaks: Public Papers and Speeches of General of the Army Douglas MacArthur*, ed. Major Vorin E. Whan Jr. (New York: Praeger, 1965), 33–35.

87. The Air Corps wanted, and the Air Corps Act called for, 1,800 serviceable aircraft, which did not count the number undergoing overhaul and major repair. Thus, MacArthur wanted to reduce the numbers available by limiting the number of Army aircraft to 1,800, including all of those that the Air Corps did not want counted. MacArthur, memorandum to the Secretary of War, "The Needs of the United States in Air Forces," 14 August 1931, 3–6, 14–15, 17, and 23, located in Cabinet Offices File, War [Department] Correspondence, Box 54L, Presidential Papers, HHPL. MacArthur also wanted the Navy to justify an air fleet of "great excess" and recommended elimination of aircraft carriers and obtaining other nations' agreement in the forthcoming disarmament conference. This idea also appealed to Hoover.

88. Memorandum for the Budget Officer for the War Department, "Estimates for Military Activities, Fiscal Year 1933," from MacArthur, 6 November 1931, Subject File, Box 19, Foulois Papers. The Air Corps reductions cuts came from reducing National Guard flying hours, cutting travel expenses and per diems, and unspecified cuts in aircraft and equipment purchases (memo, p. 5). The Air Corps' part of the cuts represented by far the largest cut within the Army, with the next-largest reductions being for construction at Army posts ($1.73 million) and arming the National Guard ($1.27 million). MacArthur also asked all organizations to be ready to rehearse their congressional presentations in his office "with a view to insuring a proper presentation" (memo, p. 10).

89. Acting Secretary Frederick H. Payne to Hoover, 15 September 1930, and Office of the Assistant Secretary of War, *Current Procurement News Digest*, vol. 5, no. 7 (7 January 1931), both in Cabinet Offices File, War [Department] Correspondence, Box 54G, Presidential Papers, HHPL.

90. "Annual Report of the Assistant Secretary of War," *Report of the Secretary of War to the President* (Washington DC: Government Printing Office, 1931), 29, accessed at AFHSO.

91. The Air Corps saved almost $2 million and returned it to the Treasury. "Annual Report of the Assistant Secretary of War," *Report of the Secretary of War to the President* (Washington DC: Government Printing Office, 1932), 39, accessed at AFHSO.

92. Peter R. Faber, "Interwar US Army Aviation and the Air Corps Tactical School: Incubators of American Airpower," in *The Paths of Heaven: The Evolution of Airpower Theory*, ed. Phillip S. Meilinger, The School of Advanced Airpower

Studies, with a foreword by General Ronald R. Fogleman, Chief of Staff, U.S. Air Force (Maxwell Air Force Base AL: Air University Press, 1997), 205.

93. Davison, oral history interview by Murray Green, MS 33, Box 62, p. 11, Green Collection, USAFAL.

94. Faber, "Interwar U.S. Army Aviation and the Air Corps Tactical School," 188–90.

95. "The Assistant Secretary of War for Aeronautics, F. Trubee Davison," 7, in James Spear Taylor and Robert M. Gates Collection, Box 1, HHPL.

96. Hoover to Davison, 25 February 1933, RG 18, General Correspondence, 1926–1933, Assistant Secretary of War (Air), Box 6, NARA II.

97. For example, note the previously mentioned Fechet reply to Summerall's survey on the need to increase, and not cut, the funding for the Air Corps.

98. *New York Times*, 9 September 1931, 3:4.

99. Fechet to Davison, 20 October 1931, in RG 18, General Correspondence, 1926–1933, Assistant Secretary of War (Air), Box 28, NARA II. Fechet did not make the official announcement until after he notified Davison, and the *New York Times* reported this episode on 28 October 1931, 11:1.

100. The newspapers rumored Foulois' probable ascent earlier in the summer. *New York Times*, 6 June 1931, 5:3. Foulois officially took the reins of the office he had long coveted on 22 December 1932. *Air Corps News Letter*, 25 January 1932, 5.

101. *New York Times*, 9 September 1931, 3:4.

102. *New York Times*, 2 January 1932, 1:4.

103. *New York Times*, 2 January 1932, 1:4. Tichenor sent Davison a copy of the same statement, with only slightly different wording, as an outprint from an upcoming *Aero Digest*. Tichenor to Davison (and attachments), 29 February 1932, in RG 18, General Correspondence, 1926–1933, Assistant Secretary of War (Air), Box 28, NARA II. Fechet's title with the publication became "National Defense Editor."

104. *New York Times*, 2 January 1932, 1:4.

105. *Aero Digest*, January 1932, 27, February 1932, 27, and March 1932, 27. The first issue without a Fechet article was September 1933. After that date he contributed regularly but not always monthly.

106. *Aero Digest*, April 1932 to December 1933.

107. James E. Fechet, *Flying* (Baltimore: Williams and Wilkins, 1933), 132–38.

108. *New York American*, 7 February 1932, 2-E.

109. *New York American*, 7 February 1932, 2-E.

110. As noted earlier, Pershing commented regularly on military issues and often wrote influential letters to sway Congress toward military policies (especially during debate on air legislation). Pershing also corresponded regularly with Republican insider John Callan O'Laughlin and submitted several open letters for publication in O'Laughlin's *Army and Navy Journal*. For example, see Persh-

ing to O'Laughlin, 15 October 1925, in Box 27, John Callan O'Laughlin Papers, LOC-MD. Wood's proposals for expanded military training in the United States countered those of President Wilson. Both Wood and Pershing were quite overt in their support for the Republican Party, and the former failed in a bid to win the 1920 GOP presidential nomination. See Chase C. Mooney and Martha E. Layman, "Some Phases of the Compulsory Military Training Movement, 1914–1920," *Mississippi Valley Historical Review* 38, no. 4 (1952): 634, 655. See also Robert D. Ward, "The Origin and Activities of the National Security League, 1914–1919," *Mississippi Valley Historical Review* 47, no. 1 (1960): 55–56.

5. CONFLICT WITHIN AND WITHOUT

1. Roosevelt cut the Army budget and tried to keep defense budgets low (at least until 1938). However, he did divert Public Works Administration money to some military building programs, and he wanted to commit more than $1 billion over a ten-year period to naval construction (which resulted in the aircraft carriers *Yorktown* and *Enterprise* and numerous other ships used during World War II). As some Army leaders feared, FDR, who had served as Assistant Navy Secretary, favored his former service during his first term and most of his second. See Frank Burt Freidel, *Franklin D. Roosevelt: A Rendezvous with Destiny* (Boston: Little, Brown, 1952), 179–81; Harold L. Ickes, *The Secret Diary of Harold L. Ickes*, vol. 1, *The First Thousand Days, 1933–1936* (New York: Simon and Schuster, 1953), 216–17; William E. Leuchtenburg, *Franklin D. Roosevelt and the New Deal: 1932–1940* (New York: Harper and Row, 1963), 133.

2. Roosevelt polled over 57 percent of the vote and garnered 472 electoral votes, compared to 59 for Hoover. Fred I. Greenstein, *The Presidential Difference: Leadership Style from Roosevelt to Clinton* (New York: Free Press, 2000), 202; James MacGregor Burns, *Roosevelt: The Lion and the Fox* (New York: Harcourt, Brace and World, 1956), 144.

3. Burns, *Roosevelt*, 149–50.

4. Foulois and Glines, *Wright Brothers to Astronauts*, 226.

5. Burns, *Roosevelt*, 139; Underwood, *The Wings of Democracy*, 27–29. As the latter explained, commercial aviation companies acted dissimilarly toward Roosevelt. Boeing refused to admit FDR to the plant during his campaign, but it allowed Republicans to enter. On the other hand, East Coast aviation companies courted the New Yorker and touted aviation projects as effective work programs while improving national defenses.

6. For information on Douglas, who counted among those the president trusted, see Burns, *Roosevelt*, 172; Arthur M. Schlesinger Jr., *The Age of Roosevelt*, vol. 2, *The Coming of the New Deal* (1958; reprint, Boston: Houghton Mifflin, 1988), 218, 289–95. Douglas's policies would soon conflict with the president's, and he would resign in mid-1934, but there remained among the advisers those

who wanted inflationary spending and those who wanted balanced budgets. Either way, the military could not look forward to receiving funding increases — quite the opposite.

7. Killigrew, "Impact of the Great Depression," 223–24.

8. *Army and Navy Journal*, 18 March 1933, 576. The *Journal* noted that Congress's "hasty response . . . to his legislative demands" indicated that "for the moment he is in the saddle," a position that allowed him more leeway to enact his policies, including military ones, with relative ease and freedom.

9. Burns, *Roosevelt*, 149–50. Burns called Roosevelt's cabinet a "strange assortment" lacking any essential principle, a body the president could dominate and force his will upon. For a short analysis on Dern's lack of closeness with Roosevelt see Killigrew, "Impact of the Great Depression," 225.

10. O'Laughlin to Major General Van Horn Moseley, 16 November 1934, Box 53, O'Laughlin Papers. O'Laughlin corresponded with the major military leaders of the day and also gleaned information from inside the Roosevelt circle from unnamed sources. He also remained a staunch Republican and a member of the Republican National Committee. After Hoover left office, O'Laughlin provided the former president with a weekly update on the inner workings of FDR's administration. Hoover thanked him for the reports and O'Laughlin noted, "I am glad to think that my reports are proving of service. They will be kept up, for I want you to be thoroughly informed . . . of what is transpiring in connection with the Government." O'Laughlin to Hoover, 3 April 1933. For an example of Hoover's letters of appreciation for the information, see Hoover to O'Laughlin, 25 March 1933. Both letters in Individual Correspondence File, Box 166, Post-Presidential Period, HHPL.

11. Killigrew, "Impact of the Great Depression," 224–25; *Army and Navy Journal*, 11 March 1933, 555. In an interview shortly after arriving, Dern admitted, "I don't know very much about this job." *Army and Navy Journal*, 11 March 1933, 549. Dern's assistant secretary did not offer much more in the way of experience. Kansas governor Harry H. Woodring vigorously pressed Roosevelt and his aides for his appointment as Secretary of Agriculture. After turning down a tentative offer of treasury secretary, he accepted the second chair in the War Department, thinking it might afford his last best chance to have any reasonable opportunity and consoling himself with the fact of at least being in the "Little Cabinet." Keith D. McFarland, *Harry H. Woodring: A Political Biography of FDR's Controversial Secretary of War* (Lawrence: University of Kansas Press, 1975), 78–79.

12. Killigrew, "Impact of the Great Depression," 224–25. Dern did not even submit a complete statement for the 1933 annual report. Instead, MacArthur wrote the War Department's recapitulation of the previous year even though Dern had been around for one-quarter of the time. *Report of the Secretary of War to the President* (Washington DC: Government Printing Office, 1933), 1–49, accessed at AFHSO.

13. D. C. James, *The Years of MacArthur*, 1:417. Roosevelt also confided to Rexford Tugwell that he considered MacArthur one of the two most dangerous men in America (the other being Louisiana senator Huey Long) because of his strong presence and leadership. D. C. James, *The Years of MacArthur*, 1:411; Leuchtenburg, *Roosevelt and the New Deal*, 96 n. 3.

14. Kilbourne to members of the General Staff, 14 February 1934, Box 56, RG 165, Records of the War Department General and Special Staffs, NARA II. For another example see "Notes on Military Aviation and the Army Air Corps," 10 April 1934, Box 55, RG 165, Records of the War Department General and Special Staffs, NARA II.

15. For congressional statistics, see http://clerkweb.house.gov/histrecs/house his/lists/divisionh.htm and http://www.senate.gov/learning/stat_13.html, both accessed 17 September 2001.

16. In addition to the latest losses, some of the earlier aviation proponents no longer roamed Capitol Hill: Florian Lampert and Charles Curry both died in 1930 while still in office; John Morin, chairman of the House Military Affairs Committee (1925–29), lost in the 1928 elections; and his Maryland district ousted John P. Hill in 1926, and he failed in election attempts in 1928 and 1936.

17. All information on the congressmen obtained from *Biographical Directory of the United States Congress*, http://bioguide.congress.gov, accessed 17–19 September 2001.

18. Sheppard entered Congress in 1902 and served in the House for eleven years, whereupon he resigned due to his election to the Senate, where he served until his 1941 death. McSwain, a World War I infantry officer, came to the House in 1921. For information on both see http://bioguide.congress.gov, accessed 1 October 2001.

19. Speech contained no date, but, due to mention of the death of Will Rogers, would probably have been in 1935. Notes to speech, Folder 3, Military affairs, 1922–1939 and undated, Box 3N188, Morris Sheppard Papers, Center for American History, University of Texas at Austin.

20. Sheppard backed legislation to relieve the "war hump" mentioned in chapter 4. For his efforts to help the Army, see General Malin Craig to Sheppard, 5 June 1936, Folder 2, Letters Received, June 1936, Box 2G195, Sheppard Papers. See also Escal Franklin Duke, "The Political Career of Morris Sheppard, 1875–1941" (Ph.D. diss., University of Texas, 1958), 431–35.

21. One newspaper called him a "principal leader" for the president's program and one who enjoyed FDR's confidence. Still, during his 1936 reelection campaign, Texas papers called him a moderate Democrat whom the country needed to ensure that "'wild-eyed' radical liberals" did not impose their will upon the country. *Paris (TX) Morning News*, 29 June 1936, and *Valley Mills (TX) Tribune*, 17 July 1936, clippings in Folder 7, Letters Written, July–December 1936, Box 2G193, Sheppard Papers. Sheppard supported Roosevelt and the New Deal even though

he did not like the administration's stand on Prohibition (Sheppard sponsored the Eighteenth Amendment and opened each session of Congress with a "dry" speech to commemorate the event—even after its 1933 repeal). Duke, "Political Career of Morris Sheppard," 421–25.

22. McSwain statement to members of the Military Affairs Committee upon his announcement of not running for reelection, June 1936, 3, Box 15, John J. McSwain Papers, Rare Book, Manuscript, and Special Collections Branch, Duke University Library, Durham NC.

23. McSwain statement to members of the Military Affairs Committee, 1.

24. *Army and Navy Journal*, 13 February 1932, 576.

25. McSwain to Roosevelt, 1 May and 25 May 1933, and McSwain to Dern, 1 May 1933, all in Official File 25, Box 31, FDRL.

26. Shiner, *Foulois and the U.S. Army Air Corps*, 79. Foulois' lack of political savvy and, as will soon be shown, his problems with a congressional subcommittee and the air mail made him a liability, and McSwain probably resented Foulois' reemerging extremism and his dishonest statements to Congress. To his credit, McSwain did not let Foulois' problems decrease his support for aviation.

27. Mitchell helped line up the Virginia Democrats during the convention, and he later played a large role in the inauguration preparations. Democrats in Virginia, his home state of Wisconsin, and Michigan supported Mitchell to replace Davison, but Mitchell, still burning to reorganize national defense, wanted a more important post, calling the assistant secretary post a "stopgap" position. Hurley, *Billy Mitchell*, 122–26; Killigrew, "Impact of the Great Depression," 222–23.

28. *Aero Digest*, January 1933, 17.

29. D. C. James, *The Years of MacArthur*, 1:437.

30. Underwood, *The Wings of Democracy*, 30.

31. Davison interview by Murray Green, Locust Valley NY, 17 April 1970, MS 33, Box 62, pp. 8–12, Green Collection, USAFAL. On two different occasions Green specifically asked Davison if economy could have played a role in the office's elimination.

32. *New York Times*, 25 March 1933, 4:5.

33. Major General R. E. Callan, Assistant Chief of Staff, G-4, to the Chief of Staff, "Reorganization of the Office of the Assistant Secretary of War (Air)," 21 July 1933, 1, in Box 56, RG 165, Records of the War Department General and Special Staffs, NARA II. Of note, the memo was endorsed at the bottom as "APPROVED: by order of the Secretary of War" and signed by General MacArthur.

34. Callan to Chief of Staff, 21 July 1933, 2.

35. Spaatz to Arnold, 29 September 1933, Box 6: Diaries and Notebooks, Spaatz Papers.

36. Foulois to Deputy Chief of Staff, "Reorganization of Office, Assistant Secretary of War (Air)," 25 July 1933, 1–2, in Box 56, RG 165, Records of the War Department General and Special Staffs, NARA II.

37. Foulois to Deputy Chief of Staff, 25 July 1933, 3.

38. Foulois to Deputy Chief of Staff, 25 July 1933, 3–4.

39. *New York Times*, 9 April 1933, 8:6.

40. "Comments on the Re-Establishment of the Office of the Assistant Secretary of War (Air)," prepared by Lieutenant Colonel J. E. Chaney, Chief of Air Corps Plans Division, 10 January 1934, located in Box 18, Subject Files, Foulois Papers.

41. Foulois statement, in response for an article request, to LeRoy Whitman, editor of the *Army and Navy Journal*, n.d., and Whitman request, 11 June 1936, 6, located in Box 5, Personal Correspondence, Foulois Papers.

42. *New York Times*, 8 June 1933, 3:5. Woodring's assumption of Davison's old duties followed naturally, since Woodring's primary function involved procurement. Recalling the definition of the air secretary's duties after the first filling of the position, the agreement with the Secretary of War had allowed Davison authority overseeing the procurement of aircraft. Thus a large portion of the air secretary's duties had been returned to the sole remaining assistant secretary.

43. *Report of the Secretary of War to the President* (Washington DC: Government Printing Office, 1934), 4.

44. For an excellent summation of Hoover's ideas and recovery actions and a comparison with those of Roosevelt, see William E. Leuchtenburg, *The Perils of Prosperity: 1914–1932*, 2nd ed. (Chicago: University of Chicago Press, 1993), 250–56, 261–64. For more on Roosevelt's program, and especially his leadership, see Leuchtenburg, *Roosevelt and the New Deal*, 42–62 passim; see also Burns, *Roosevelt*, 203–5.

45. The exceptions were monies for the Works Progress Administration and a naval building program.

46. *Army and Navy Journal*, 18 March 1933, 576. See also West, "Laying the Legislative Foundation," 279–87.

47. Quote from West, "Laying the Legislative Foundation," 278. See Paul K. Conkin, *The New Deal* (New York: Crowell, 1967), 16, for his beliefs on FDR's charisma. For more on the complex Roosevelt personality, and his use of charm and deception, see Patrick J. Maney, *The Roosevelt Presence: A Biography of Franklin Delano Roosevelt* (New York: Twayne, 1992), 78–82; Burns, *Roosevelt*, 203–5. For example, Roosevelt for a time caused Carl Vinson, the foremost advocate of the Navy in Congress, to reverse his position and support the president. See West, "Laying the Legislative Foundation," 277–81.

48. Foulois and Glines, *Wright Brothers to Astronauts*, 188.

49. "America First," speech before the National Aeronautical Association convention, 18 August 1932, MS 17, Box 10, Series 7, Folder 3, Foulois Collection, USAFAL.

50. Foulois, "What Is an Adequate Defense?" 1933, MS 17, Box 6, Series 2, Folder 2, Foulois Collection, USAFAL.

51. McSwain had introduced a similar bill during the previous year and session of Congress, and Foulois testified in February 1932. McSwain praised "the dignity and force" of Foulois' 1932 stand against the "blue bloods" of the Army. Foulois replied that he would always be glad to work with McSwain. McSwain to Foulois, 5 March 1932, and Foulois to McSwain, 8 March 1932, both in Box 26, Foulois Papers.

52. Foulois testimony available in Box 27, Foulois Papers.

53. Foulois testimony, Box 27, Foulois Papers; Shiner, *Foulois and the U.S. Army Air Corps*, 83.

54. *New York Times*, 1 April 1933, 8:3.

55. *Report of the Secretary of War to the President*, for fiscal year ending 30 June 1932, 42, accessed at AFHSO.

56. Only a few months later, in the 1933 annual report, Foulois did not request additional aircraft, and, breaking with tradition, he did not report on the numbers of aircraft in the inventory. Of his primary recommendations and needs, he requested funds for housing and facilities at the Air Corps Tactical School and four airfields—nowhere did he mention such a drastic shortage of aircraft. *Annual Report of the Chief of the Air Corps, 1933*, 66, AFHSO. Additionally, MacArthur—who would, of course, want to place the Air Corps in a better light to obtain more funds for other branches and his mechanization goals—ranked the air component as the second or third best in the world and the best-prepared branch in the Army. *Report of the Secretary of War to the President*, for fiscal year ending 30 June 1932, 33, accessed at AFHSO.

57. Westover to Kilbourne, 13 April 1933, and Drum to Kilbourne, 10 April 1933, both in Box 27, Foulois Papers.

58. Drum to Westover, 13 April 1933, Box 27, Foulois Papers. The Drum memo provided an official record for the clarification, following Westover's meeting with Drum on the same date, and reasserted the Air Corps' right to present testimony to Congress.

59. Foulois to Barbour, 14 February 1933, Box 5, Foulois Papers. Barbour resumed his law practice and never returned to Congress. See http://bioguide.con gress.gov, accessed 20 October 2001.

60. Foulois to Willford, 23 February 1934, and Willford to Foulois, 26 February 1934, both in Box 5, Foulois Papers.

61. D. C. James, *The Years of MacArthur*, 1:356–62, 378–81, 426–27; Tate, *The Army and Its Air Corps*, 95–96; Killigrew, "Impact of the Great Depression," 115–22; Wilson, "The Quaker and the Sword," 45–46. In the most famous incident, MacArthur defeated Collins's proposal to slash the officer corps in 1931–32. Collins's measure won in the House, but MacArthur's influence helped defeat it in the Senate and Collins could not resurrect the measure. This incident prompted the famous telegram from MacArthur to Assistant Secretary of War Frederick Payne: "Just hog-tied a Mississippi cracker."

62. Shiner, *Foulois and the U.S. Army Air Corps*, 84. Shiner felt that Foulois did not sway Collins but simply reinforced the congressman's existing attitudes.

63. Quesada oral history interview by DeWitt S. Copp, 7 December 1976, MS 46, Box 3, Folder 15, DeWitt S. Copp Collection, USAFAL.

64. Davison oral history interview by Murray Green, 17 April 1970, MS 33, Box 62, Green Collection, USAFAL.

65. *New York Times*, 8 January 1933, VIII, 8:1.

66. *Annual Report of the Chief of the Air Corps, 1933*, 56, AFHSO. The Air Corps also cut other expenses as much as possible to keep aircraft flying. Two examples of such frugality included eliminating the Air Corps Band and halting publication of the *Air Corps News Letter* for more than a year. U.S. Congress, House, Subcommittee of the House Committee on Appropriations, *War Department Appropriation Bill for 1934: Hearing before the Subcommittee of the Committee on Appropriations*, 72nd Cong., 2nd sess., 1932–33, 581; The cover page of the *Air Corps News Letter*, 31 October 1933, announced the last issue, but publication resumed with the *Air Corps News Letter*, 15 January 1935.

67. Lewis Douglas recommended eliminating $90 million of the budget's $277.1 million for military activities and reducing the overall War Department budget by some $144 million. *New York Times*, 18 April 1933, 1:6; see also Shiner, *Foulois and the U.S. Army Air Corps*, 115.

68. See "Army Officer Reductions and Other Defense Cuts Protested by Nation's Press," *Army and Navy Journal*, 29 April 1933, 690.

69. MacArthur, *Reminiscences*, 100–101; D. C. James, *The Years of MacArthur*, 1:428–29.

70. J. Britt McCarley, "General Nathan Farragut Twining: The Making of a Disciple of American Strategic Air Power, 1897–1953" (Ph.D. diss., Temple University, 1989), 31.

71. McFarland, *Harry H. Woodring*, 97–98; Futrell, *Ideas, Concepts, Doctrine*, 69–70; Johnson, *Fast Tanks and Heavy Bombers*, 157–58.

72. Michael S. Sherry, *The Rise of American Air Power: The Creation of Armageddon* (New Haven: Yale University Press, 1987), 51–53. See also Robert W. Krauskopf, "The Army and the Strategic Bomber, 1930–1939," *Military Affairs* 22, no. 2 (1958): 86–87.

73. D. C. James, *The Years of MacArthur*, 1:367–68.

74. "Air Plan for the Defense of the United States," Box 52, RG 165, Records of the War Department General and Special Staffs, NARA II.

75. Kilbourne to MacArthur, 25 July 1933, Box 55, RG 165, Records of the War Department General and Special Staffs, NARA II. Assistant Chief of Staff Drum also agreed with Kilbourne, calling the plan, and Foulois' actions, "based primarily in the getting up an argument for an increase in the Air Corps." Drum to MacArthur, 3 August 1933, Box 55, RG 165, Records of the War Department General and Special Staffs, NARA II.

76. Foulois, in an endorsement to the original report, 28 July 1933, Box 52, RG 165, Records of the War Department General and Special Staffs, NARA II.

77. Ordered by Secretary of War directive, 11 August 1933, Box 52, RG 165, Records of the War Department General and Special Staffs, NARA II. The other members included Kilbourne; Commandant of the Army War College, Major General George S. Simonds; and Chief of Coast Artillery, Major General John W. Gulick.

78. "Report of the Special Committee, General Council, on Employment of Army Air Corps under Certain Strategic Plans," 4, Box 52, RG 165, Records of the War Department General and Special Staffs, NARA II.

79. "Report of the Special Committee, General Council, on Employment of Army Air Corps under Certain Strategic Plans," 9, 22.

80. Holley, *Buying Aircraft*, 55. The Baker Board findings would allude to the earlier Drum Board and thus made it available to the public. Congress found out about the Drum Board during secret testimony before the Military Affairs Committee by MacArthur and Foulois in February 1934. U.S. Congress, House, Committee on Military Affairs, *Statement of Maj. Gen. B. D. Foulois, Chief of the Air Corps, before the Committee on Military Affairs*, 73rd Cong., 2nd sess., 1 February 1934, Committee Print, 27–28. The committee then requested a copy of that report and ordered General Drum to testify (39).

81. Foulois' biographer believed that the Drum Board made its report in a "relatively cooperative spirit" and that Foulois thought a more accommodating temperament would work at this time in his dealings with the General Staff. Shiner, *Foulois and the U.S. Army Air Corps*, 96. The aviation officers had long wanted a GHQ Air Force, since it would consolidate aviation's striking power and place it under the command of a pilot.

82. Testimony before Subcommittee No. 3 (also known as the Rogers Subcommittee) of the House Military Affairs Committee: Drum, 5 June 1934, 8–32 passim; Kilbourne, Gulick, and Simonds, 6 June 1934, 9–11; Box 23, War Department, Case of Major-General Benjamin D. Foulois, Air Corps, RG 159, Records of the Office of the Inspector General, NARA II. Events surrounding the entire Rogers Subcommittee, its investigation, and the subsequent and related investigation of Foulois by the Inspector General's office will be covered later in this chapter.

83. *Army and Navy Journal*, 6 January 1934, 365.

84. *New York Times*, 7 January 1934, 10:1.

85. Copy of H.R. 7601, "To Provide more effectively for the National Defense by further increasing the effectiveness and efficiency of the Air Corps of the Army," contained in Box 28, Subject File, Foulois Papers. See also *Army and Navy Journal*, 3 February 1934, 445, 455.

86. Foulois to MacArthur, 20 May 1935, Box 28, Subject File, Foulois Papers.

87. "Final Statement of General Foulois," n.d., Box 47, Subject File, Foulois Papers.

88. Memorandum for General Kilbourne, "Preparation of Bills for Separate Air Corps, 1933–1934," from Major Follett Bradley, 14 January 1935, in Box 28, War Department, Case of Major-General Benjamin D. Foulois, Air Corps, RG 159, Records of the Office of the Inspector General, NARA II. Of note, Major Carl Spaatz participated in the committees preparing these bills. When Foulois changed his position in advocating these bills, Colonel Weaver, upset that Foulois seemed disloyal to Weaver, requested his release from the Office of the Chief of the Air Corps. One General Staff officer requested that actions be taken to remove the "stigma" from Weaver's name and to reward the officer with an assignment to Army War College. Memorandum for General Kilbourne, "Preparation of Bills for Separate Air Corps, 1933–1934," from Major Follett Bradley, 14 January 1935; Bradley to Kilbourne, 18 January 1934, Box 28, RG 159, NARA II.

89. Bradley to Kilbourne, 18 January 1934, Box 28, RG 159, NARA II.

90. For examples of his interoffice memoranda of support, see Kilbourne to MacArthur, "General Staff Supervision of the Air Corps," 19 February 1934, and memorandum for General Drum, 14 February 1934, both in Box 56, RG 165, Records of the War Department General and Special Staffs, NARA II.

91. One release contained a handwritten note by Kilbourne, "prepared for release[,] but C. S. [MacArthur]" did not acquiesce to send it out as an official release. Kilbourne, "Notes on Military Aviation and the Army Air Corps," 10 April 1934, Box 55, RG 165, Records of the War Department General and Special Staffs, NARA II. Kilbourne sent a proposed press release and an accompanying letter to Congressman James W. Wadsworth, asking him to "perform a real service to the Army and to the country." The press release contained a note "suggested for the press." Kilbourne to Wadsworth, 20 March 1934, Box 55, RG 165, Records of the War Department General and Special Staffs, NARA II.

92. Kilbourne to McSwain, Sheppard, et al., 16 and 20 February 1934, Box 56, RG 165, Records of the War Department General and Special Staffs, NARA II.

93. Kilbourne to the Assistant Chiefs of Staff, War Department General Staff, 9 February 1934, Box 56, RG 165, Records of the War Department General and Special Staffs, NARA II.

94. The Air Corps may not have been where its officers wanted or at the numbers of the Morrow Board and Air Corps Act, but it remained one of the best branches in the Army. Foulois' actions, therefore, incited the General Staff even more. Shiner, *Foulois and the U.S. Army Air Corps*, 117. See also Statement of Lieutenant Colonel L. P. Collins, General Staff, WPD, "Memorandum for Record with the Proceedings of the Special Committee [Drum Board]," 3, Box 56, RG 165, NARA II.

95. In the 1934 budget hearings (for the FY 1935 budget), Foulois promised

congressmen that all Liberty engines would be discontinued before the end of 1934. U.S. Congress, House, Subcommittee of the House Committee on Appropriations, *War Department Appropriation Bill for 1935, Military Activities: Hearing before the Subcommittee of the Committee on Appropriations*, 73rd Cong., 2nd sess., 1934, 567.

96. McSwain to Dern, 21 February 1934, 1, Box 8, McSwain Papers.

97. McSwain to Dern, 21 February 1934, 3.

98. McSwain to Dern, 21 February 1934, 2.

99. McSwain to Dern, 21 February 1934, 2; Dern to Stephen Early (secretary to the president), 6 March 1934, Official File 25, Box 1, FDRL.

100. *New York Times*, 2 March 1934, 6:2.

101. "Statement of the Honorable George H. Dern, Secretary of War: To the Committee on Military Affairs, House of Representatives, Legislative Proposals for Reorganization of Army Air Corps," 3, Official File 25, Box 1, FDRL.

102. "Statement of the Honorable George H. Dern," 6.

103. "Statement of the Honorable George H. Dern," 6.

104. McSwain to Roosevelt, 14 March 1934, Box 19, Mitchell Papers.

105. John Callan O'Laughlin to Hoover, 10 and 17 February 1934, Individual Correspondence File, Box 167, Post-Presidential Period, HHPL. For additional information on the background and a complete timeline of the air mail developments, see Paul D. Tillett, *The Army Flies the Mails*, Inter-University Case Program No. 24 (n.p.: Published for the Inter-University Case Program by The Bobbs-Merrill Company, Inc., n.d.), 3–5. For more information on this episode, see Shiner, *Foulois and the U.S. Army Air Corps*, 125–49; Carroll V. Glines, *The Saga of the Air Mail* (New York, Arno Press, 1968), 127–41.

106. Shiner, *Foulois and the U.S. Army Air Corps*, 125–27; Foulois and Glines, *Wright Brothers to Astronauts*, 236–39; Tillett, *The Army Flies the Mails*, 28–30. During the subsequent investigations, Representative James asked Westover about the decision. Westover admitted to having doubts about the Air Corps' ability to fly the mail, but in either case, had he been in charge, he agreed that he would have consulted the War Department before giving an answer. Kilbourne to MacArthur, "Hearings before the Rogers Sub-committee of the Military Affairs Committee of the House of Representatives (morning session) March 16, 1934," [a memorandum to inform MacArthur about the ongoing investigation], Box 55, RG 165, NARA II.

107. U.S. Congress, House, Committee on the Post Office and Post Roads, *Air Mail: Hearings before the Committee on the Post Office and Post Roads*, 73rd Cong., 2nd sess., 16 February 1934, 104 [hereinafter cited as *Air Mail Hearings*].

108. Shiner wrote that these navigational items "were absolutely a must for air mail operations." Shiner, *Foulois and the U.S. Army Air Corps*, 131. See also John F. Shiner, "The General and the Subcommittee: Congress and U.S. Army Air

Corps Chief Benjamin D. Foulois, 1934–1935," *Journal of Military History* 55, no. 4 (1991): 490.

109. *Air Mail Hearings*, 103.

110. Shiner, *Foulois and the U.S. Army Air Corps*, 132–33; *Air Mail Hearings*, 98.

111. Arnold, *Global Mission*, 143. Quesada oral history interview by Copp, 18 November 1976, MS 46, Box 3, Folder 15, Copp Collection, USAFAL.

112. Quoted in Tillett, *The Army Flies the Mails*, 30. Even after some flyers had died, Foulois did a Sunday feature article and called the carrying of the mail a wonderful opportunity to build up the service. *New York Times*, 11 March 1934, VI, 3.

113. Quoted in Foulois and Glines, *Wright Brothers to Astronauts*, 243.

114. Tillett, *The Army Flies the Mails*, 42; Foulois to Congressman William N. Rogers, 9 March 1934, Box 14, Foulois Papers.

115. *New York Times*, 26 February 1934, 5:5; *Congressional Record*, 73rd Cong., 2nd sess., vol. 78, pt. 4, 3616.

116. *New York Times*, 10 March 1934, 2:1, 11 March 1934, 1:8 and 3:4. See also *Washington Post*, 10 March 1934, 1; Roosevelt to Dern, 10 March 1934, Official File 25, Box 1, FDRL.

117. Schlesinger, *The Coming of the New Deal*, 455.

118. *Congressional Record*, 73rd Cong., 2nd sess., vol. 78, pt. 4, 3614.

119. *New York Times*, 1 March 1934, 14:2; see also *Congressional Record*, 73rd Cong., 2nd sess., vol. 78, pt. 4, 3614–15.

120. O'Laughlin to Hoover, 10 March 1934, Individual Correspondence File, Box 167, Post-Presidential Period, HHPL.

121. Roosevelt to Dern, 10 March 1934, Official File 25, Box 1, FDRL.

122. Foulois and Glines, *Wright Brothers to Astronauts*, 253–54.

123. Foulois and Glines, *Wright Brothers to Astronauts*, 255.

124. Quoted in Foulois and Glines, *Wright Brothers to Astronauts*, 256.

125. O'Laughlin to Hoover, 17 March 1934, Individual Correspondence File, Box 167, Post-Presidential Period, HHPL. Neither MacArthur nor his biographer mentioned the incident, nor did the President's Secretary's Files or Foulois' files and memoirs (Foulois supposedly witnessed MacArthur's call).

126. Section 10, Air Corps Act, *Statutes at Large*, 44:784–89.

127. Shiner, *Foulois and the U.S. Army Air Corps*, 150–51.

128. House Subcommittee of the House Committee on Appropriations, *War Department Appropriation Bill for 1935, Military Activities*, 486–91.

129. House Subcommittee of the House Committee on Appropriations, *War Department Appropriation Bill for 1935, Military Activities*, 515–28.

130. *Army and Navy Journal*, 17 February 1934, 488; *New York Times*, 10 February 1934, 3:1; Shiner, *Foulois and the U.S. Army Air Corps*, 157. O'Laughlin reported to Hoover on James's working behind the scenes and supporting

McSwain. He also characterized the McSwain-Dern relationship as "tense." O'Laughlin to Hoover, 3 March 1934, Individual Correspondence File, Box 167, Post-Presidential Period, HHPL.

131. U.S. Congress, House, Committee on Military Affairs, *Investigation under House Resolution 275*, Report No. 1506, 73rd Cong., 2nd sess., 1934. Parts of the report quoted sections of the 1934 budget testimony, unavailable in printed form elsewhere.

132. *New York Times*, 3 March 1934, 1:7.

133. Other subcommittee members included Democrats Lister Hill of Alabama, Numa F. Montet of Louisiana, and Dow W. Harter of Ohio; Republicans W. Frank James of Michigan, Edward Goss of Connecticut, and Charles Plumley of Vermont; and Paul Kvale of Minnesota, a Farm-Labor representative.

134. House Committee on Military Affairs, *Investigation under House Resolution 275*, 26–32.

135. *Washington Post*, 6 March 1934, 1, and 7 March 1934, 6.

136. *New York Times*, 8 March 1934, 7:1; *Washington Post*, 8 March 1934, 1–2; House Committee on Military Affairs, *Investigation under House Resolution 275*, 32; copies of Foulois testimony of 7 March 1934 in Box 45, Foulois Papers. See also Rutkowski, *Politics of Military Aviation Procurement*, 94–105; Shiner, *Foulois and the U.S. Army Air Corps*, 160–62.

137. Copies of Foulois testimony of 7 March 1934 in Box 45, Foulois Papers.

138. Foulois testimony, Exhibit H, 13, in Box 23, RG 159, NARA II.

139. House Committee on Military Affairs, *Statement of Maj. Gen. B. D. Foulois, Chief of the Air Corps, before the Committee on Military Affairs*, Committee Print, obtained from Box 23, RG 159, NARA II.

140. O'Laughlin to Hoover, 31 March 1934, Individual Correspondence File, Box 167, Post-Presidential Period, HHPL.

141. *Army and Navy Journal*, 14 April 1934, 645.

142. Quoted in *Army and Navy Journal*, 14 April 1934, 645.

143. U.S. Congress, House, Committee on Military Affairs, *Investigation of Profiteering in Military Aircraft, under H. Res. 275*, Report No. 2060, 73rd Cong., 2nd sess., 15 June 1934, 3.

144. House Committee on Military Affairs, *Investigation of Profiteering in Military Aircraft*, under H. Res. 275, 11–12.

145. House Committee on Military Affairs, *Investigation of Profiteering in Military Aircraft*, under H. Res. 275, 14.

146. *New York Times*, 11 March 1934, VI, 3; *Army and Navy Journal*, 17 March 1934, 566; *Wichita Beacon*, 28 March 1934, clipping obtained from Box 54, William P. MacCracken Jr. Papers, HHPL.

147. McSwain to Hearst, telegram, 5 March 1934, and McSwain to Hearst, letter, 6 March 1934, both in Box 8, McSwain Papers.

148. Copy of editorial from the Hearst paper *San Antonio Light*, 23 March 1934, Box 14, McSwain Papers.

149. A sampling of newspaper comments collected in "Editors Give Views Pro and Con on Accusations against General Foulois," *Army and Navy Journal*, 23 June 1934, 870.

150. "Newspaper Editors Comment on Subcommittee's Attack on General Foulois," *Army and Navy Journal*, 7 July 1934, 910. I also examined the editorials from various editions from May to July 1934 from the *New York Times* and various Washington DC papers. Their comments follow the divergence of opinions noted in the text, with the *Washington Herald* providing the most support for Foulois.

151. *New York Times*, 11 May 1934, 2:6, and 19 June 1934, 3:5; *Washington Post*, 6 March 1934, 1, and 7 March 1934, 6.

152. House Committee on Military Affairs, *Statement of Maj. Gen. B. D. Foulois, Chief of the Air Corps, before the Committee on Military Affairs*, Committee Print, 1; "Statement: Regarding Accusations Made Against the Chief of the Army Air Corps (Major General B.D. Foulois) and Other Responsible Officers of the Air Corps," 5, in Box 46, Foulois Papers.

153. Foulois to McSwain, 14 July 1934, as Exhibit 17, Box 23, RG 159, NARA II.

154. McSwain to Foulois, 22 June 1934, and Rogers to Foulois, 3 July 1934, both in Box 23, RG 159, NARA II.

155. Foulois to McSwain et al., 14 July 1934, Harter to Foulois, 7 July 1934, Hill to Foulois, 13 July 1934, Kvale to Foulois, 7 July 1934, Ransley to Foulois, 23 July 1934, and Thompson to Foulois, 19 July 1934, all in Box 23, RG 159, NARA II.

156. Rogers to Roosevelt, 18 June 1934, Box 46, Foulois Papers; Dern to Rogers, 21 August 1934, Box 23, RG 159, NARA II.

157. Dern to Rogers, 21 August 1934, Box 23, RG 159, NARA II.

158. "Newspapers Commend Secretary Dern's Action in General Foulois Case," *Army and Navy Journal*, 8 September 1934, 26.

159. O'Laughlin to Hoover, 20 June 1934, Individual Correspondence File, Box 167, Post-Presidential Period, HHPL.

160. Dern to Inspector General, 13 December 1934, Box 23, RG 159, NARA II; see also Shiner, *Foulois and the U.S. Army Air Corps*, 188.

161. Colonel Thorne Strayer, who had led the Inspector General investigation of the Arnold-Dargue affair, began the investigation. However, Drum protested to McSwain that Strayer seemed pro-Foulois, and the chairman applied political pressure to have the lead inspector changed. The Inspector General assigned Reed, who began the investigation anew. The change angered Foulois. Shiner, *Foulois and the U.S. Army Air Corps*, 188–89.

162. Interviews by Reed with Drum (Exhibit "L") and Kilbourne (Exhibit "O") in Box 23, RG 159, NARA II.

163. Reed interview of Cheney (Exhibit "N"), 12–13, 15, 24, in Box 23, RG 159, NARA II.

164. Reed interview of Cheney, 61.

165. Reed interview of Westover, 17, Box 23, RG 159, NARA II.

166. Reed interview of Foulois, 372, 388–89, Box 23, RG 159, NARA II.

167. Shiner, *Foulois and the U.S. Army Air Corps*, 190–91.

168. Dern to McSwain, 14 June 1935, Box 11, McSwain Papers.

169. Foulois and Glines, *Wright Brothers to Astronauts*, 273; *Congressional Record*, 74th Cong., 1st sess., vol. 79, pt. 9, 9381.

170. Shiner, *Foulois and the U.S. Army Air Corps*, 192; Foulois and Glines, *Wright Brothers to Astronauts*, 274; *Congressional Record*, 74th Cong., 1st sess., vol. 79, pt. 9, 9381–93. Rogers actually interrupted an unrelated bill under consideration with an objection, and then he used parliamentary rules and control of the floor to expand his comments and fight against Foulois and read part of Dern's report into the record. Foulois' formal retirement occurred in December 1935, but he began his terminal leave, and ceased performing his duties, on 25 September 1935.

171. *New York Times*, 15 March 1934, 19:2.

172. O'Laughlin to Hoover, 21 April 1934, Individual Correspondence File, Box 167, Post-Presidential Period, HHPL. Foulois also worried that another flurry of boards and investigations would only delay congressional budget actions and procurement. Foulois and Glines, *Wright Brothers to Astronauts*, 271–72.

173. *New York Times*, 14 March 1934, 1:6.

174. *New York Times*, 15 March 1934, 19:2; *Army and Navy Journal*, 31 March 1934, 606. The press took divergent opinions on Lindbergh. Some lauded him for not taking part in another government "whitewash" of aviation, while others lambasted him for ignoring his civic duty and providing needed expertise.

175. Dern to McIntyre, telegram, 4 April 1934, Official File 25, Box 31, FDRL.

176. *New York Times*, 11 April 1935, 18:2; War Department Special Committee on Army Air Corps, *Final Report of War Department Special Committee on Army Air Corps*, 18 July 1934, 1 (henceforth referred to as the Baker Board Report). The other civilians on the board included Chamberlin, James H. Doolittle (former Air Corps officer, noted pilot, and company executive), Dr. Karl T. Compton (president of the Massachusetts Institute of Technology), Edgar S. Gorrell (World War I aviator and president of Stutz Motor Company), and Dr. G. W. Lewis (research director of the National Advisory Committee for Aeronautics). The military members remained the same as the Drum Board: Generals Foulois, Drum, Kilbourne, Simonds, and Gulick.

177. Beaver, *Baker and the American War Effort*, 169.

178. The War Department sent a telegram to all stations asking officers to submit "any constructive suggestions they may desire" through the Adjutant General. For an example, see telegram to Commanding Officer, Bolling Field DC, 19 April 1934, Box 529, RG 18, NARA II.

179. Underwood rightly characterizes the report as something seemingly written by the General Staff. Underwood, *The Wings of Democracy*, 51.

180. Baker Board Report, 61–75.

181. Baker Board Report, 63.

182. Baker Board Report, 62.

183. Foulois and Glines, *Wright Brothers to Astronauts*, 260.

184. John L. Frisbee, *Makers of the United States Air Force* (1987; reprint, Washington DC: Air Force History and Museums Program, 1996), 32.

185. Baker Board Report, 64–65.

186. Arnold, *Global Mission*, 145.

187. *New York Times*, 22 September 1934, 16:5.

188. Part of the Navy's success came from a pro-Navy president and strong congressional supporters, especially Carl Vinson. Yet the Navy's air expansion program, like the Army's, competed with other expenditures and desires and was still completed on time.

189. Layman, *Air Corps Personnel and Training Programs*, 108.

190. Layman, *Air Corps Personnel and Training Programs*, 110–17.

6. MODERATES AND MONEY

1. Layman, *Air Corps Personnel and Training Programs*, 40–43, 114–17. The only notable air independence legislation came from recently elected (and air-friendly) Florida Democrat James Mark Wilcox, in H.R. 3151.

2. The Air Corps Tactical School formulated the concept of high-altitude, daylight bombing raids against the enemy with a high-speed bomber that pursuit fighters could not catch. Prior to the arrival of the B-17 they could not test this theory, and the technology lagged behind the concept. All of that changed with the 1935 arrival of the Flying Fortress, the plane that they believed fulfilled their doctrine. Robert T. Finney, *Air Force Historical Studies: No. 100: History of the Air Corps Tactical School, 1920–1940* (1955; reprint, Washington DC: Center for Air Force History, 1992), 66–68.

3. Quoted in "Air Force History Notes," n.d., Reel 197: Subject File, Henry Harley Arnold Papers, LOC-MD.

4. Frisbee, *Makers of the United States Air Force*, 33.

5. Quoted in McClendon, *Autonomy of the Air Arm*, 74.

6. Special Order No. 296, 14 December 1934, Official File 25, Box 30, FDRL. See also *New York Times*, 15 November 1934, 4:4, and 13 December 1934, 30:8.

7. O'Laughlin to Hoover, 22 May and 22 June 1934, Individual Correspondence File, Box 167, Post-Presidential Period, HHPL.

8. Even before FDR announced the final decision, two Democrats—Senator Kenneth McKellar of Tennessee and Representative Arthur Lamneck of Ohio—wrote the president to oppose the extension, and the former called the action

illegal. McKellar to Roosevelt, telegram, 2 October 1934, and Lamneck to Roosevelt, n.d., both in Official File 25, Box 30, FDRL. On the other hand, prominent congressmen such as Majority Leader Joseph Byrns wanted to retain MacArthur. Byrns to Roosevelt, 6 July 1934, Official File 25, Box 30, FDRL.

9. FDR to Dern, 13 July 1935, Official File 25, Box 1, FDRL.

10. FDR, on a handwritten, rank-ordered list (with ages listed), preferred General Dennis E. Nolan over Craig, but Craig, at fifty-nine, was three years younger and thus met the president's criteria. The president's list rejected outright Generals Fox Conner and George Van Horne Moseley, and FDR listed General Simonds sixth (MacArthur and Dern ranked Simonds second). All of these generals had at one time or another expressed anti–Air Corps sentiments.

11. Dern to FDR, 23 December 1935, Official File 25, Box 1, FDRL.

12. Dern to FDR, 23 December 1935.

13. Ira Eaker, "Memories of Six Air Chiefs," n.d., 4, Series 1, Box 3, Folder 26, Copp Collection, MS 46, USAFAL.

14. Underwood, *The Wings of Democracy*, 23.

15. Oscar Westover, "Air Corps Policies," 6 November 1935, Box 16, Foulois Papers.

16. Westover, "Air Corps Policies."

17. McClendon, *Autonomy of the Air Arm*, 84–85. Westover's opposition to immediate independence derived from his disciplinarian nature and deference to higher authorities. Although he opposed confrontation as a means to gain autonomy, he remained an advocate of air power. See also Nalty, *Winged Shield, Winged Sword*, 1:131–33.

18. Quoted in Arnold, *Global Mission*, 162.

19. Oscar Westover, "Air Corps Policies," 6 November 1935, in Box 16, Foulois Papers. For more on Arnold's understanding of—and playing to—the press, see Daso, *Hap Arnold*, 230.

20. Dern to FDR, 23 December 1935, Official File 25, Box 1, FDRL.

21. Dern to FDR, 23 December 1935.

22. Dern to FDR, 23 December 1935.

23. Underwood, *The Wings of Democracy*, 42–45.

24. Arnold to McSwain, 20 March 1935, Box 11, McSwain Papers.

25. Ira Eaker, "Memories of Six Air Chiefs," n.d., 4, Series 1, Box 3, Folder 26, Copp Collection, MS 46, USAFAL. Eaker joined the air staff in 1936.

26. Andrews became a brigadier general on 1 March 1935 with the activation of GHQ Air Force, and on 1 December he added another star (both temporary ranks). "Andrews, Frank Maxwell," in Fogerty, *Biographical Study*. When Andrews first set out to form his headquarters (prior to full activation at Langley) he needed office space. Lieutenant Elwood Quesada (temporarily assigned to GHQ Air Force) knew that Davison's old offices sat empty, so they set up shop there. Quesada

oral history interview by Copp, 18 November 1976, MS 46, Box 3, Folder 15, Copp Collection, USAFAL.

27. Mark Clodfelter described Andrews as "an airpower disciple who relentlessly spouted Mitchellese to both the War Department and the public." Clodfelter, "Molding Airpower Convictions," in *Paths of Heaven*, 108.

28. Quesada oral history interview by Copp, 18 November 1976.

29. Colonel G. C. Brant to Andrews, 6 January 1936, Box 1: General Correspondence File, Lieutenant General Frank M. Andrews Papers, LOC-MD.

30. McIntyre to Early, telegram, 25 September 1936, Official File 25, Box 2, FDRL.

31. Roosevelt to Woodring, telegram, 25 September 1936, Official File 25, Box 2, FDRL.

32. O'Laughlin to Hoover, 29 August 1936, Individual Correspondence File, Box 168, Post-Presidential Period, HHPL. Roosevelt did carry Kansas in the 1936 election. Landon won only Vermont and Maine.

33. On the announcement of the new cabinet to those members, see O'Laughlin to Hoover, 23 January and 23 March 1937, Individual Correspondence File, Box 168, Post-Presidential Files, HHPL. FDR sent the nomination of Woodring as Secretary of War to the Senate on 27 April 1937, eight months after Dern's death. Press release by Early, Official File 25, Box 2, FDRL.

34. On Woodring's recommendation, Louis Johnson became assistant secretary on 28 June 1937. During the eleven months of there being only one secretary, General Craig filled in as Acting Secretary of War during Woodring's absence from the office and signed various memoranda and letters as such. See Official File 25, Box 2, FDRL.

35. Of the 14 Republican House losses, Democrats gained 9 and other parties 5 (103 Republicans, 322 Democrats, 10 from other parties). Democrats gained 10 of the 11 Senate seats lost (25 Republicans, 69 Democrats, and 2 from other parties). See http://www.senate.gov/learning/stat_13.html and http://clerkweb.house.gov/histrecs/househis/lists/divisionh.htm, both accessed 8 November 2001.

36. Craig, "Notes on Wilcox Bill (H.R. 3151)," 23 April 1937, in Official File 25, Box 32, FDRL.

37. Woodring to Roosevelt, 9 June 1937, Official File 25, Box 32, FDRL.

38. Copp, *A Few Great Captains*, 374–79, 386–88; Underwood, *The Wings of Democracy*, 101. Because the bill was never heard, congressional records provide only slim information. Copp's narrative contains the best iteration of the complete story. For information on Knerr's assignment see "Knerr, Hugh Johnson," in Fogerty, *Biographical Study*.

39. Air Corps Act, *Statutes at Large*, 44:784.

40. Holley, *Buying Aircraft*, 49–50.

41. Holley, *Buying Aircraft*, 54–58.

42. Holley, *Buying Aircraft*, 43.

43. U.S. Congress, Senate, Committee on Military Affairs, *Report to Accompany H.R. 11140, To Provide More Effectively for the National Defense by Further Increasing the Effectiveness and Efficiency of the Air Corps of the Army of the United States*, 74th Cong., 2nd sess., 12 May 1936, S. Report 2131, Serial 9989, 3.

44. *Report to Accompany H.R. 11140*, S. Report 2131, 1.

45. U.S. Congress, Senate, Committee on Military Affairs, *Efficiency of the Air Corps of the Army: Hearings before the Committee on Military Affairs on H.R. 11140, H.R. 11920 and H.R. 11969*, 74th Cong., 2nd sess., 15 May 1936, 19.

46. Holley, *Buying Aircraft*, 60. Roosevelt used the same justification that Coolidge had employed in 1926 against the Five-Year Plan of the original Air Corps Act.

47. U.S. Congress, House, *Conference Report to Accompany H.R. 11140, Increase Effectiveness and Efficiency of the Air Corps*, 74th Cong., 2nd sess., 15 June 1936, Report 2994, Serial 9994, 2.

48. Roosevelt called any radical increase in the numbers of aircraft and budgets "unwise," but he wanted the military to lay the foundations for increased production and improve the manufacturers' ability to deliver aircraft more quickly. Roosevelt to Dern, 15 January 1936, Box 81, President's Secretary's File, FDRL.

49. Underwood, *The Wings of Democracy*, 74, 77.

50. McFarland, *Harry H. Woodring*, 99.

51. For a good summary of budget numbers and increases, see the chart "Appropriations for the Military Establishment Showing Approximate Breakdown into Major Functions, FY 1925–1940 inclusive" in Box 14, McSwain Papers. Even with the increased budgets, the War Department still needed every penny. When Roosevelt asked Woodring to cut back on money already appropriated during FY 1938, the secretary replied that the service needed all the money, especially since the Budget Bureau (to $410 million) and Congress (over $409 million) both reduced the initial request for over $590 million. Woodring specifically mentioned the Air Corps needing its funds for expansion and procurement. Woodring to Roosevelt, 7 July 1937, Official File 25, Box 3, FDRL.

52. Those numbers reflect total aircraft production for both the Army and Navy. G. R. Simonson, "The Demand for Aircraft and the Aircraft Industry, 1907–1958," *Journal of Economic History* 20, no. 3 (1960): 370.

53. For example, during the 1936 budget process the Air Corps initially asked for almost $53 million. The Budget Bureau cut that to just over $45 million, but Congress restored a little over half of the administration's cut and appropriated $50 million. U.S. Congress, House, Subcommittee of the House Committee on Appropriations, *War Department Appropriations Bill for 1937: Military Activities: Hearings before the Subcommittee of the House Committee on Appropriations*, Pt. 1, 74th Cong., 2nd sess., 1936, 308.

54. In the year between the 1936 and 1937 budgets, the Air Corps received

an increase of $15 million—a 23 percent increase. *War Department Appropriations Bill for 1937*, 319. See also *Annual Report of the Chief of the Air Corps, 1936*, 69–71; *Annual Report of the Chief of the Air Corps, 1937*, 76–78; and *Annual Report of the Chief of the Air Corps, 1938*, 78–80. All reports accessed at AFHSO.

55. Krauskopf, "The Army and the Strategic Bomber," 85–86.

56. Thomas H. Greer, *Army Air Forces Historical Studies: No. 89: The Development of Air Doctrine in the Army Air Arm, 1917–1941* (Maxwell Air Force Base AL: Air University USAF Historical Division, 1955), 51. The Air Corps Tactical School understood the divergence of its doctrine from the stated national strategy, and the school instructed its officers that the stated national strategy remained defensive. According to Greer, the instructors often ignored the actual strategic situation and policy and taught offensive air power in hypothetical terms (53).

57. Upon the retirement of Admiral Pratt (1933), the Navy renounced the MacArthur-Pratt agreement, and the debate continued until World War II. However, displaying the new moderate and Army-team attitude, the air leaders worked with the Chief of Staff to keep the disagreements out of the press. Underwood, *The Wings of Democracy*, 25, 78–79.

58. Many sources cover the United States' isolationist attitude between the world wars. For citations on specific arguments for the bomber as a defensive weapon, see Underwood, *The Wings of Democracy*, 56–58, 81–83; Sherry, *The Rise of American Air Power*, 52–53.

59. G-4 memorandum for Chief of Staff, 8 August 1936, quoted in Krauskopf, "The Army and the Strategic Bomber," 85.

60. *New York Times*, 28 April 1935, 28:3.

61. *New York Times*, 2 May 1935, 11:1.

62. Copp, *A Few Great Captains*, 326. The first B-17 outperformed the minimum standards. It could fly at 250 miles per hour and carry 2,500 pounds of bombs 2,260 miles (or 5,000 pounds 1,700 miles). Greer, *Development of Air Doctrine*, 46–47.

63. "Development of 4-Engine Bombers, 1933–1939," memorandum to General Arnold from Brigadier General L. W. Miller, U.S. Army Budget and Fiscal Officer, 15 April 1943, Box 166: Official File, Arnold Papers.

64. "Development of 4-Engine Bombers, 1933–1939."

65. Underwood, *The Wings of Democracy*, 66.

66. "Development of 4-Engine Bombers, 1933–1939."

67. Underwood, *The Wings of Democracy*, 66. For costs of producing the bomber, see Copp, *A Few Great Captains*, 326. For contract date and other information see "Development of 4-Engine Bombers, 1933–1939."

68. Arnold, *Global Mission*, 156–57.

69. Arnold, *Global Mission*, 154.

70. "Development of 4-Engine Bombers, 1933–1939."

71. The Project A aircraft also served as a blueprint or "parent" experiment

for heavy-bombardment aircraft that followed, including the B-29 and the B-24, the latter of which carried the weight of the World War II bombing campaigns and was produced in higher numbers than any other bomber aircraft of the war (18,000 as compared to 12,600 B-17s). See Craven and Cate, *Plans and Early Operations*, 66; Nalty, *Winged Shield, Winged Sword*, 1:243.

72. "Augmentation in Aircraft to be included in F.Y. 1938 Estimates," Brigadier General George R. Spalding to the Chief of Staff, n.d. (endorsements dated 2 July 1936), Box 166: Official File, Arnold Papers.

73. "Augmentation in Aircraft to be included in F.Y. 1938 Estimates."

74. The War Department worked concurrently on the budgets for three fiscal years: spending the current year's money, finalizing the next year's budget, and initial submissions for the budget two years away. Thus, for planning purposes, at this time in 1936, the War Department tried to finalize the numbers for the FY 1937 budget while planning the FY 1938 budget. They took these steps in conjunction with the administration, the Budget Bureau, and Congress.

75. "Augmentation in Aircraft to be included in F.Y. 1938 Estimates."

76. Arnold, *Global Mission*, 167. When Congress finalized the 1938 budget the lawmakers added almost $7 million to the War Department's requested amount for the aircraft procurement, but the Woodring-induced savings remained. Westover calculated that the War Department substituted forty-four two-engine bombers for the twenty four-engine models requested by the Air Corps. *Annual Report of the Chief of the Air Corps, 1938*, 80, AFHSO.

77. "Procurement of Four-Engine Airplanes," General Andrews to the Adjutant General, 17 June 1937, Box 9: Official Papers File, Andrews Papers.

78. For chains-of-command wiring diagrams, see Holley, *Buying Aircraft*, 95, 96, and 103. Of note, the vacant office of the Assistant Secretary of War for Air still represented a step in the official chain, coming between the Chief of the Air Corps and the Secretary of War.

79. Daily Record of Events, 16 June 1937, Box 178: Official File, Arnold Papers.

80. Daily Record, 16 June 1937.

81. "Procurement of Four-Engine Airplanes."

82. "Memorandum for Chief of Staff, U.S. Army," from General Andrews, 24 November 1937, Box 2: General Correspondence File, Andrews Papers.

83. Johnson took office as Assistant Secretary of War on 29 June 1937. He was an Army veteran of World War I and became national commander of the American Legion in 1932. A staunch Democrat, Johnson organized the Veterans Division of the Democratic National Committee in 1936 and was rewarded with the second chair in the War Department. He initially turned down the job, as he and Woodring did not get along, and Johnson wanted the secretary's position. However, after allegedly receiving assurances from insiders that Roosevelt would relieve Woodring and make Johnson secretary, the Virginian accepted the lower

position. For more information on the Woodring-Johnson disagreements and political intrigue, see McFarland, *Harry H. Woodring*, 144–50.

84. Background on the plans and the submissions in "Five-Year Program for the Air Corps," Assistant Chief of Staff George R. Spalding to Chief of Staff, 22 January 1938, Box 11: Official Papers File, Andrews Papers.

85. "Procurement Program for the Air Corps from 1940–1945," General Andrews to the Assistant Secretary of War, 24 November 1937, Box 11: Official Papers File, Andrews Papers.

86. General Andrews to General Westover, 27 September 1937, Box 7: General Correspondence File, Andrews Papers.

87. Andrews to Westover, 27 September 1937.

88. Andrews to Westover, 27 September 1937.

89. Eric Larrabee, *Commander in Chief: Franklin Delano Roosevelt, His Lieutenants, and Their War* (New York: Harper and Row, 1987), 26–27.

90. General Andrews to Colonel Watson, n.d., Box 11: Official Papers File, Andrews Papers. Andrews's memorandum is an undated draft. However, from the information contained therein and its location in his official papers, the date is interpreted to be 1937.

91. "Five-Year Program for the Air Corps," emphasis in original.

92. "Five-Year Program for the Air Corps."

93. "Five-Year Program for the Air Corps."

94. General Andrews to General Spalding, 29 January 1938, Box 11: Official Papers File, Andrews Papers.

95. Millett and Maslowski, *For the Common Defense*, 387, 397.

96. Confidential Record of Events, 3 June 1938, Box 180: Official File, Arnold Papers.

97. Daily Record (Secret Items) for 30 June 1938, Box 180: Official File, Arnold Papers.

98. General Arnold to General Westover, 7 June 1938, Box 178: Official File, and Box 180: Daily Record (Secret Items) for 30 June 1938, both in Arnold Papers.

99. Quote is Arnold's from Johnson's meeting, as annotated in Daily Record (Secret Items) for 30 June 1938, Box 180: Official File, Arnold Papers.

100. Daily Record (Secret Items) for 30 June 1938.

101. "Procurement of Bombardment Aircraft," General Andrews to the Adjutant General, 25 June 1938, Box 9: Official Papers File, Andrews Papers.

102. "Procurement of Bombardment Aircraft."

103. Daily Record of Events, 29 September 1938, Box 180: Official File, Arnold Papers.

104. Daily Record of Events, 29 September 1938.

105. Daily Record of Events, 7 October 1938, Box 180: Official File, Arnold Papers.

106. Daily Record of Events, 15 October 1938, Box 180: Official File, Arnold Papers.

107. Daily Record of Events, 18 October 1938, Box 180: Official File, Arnold Papers.

108. Daily Record of Events, 18 October 1938.

109. Jean H. Dubuque and Robert F. Gleckner, *Air Force Historical Studies: No. 6: The Development of the Heavy Bomber, 1918 to 1944* (Maxwell Air Force Base AL: Air University USAF Historical Division, 1951), 33.

110. For a detailed explanation of the crash, see Copp, *A Few Great Captains*, 438–40; see also Arnold, *Global Mission*, 169.

111. Arnold did not know who started those rumors, but Copp suspected Louis Johnson, who backed Andrews, or John Callan O'Laughlin. For the full story and intrigue, see Copp, *A Few Great Captains*, 441–44. Woodring supported Arnold's ascendancy to the Air Corps' top office and also recommended Lieutenant Colonel Walter G. Kilner for the second position. Woodring to Roosevelt, 26 September 1938, Official File 25, Box 32, FDRL.

112. McFarland, *Harry H. Woodring*, 164.

113. Underwood, *The Wings of Democracy*, 132–33.

114. Arnold's handwritten notes from the 14 November meeting is contained on Reel 4: Correspondence, Arnold Papers. Arnold later organized and typed up the meeting notes and sent them to General Craig, available in Official File 25, Box 30, FDRL.

115. Arnold, *Global Mission*, 177. Reflecting the morals of the printed word in 1949, Arnold's quote deleted Roosevelt's expletives.

116. Arnold, *Global Mission*, 179.

117. Arnold, *Global Mission*, 179. For secondary source coverage of the meeting, see Underwood, *The Wings of Democracy*, 131–35; McFarland, *Harry H. Woodring*, 165–66.

118. McFarland, *Harry H. Woodring*, 166–67. Arnold's book (*Global Mission*) mistakenly included Woodring among those present, but McFarland and other sources note Woodring's exclusion. Also, Arnold's typed notes from the meeting did not list Woodring. Memorandum for the Chief of Staff, 15 November 1938, Official File 25, Box 30, FDRL.

119. Arnold, *Global Mission*, 179.

120. A good example of how they publicized the bomber and ever-improving Air Corps capabilities (and the defensive mission) occurred in May 1938, when B-17s on a navigational exercise intercepted the Italian liner *Rex* six hundred miles off the coast. The Air Corps leaders trumpeted the success in the press, which gave the public, and the national leaders, increased confidence in the Air Corps — and in its ability to *defend* the United States. See Krauskopf, "The Army and the Strategic Bomber," 88; Underwood, *The Wings of Democracy*, 113–14.

121. As the number of old aviation activists dwindled, not needing to rely

on Congress probably served the Air Corps better. Added to those lost (listed in the previous chapter), Frank James failed in 1934 and 1936 election bids, and McSwain decided not to stand for reelection in 1936 due to health reasons (he died that September).

7. THE POLITICS OF AIR CORPS EXPANSION

1. Arnold, *Global Mission*, 177.

2. Arnold, *Global Mission*, 179.

3. Underwood, *The Wings of Democracy*, 155. See also Franklin D. Roosevelt, "The President Again Seeks a Way to Peace. A Message to Chancellor Adolf Hitler and Premier Benito Mussolini. April 14, 1939," in Franklin D. Roosevelt, *The Public Papers and Addresses of Franklin D. Roosevelt*, ed. Samuel Irving Rosenman, vol. 8, *1939: War—And Neutrality* (New York: Macmillan, 1941), 202.

4. For an analysis of Roosevelt pitting subordinates against one another, see Burns, *Roosevelt*, 372–73; Conkin, *The New Deal*, 90.

5. Harold Ickes observed how the president acted like a "beaten man" and had lost confidence and courage. Ickes, *The Secret Diary of Harold L. Ickes*, vol. 2, *The Inside Struggle, 1936–1939* (New York: Simon and Schuster, 1954), 326. For updated and expanded coverage of the "court-packing" scheme, see William E. Leuchtenburg, *The Supreme Court Reborn: The Constitutional Revolution in the Age of Roosevelt* (New York: Oxford University Press, 1995), 82–162.

6. D. M. Kennedy, *Freedom from Fear*, 337–44; Leuchtenburg, *Roosevelt and the New Deal*, 231–39, 252–54.

7. See http://clerkweb.house.gov/histrecs/househis/lists/divisionh.htm and http://www.senate.gov/learning/stat_13.html, both accessed 19 November 2001. The Democrats still controlled the House by a 262–169 margin, and the Senate with 69 of the 96 seats.

8. Leuchtenburg, *Roosevelt and the New Deal*, 271–72.

9. Burns, *Roosevelt*, 254.

10. Burns, *Roosevelt*, 255; D. M. Kennedy, *Freedom from Fear*, 394–95. The 1937 Neutrality Act replaced time limits on the 1935 law with permanent assurances of American nonintervention.

11. Arnold, *Global Mission*, 177.

12. Franklin D. Roosevelt, "The Four Hundred and Ninety-first Press Conference (Excerpts). October 14, 1938," in Franklin D. Roosevelt, *The Public Papers and Addresses of Franklin D. Roosevelt*, ed. Samuel Irving Rosenman, vol. 7, *1938: The Continuing Struggle for Liberalism* (New York: Macmillan, 1941), 548; *New York Times*, 15 October 1938, 1:4.

13. Johnson to Craig, 15 October 1938, and sent to Roosevelt on same date with a cover note, Box 83, President's Secretary's File, FDRL. By comparison, England, under the pressure of imminent war, still produced only 3,000 aircraft per year.

14. Underwood, *The Wings of Democracy*, 132–33.

15. "Strength of Army Air Corps," Arnold to Johnson, 10 November 1938, Reel 4: Correspondence File, Arnold Papers.

16. Arnold's numbers came close to those of another study, conducted by Air Corps Major Alfred Lyon and sent to Harry L. Hopkins of the Works Progress Administration. Lyon estimated the maximum production with available facilities at 7,500 aircraft per year, which meant a sixfold increase over 1938 production. The maximum production rate would also require the government to contract for large orders. Any higher production rate required building new plants, probably at government expense. Lyon to Hopkins, 3 November 1938, Box 81, President's Secretary's File, FDRL.

17. "Strength of Army Air Corps."

18. Captain Park Holland, at a meeting with Johnson, claimed that Johnson wanted to "buy the maximum number of bombers," and "All we [the Army] needed was a mass of bombers." Quoted in Daily Record (Secret Items) for 30 June 1938, Box 180: Official File, Arnold Papers. Quote is not a direct quote from Johnson but is what Holland recorded (paraphrased). For Underwood's belief in Johnson's being swayed by New Dealers, see Underwood, *The Wings of Democracy*, 133.

19. Arnold, *Global Mission*, 172–73.

20. Arnold, *Global Mission*, 177–79.

21. Arnold, *Global Mission*, 172. The other two were Harry Hopkins and Robert Lovett, a future Assistant Secretary of War for Air (1941–45).

22. Forrest C. Pogue, *George C. Marshall*, vol. 1, *Education of a General, 1880–1939* (New York: Viking Press, 1963), 334. Marshall believed the president's 10,000-plane program was unsound and even drafted a letter from General of the Armies John J. Pershing to Roosevelt (in November 1938) pushing for additional ground forces. See George C. Marshall, *The Papers of George Catlett Marshall*, ed. Larry I. Bland and Sharon R. Ritenour, vol. 1, *"The Soldierly Spirit," December 1880–June 1939* (Baltimore: Johns Hopkins University Press, 1981), 654–55, 680.

23. Marshall to Andrews, 1 April 1939, Box 12: Personal File, Andrews Papers.

24. Of note, after Marshall became the Chief of Staff in September 1939, he appointed Andrews his Assistant Chief of Staff, G-3. See Pogue, *Education of a General*, 409 n. 33.

25. Underwood, *The Wings of Democracy*, 136–37. Underwood wrote that Andrews believed Roosevelt had the interest of the Air Corps in mind. However, Underwood himself asserts, here and at various other places in his book, that Roosevelt used the situation primarily to assist Britain and France, and the by-product of American production and purchase money from those countries only added to the deal. John Haight argued that Roosevelt wanted the planes for Britain and France but that Chamberlain wanted only raw materials, the British staff wanted America to build a war reserve, and that only France desperately

wanted American bombers. John McVickar Haight Jr., *American Aid to France, 1938–1940* (New York: Atheneum, 1970), 62–63, 67–68.

26. Marshall to Arnold, 15 December 1938, microfilm, 3:898, General of the Armies George C. Marshall Papers, LOC-MD.

27. *New York Times*, 1 December 1938, 1:1.

28. *New York Times*, 1 December 1938, 1:1 (and continuation on 16), and 4 December, 1:6.

29. *New York Times*, 1 December 1938, 16. Biographical information from http://bioguide.congress.gov/biosearch/biosearch.asp, accessed 20 November 2001.

30. *New York Times*, 1 December 1938, 1:1 (and continuation on 16).

31. *New York Times*, 1 December 1938, 1:1 (and continuation on 16), 1:6.

32. *New York Times*, 4 December 1938, 1:4.

33. *New York Times*, 12 December 1938, 18:1.

34. *New York Times*, 15 December 1938, 26:3.

35. *New York Times*, 28 December 1938, 3:4.

36. Daily Record of Events, 11–12 January 1939, Reel 180: Official File, Arnold Papers. Woodring and Craig both let their deputies take the lead in formulating a program for the president. Woodring let Johnson handle the proposal because Johnson was responsible for procurement. Because Marshall had a better understanding of aviation, Craig probably wanted Marshall to work with Arnold.

37. For comparison, the additional money would almost double the original budgeted amount and would bring the 1940 defense request to $1.126 billion. The total expenditure for national defense for the previous two completed fiscal years totaled $980 million in 1938 and $895 million in 1937. The first New Deal budget of 1933 gave the least amount of money to national defense during the 1930s, $494 million. Figures from Annual Budget Message Chart, Roosevelt, *1939: War—And Neutrality*, 39.

38. Roosevelt, *1939: War—And Neutrality*, 71–72. For an explanation of "educational orders," see Wesley Frank Craven and James Lea Cate, eds., *The Army Air Forces in World War II*, vol. 6, *Men and Planes* (1955; reprint, Washington DC: Office of Air Force History, 1983), 300–301. These orders were a measure, in practice since 1927, to allow the War Department to "test" the company's ability to produce a critical item during a war and allow the military to secure production data.

39. U.S. Congress, Senate, Committee on Military Affairs, *National Defense: Hearings before the Committee on Military Affairs on H.R. 3791, An Act to Provide More Effectively for the National Defense by Carrying Out the Recommendations of the President in his Message of January 12, 1939, to the Congress*, 76th Cong., 1st sess., 17 January to 22 February 1939, 5.

40. *Hearings before the Committee on Military Affairs on H.R. 3791*, 18.

41. *Hearings before the Committee on Military Affairs on H.R. 3791*, 36–41.

42. *Hearings before the Committee on Military Affairs on H.R. 3791*, 31.

43. Information from the Associated Press, quoted in *Hearings before the Committee on Military Affairs on H.R. 3791*, 64. Information about "Smithson" is on page 66.

44. For information on Clark's foreign policy beliefs, see Robert Dallek, *Franklin D. Roosevelt and American Foreign Policy, 1932–1945* (New York: Oxford University Press, 1979), 103–4, 108, 119. Arnold believed that most members of the Committee were outstanding isolationists. Arnold, *Global Mission*, 185.

45. Arnold, *Global Mission*, 186.

46. *Hearings before the Committee on Military Affairs on H.R. 3791*, 94.

47. Arnold to Wolfe, 19 January 1939, quoted in Hearings before the Committee on Military Affairs on H.R. 3791, 94.

48. Morgenthau Diary Entry 0047, 31 January 1939, Box 514: Presidential Diaries, Henry Morgenthau Jr. Papers, FDRL.

49. Arnold, *Global Mission*, 185–86.

50. *Hearings before the Committee on Military Affairs on H.R. 3791*, 192–96.

51. *Hearings before the Committee on Military Affairs on H.R. 3791*, 126. Here, Craig received the same treatment from the Committee as Arnold, asking "What right has the Treasury Department of the United States to give Orders to the Army? What has the Treasury Department to do with the Army insofar as orders of that sort are issued?" To these, Craig replied "None ordinarily, sir."

52. *Hearings before the Committee on Military Affairs on H.R. 3791*, 177–80.

53. U.S. Congress, House, Subcommittee of the Committee on Appropriations, *Hearings on the Military Establishment Appropriation Bill for 1940*, 76th Cong., 1st sess., 24 January to 15 February 1939, 318–9. Of note, at this encounter Engel sarcastically asked Arnold if he took orders from the Treasury Department.

54. See Morgenthau Diaries, microfilm, Book 172, p. 78, 30 December 1938, and pp. 80–81, 31 December 1938, Morgenthau Papers.

55. Morgenthau Diaries, microfilm, Book 173, p. 70, 27 January 1939.

56. Morgenthau Diaries, microfilm, Book 173, pp. 72–75, 27 January 1939.

57. Morgenthau Diaries, microfilm, Book 173, pp. 170–71, 27 January 1939.

58. Morgenthau Diaries, microfilm, Book 173, pp. 149–51, 27 January 1939. For a thorough analysis of Roosevelt's desires to help the French, see Haight, *American Aid to France;* Dallek, *Roosevelt and American Foreign Policy*, 171–232 passim.

59. Arnold, *Global Mission*, 186.

60. Arnold, *Global Mission*, 186–87.

61. Arnold, *Global Mission*, 194.

62. *Hearings before the Committee on Military Affairs on H.R. 3791*, 229.

63. *Hearings on the Military Establishment Appropriation Bill for 1940*, 10–11.

64. Holley, *Buying Aircraft*, 174.

65. Craven and Cate, *Men and Planes*, 301–2.

66. *Hearings on the Military Establishment Appropriation Bill for 1940*, 318–19.

67. Craven and Cate, *Men and Planes*, 301–2.

68. Craven and Cate, *Men and Planes*, 301–2.

69. Daily Record for 8 and 9 January 1940, Box 180: Official File, Arnold Papers.

70. Daily Record for 19 March 1940, Box 180: Official File, Arnold Papers.

71. Roosevelt, *1940: War—And Aid to Democracies*, 198–202.

72. Underwood, *The Wings of Democracy*, 155.

73. Underwood, *The Wings of Democracy*, 155. For the specific breakdown of dollars requested and authorized, see Roosevelt, *1940: War—And Aid to Democracies*, 203 and 205.

74. Arnold, *Global Mission*, 193.

EPILOGUE

1. Roosevelt to Woodring, 19 June 1940, Box 84, President's Secretary's File, FDRL.

2. Woodring to Roosevelt, 20 June 1940, Box 84, President's Secretary's File, FDRL.

3. Press Release, 20 June 1940, Box 84, President's Secretary's File, FDRL. Of note, Roosevelt also replaced the Secretary of the Navy, and both new secretaries were Republicans. See also *Kansas City Star*, 20 June 1940, 1. Newspapers postulated that Roosevelt replaced Woodring because of differences over foreign policy and Woodring's being more isolationist than interventionist. See *Kansas City Times*, 22 June 1940, in Box 84, President's Secretary's File, FDRL. Roosevelt may have viewed differing views as being disloyal—and though he allowed bickering among his staff, he did not tolerate disloyalty. See Conkin, *The New Deal*, 90.

4. Johnson to Roosevelt, 25 July 1940, Box 83, President's Secretary's File, FDRL. Stimson requested Robert P. Patterson to replace Johnson, and Roosevelt agreed.

5. Arnold, *Global Mission*, 195.

6. Oral history interview with Jacob Smart by Murray Green, 13 November 1969, Washington DC, transcript in Box 76, Green Collection, USAFAL.

7. U.S. Strategic Bombing Survey, *The United States Strategic Bombing Survey: Summary Report*, vol. 1, with an introduction by David MacIsaac (New York: Garland, 1976), 15–16. Richard Davis wrote that strategic bombing "vindicated the treasure expended on it," even though subsequent studies showed that it did not have an adverse impact on enemy morale—a goal of the Air Corps Tactical School strategy adopted from the early air power theorists. Richard Davis, *Carl A.*

Spaatz and the Air War in Europe (Washington DC: Center for Air Force History, 1993), 590.

8. Mark Clodfelter, *The Limits of Air Power: The American Bombing of North Vietnam* (New York: Free Press, 1989), 10–11.

9. Walter J. Boyne, *Beyond the Wild Blue: A History of the United States Air Force, 1947–1997* (New York: St. Martin's Press, 1997), 32. For insight on the different proposals and perceptions on creating the new military establishment, see also David R. Mets, *Master of Airpower: General Carl A. Spaatz* (Novato CA: Presidio Press, 1988), 317–22. Russell F. Weigley noted how the Army struggled after World War II because it seemed irrelevant, as the Air Force controlled an atomic monopoly that could win any war quickly and without ground forces. Weigley, *History of the United States Army* (1967; reprint, Bloomington: Indiana University Press, 1984), 501.

10. Kohn identified these actions as ways military officers have gained influence over military policy in the twentieth century, to the detriment of civilian control of the military. He proposes that civilian control is "situational, dependent on the people, issues, and political and military forces involved." Kohn, "Erosion of Civilian Control," 16.

11. Michael Desch compares the different situations of internal and external threats on civil-military relations. However, he also asserts that the ensuing civil-military conflict in an environment with low internal and external threats would generate problems of "coordination rather than insubordination." In the case of the early air proponents, insubordination emerged as the major problem. Desch, *Civilian Control of the Military*, 15–17.

12. John "Fred" Shiner identified several Mitchell shortcomings, but he did not include the two most damaging ones: insubordination and inappropriate political behavior for a military officer. Shiner, "From Air Service to Air Corps: The Era of Billy Mitchell," in Nalty, *Winged Shield, Winged Sword*, 1:100. James Cooke's more recent biography summarized Mitchell as sometimes a "shameless promoter, a difficult man to deal with," and "a brilliant fighter but also a poor subordinate." Cooke, *Billy Mitchell*, 1. At the same time, *Makers of the United States Air Force*, meant to round out air history by highlighting some of the "forgotten" air leaders, included chapters on Andrews, Knerr, and even Foulois but failed to cover Air Corps chiefs Westover or Fechet. Frisbee, *Makers of the United States Air Force*. Only recently was Robert White's biography of Patrick published, and no history of air power adequately emphasizes the importance of the two civilian aviation secretaries, Davison and Lovett.

13. See the Air Force Academy's cadet handbook (and required freshman knowledge). Mueller, *Contrails: Air Force Academy Cadet Handbook*, 61–64. The professional officers' education system also touts Mitchell's contributions, especially in the Air Force's Squadron Officer School (for captains), and less so in Air Command and Staff College (for majors), without discussing his insubor-

dination and damage done to the service. More recent works expose some of Mitchell's flaws. Mark Clodfelter admits that the "real Mitchell" lies between the renegade and the brilliant theorist. See Clodfelter, "Molding Airpower Convictions: Development and Legacy of William Mitchell's Strategic Thought," in Air Command and Staff College, vol. 4, *Military Studies* (Maxwell Air Force Base AL: Air Command and Staff College, 2000), 233.

14. Pogue, "George C. Marshall on Civil Military Relationships," in Kohn, *United States Military*, 194.

15. Morris Janowitz, *The Professional Soldier: A Social and Political Portrait* (Glencoe IL: Free Press of Glencoe, 1960), 233. Weigley asserts that the tradition became cemented in American culture due to Lincoln's actions during the Civil War. Prior to the 1860s, the American military's size and peripheral relation to society limited its ability to challenge civilian control. Russell F. Weigley, "The American Civil-Military Cultural Gap: A Historical Perspective, Colonial Times to the Present," in *Soldiers and Civilians: The Civil-Military Gap and American National Security*, ed. Peter Douglas Feaver and Richard H. Kohn (Cambridge: MIT Press, 2001), 215–16.

16. Richard C. Brown, *Social Attitudes of American Generals, 1898–1940* (New York: Arno Press, 1979), 172.

17. Janowitz, *The Professional Soldier*, 12.

18. Weigley, "The American Civil-Military Cultural Gap," 216–17. In an earlier, related article, Weigley asserted that civil-military problems occurred during World War II, and he called it "remarkable" that harmony reigned despite those differences. Russell F. Weigley, "The American Military and the Principle of Civilian Control from McClelland to Powell," *Journal of Military History* 57, no. 2 (1993): 55.

19. Weigley, "The American Civil-Military Cultural Gap," 238.

20. Peter D. Feaver and Richard H. Kohn, "Conclusion: The Gap and What it Means for American Security," in Feaver and Kohn, *Soldiers and Civilians*, 464–65. See also Feaver, *Armed Servants*, 300–301.

21. Feaver and Kohn, "Conclusion," 473.

22. Kohn, "Erosion of Civilian Control," 16. See also Huntington, *The Soldier and the State*, 80–89. Military theorist Carl von Clausewitz commented upon the need for "genius" that would allow a commander to be a statesman and understand the entire political situation. Although he emphasized the need to understand the political goals that prompt war, his assertion that a military commander must be politically savvy and work with others remains valid. Carl von Clausewitz, *On War*, ed. and trans. Michael Howard and Peter Paret (1976; reprint, Princeton: Princeton University Press, 1984), 111–12.

23. Arnold's team meant himself and George Marshall as the uniformed contingent, and Lovett, Stimson, and Assistant Secretary Robert Patterson on the civilian side of the War Department hierarchy. Arnold, *Global Mission*, 195.

24. Johnson, "From Frontier Constabulary to Modern Army," in Winton and Mets, *The Challenge of Change*, 195.

25. Johnson, "From Frontier Constabulary to Modern Army," 162–204; Posen, *Sources of Military Doctrine*, 34–80; Odom, *After the Trenches*, 3–5, 236–45.

26. One could call this a "good cop/bad cop" relationship, but there is no evidence that air leaders pursued such a strategy consciously or purposefully. The attitudes of the two different types of leaders were never knowingly coordinated to create these conditions; rather, such a scenario came about because of the styles and approaches of the different men—again highlighting the situational and personality-dependent nature of civil-military relations.

Bibliography

MANUSCRIPT COLLECTIONS

Air Force History Support Office. Bolling Air Force Base/U.S. Navy Yard Anacostia Annex, Washington DC.

> Annual Reports of the Chief of the Air Service/Corps.
>
> *Report of the Secretary of War to the President*. Washington DC: Government Printing Office, for years 1927–34.

Army War College Archives. Carlisle Barracks PA.

> W. Frank James, "Handling Military Legislation in the House of Representatives." Lecture given before the Army War College, Washington DC, 16 June 1927. File no. 340A-4.

Center for American History. University of Texas at Austin.

> Morris Sheppard Papers.

Walter Royal Davis Library. University of North Carolina at Chapel Hill.

> Calvin Coolidge Papers. Washington DC, Library of Congress, 1959. Microfilm.

Duke University Library. Rare Book, Manuscript, and Special Collections Branch. Durham NC.

> John J. McSwain Papers.

Herbert Hoover Presidential Library. West Branch IA.

> William P. MacCracken Jr. Papers.
>
> Post-Presidential Period.
>
> Presidential Period.
>
> E. French Strother Papers.
>
> James Spear Taylor and Robert M. Gates Collection.

Library of Congress, Manuscript Division. Washington DC.

> Lieutenant General Frank M. Andrews Papers.
>
> General of the Air Force Henry Harley Arnold Papers.
>
> Newton D. Baker Papers.
>
> General Ira C. Eaker Papers.
>
> Major General Benjamin D. Foulois Papers.

General of the Armies George C. Marshall Papers.
Brigadier General William Mitchell Papers.
John Callan O'Laughlin Papers.
General of the Armies John J. Pershing Papers.
General Carl A. Spaatz Papers.
General Charles P. Summerall Papers.
National Archives and Records Administration. Washington DC.
Record Group 233. Records of the Armed Services Committee (House) and Its Predecessors, 1822–1988.
Records of Select Committees. *Of Inquiry into the Operation of the U.S. Air Services*. 68th–69th Congresses, 1924–25.
National Archives and Records Administration II. College Park MD.
Record Group 18. Records of the Army Air Forces
Major General Mason M. Patrick Papers.
Records of the Headquarters, Army Air Forces, 1917–1949.
Records of the Office of the Chief of the Air Service and the Office of the Chief of the Air Corps, 1917–1944.
Record Group 107. Records of the Office of the Secretary of War.
Records of the Office of the Assistant Secretary of War for Air, 1926–1947.
Record Group 159. Records of the Office of the Inspector General.
Record Group 165. Records of the War Department General and Special Staffs.
Records of the War Plans Division.
Franklin D. Roosevelt Library. Hyde Park NY.
Henry Morgenthau Jr. Papers.
Franklin D. Roosevelt Presidential Papers
Official File 25 (War Department).
President's Secretary's File.
United States Air Force Academy Library. Special Collections Branch. United States Air Force Academy, Colorado Springs CO.
DeWitt S. Copp Collection, MS 46.
Benjamin D. Foulois Collection, MS 17.
Murray Green Collection, MS 33.

GOVERNMENT DOCUMENTS
Air Commerce Act. *Statutes at Large* 44 (1926).
Air Corps Act. *Statutes at Large* 44.
National Defense Act of 1916. *Statutes at Large* 39.
National Defense Act of 1920. *Statutes at Large* 41.
President's Aircraft Board. *Aircraft: Hearings before the President's Aircraft Board*. Vols. 1–4. Washington DC: Government Printing Office, 1925.

U.S. Congress. *War Department Appropriations Bill, 1929.* 70th Cong., 1st sess. H. Rept. 497. Serial 8835.

———. *War Department Appropriations Bill, 1931.* 71st Cong., 2nd sess. H. Rept. 97. Serial 9190.

U.S. Congress. House. Committee on Military Affairs. *Air Corps: Progress under Five-Year Program: Hearing before the Committee on Military Affairs.* 69th Cong., 2nd sess., 19 January 1927.

———. *Air Service Unification: Hearing before the Committee on Military Affairs.* 68th Cong., 2nd sess., 8 January to 17 February 1925.

———. *Army Reorganization: Hearings before the Committee on Military Affairs.* Vol. 1. 66th Cong., 1st sess., September 3, 1919 to November 12, 1919.

———. *Conference Report to Accompany H.R. 11140, Increase Effectiveness and Efficiency of the Air Corps.* 74th Cong., 2nd sess., 15 June 1936, Report 2994, Serial 9994.

———. *Department of Defense and Unification of Air Service: Hearing before the Committee on Military Affairs.* 69th Cong., 1st sess., 19 January to 9 March 1926.

———. *Investigation of Profiteering in Military Aircraft, under H. Res. 275.* Report no. 2060, 73rd Cong., 2nd sess., 15 June 1934.

———. *Investigation under House Resolution 275.* Report no. 1506, 73rd Cong., 2nd sess., 1934.

———. *Statement of Maj. Gen. B. D. Foulois, Chief of the Air Corps, Before the Committee on Military Affairs.* 73rd Cong., 2nd sess., 1 February 1934. Committee Print.

———. *United Air Service: Hearing before a Subcommittee of the Committee on Military Affairs.* 66th Cong., 2nd sess., December 1919.

U.S. Congress. House. Committee on the Post Office and Post Roads. *Air Mail: Hearings before the Committee on the Post Office and Post Roads.* 73rd Cong., 2nd sess., 20 April 1933 and 15–21 February 1934.

U.S. Congress. House. Select Committee of Inquiry. *Inquiry into Operations of the United States Air Services.* Pts. 1–6. 68th Cong., 1925. Available on microfiche under U.S. Congressional Hearings. Vols. 379–81. Senate Library, 1925.

———. *Report of the Select Committee of Inquiry into Operations of the United States Air Services.* 68th Cong., 2nd sess., 14 December 1925.

U.S. Congress. House. Subcommittee of the House Committee on Appropriations. *Hearings on the Military Establishment Appropriation Bill for 1940.* 76th Cong., 1st sess., 24 January to 15 February 1939.

———. *War Department Appropriation Bill for 1930: Hearing before the Subcommittee of the House Committee on Appropriations.* 70th Cong., 2nd sess., 1928.

———. *War Department Appropriation Bill for 1934: Hearing before the Subcommittee of the Committee on Appropriations.* 72nd Cong., 2nd sess., 1932–33.

———. *War Department Appropriation Bill for 1935, Military Activities: Hearing before the Subcommittee of the Committee on Appropriations.* Pt. 1. 73rd Cong., 2nd sess., 1934.

———. *War Department Appropriations Bill for 1937: Military Activities: Hearings before the Subcommittee of the House Committee on Appropriations.* Pt. 1. 74th Cong., 2nd sess., 1936.

U.S. Congress. Senate. Committee on Military Affairs. *Efficiency of the Air Corps of the Army: Hearings before the Committee on Military Affairs on H.R. 11140, H.R. 11920 and H.R. 11969.* 74th Cong., 2nd sess., 15 May 1936.

———. *National Defense: Hearings before the Committee on Military Affairs on H.R.3791, An Act to Provide More Effectively for the National Defense by Carrying Out the Recommendations of the President in his Message of January 12, 1939, to the Congress.* 76th Cong., 1st sess., 17 January to 22 February 1939.

———. *Reorganization of the Army Air Service: Hearing before the Committee on Military Affairs on S. 2614.* 69th Cong., 1st sess., 5 February 1926.

———. *The Army Air Service: Hearing before the Committee on Military Affairs on H.R. 10827.* 69th Cong., 1st sess., 10 May 1926.

———. *To Increase the Efficiency of the Army Air Service.* Report no. 830, to accompany H.R. 10827. 69th Cong., 1st sess., 1926.

———. *Report to Accompany H.R. 11140, To Provide More Effectively for the National Defense by Further Increasing the Effectiveness and Efficiency of the Air Corps of the Army of the United States.* 74th Cong., 2nd sess., 12 May 1936, S. Report 2131, Serial 9989.

U.S. Congress. Senate. Subcommittee of the Committee on Appropriations. *War Department Appropriation Bill, 1929: Hearing before the Subcommittee of the Committee on Appropriations.* 70th Cong., 1st sess., 1928.

———. *War Department Appropriation Bill for 1931: Hearing before the Subcommittee of the Committee on Appropriations.* 71st Cong., 2nd sess., 1930.

U.S. Strategic Bombing Survey. *The United States Strategic Bombing Survey: Summary Report.* Vol. 1. With an introduction by David MacIsaac. New York: Garland, 1976.

War Department. *General Orders and Bulletins, 1920.* Washington DC: Government Printing Office, 1921.

———. *General Orders and Bulletins, 1922.* Washington DC: Government Printing Office, 1923.

War Department. Special Committee on Army Air Corps. *Final Report of War Department Special Committee on Army Air Corps,* 18 July 1934.

BOOKS, ARTICLES, AND DISSERTATIONS

Almond, Gabriel A. "The Political Attitudes of Wealth." *Journal of Politics* 7, no. 3 (1945): 213–55.

Arnold, Henry H. *Global Mission.* New York: Harper and Brothers, 1949.

Avant, Deborah D. *Political Institutions and Military Change: Lessons from Peripheral Wars*. Ithaca: Cornell University Press, 1994.

Beaver, Daniel R. *Newton D.Baker and the American War Effort, 1917–1919*. Lincoln: University of Nebraska Press, 1966.

Ben-Meir, Yehuda. *Civil-Military Relations in Israel*. New York: Columbia University Press, 1995.

Berdahl, Clarence A. "American Government and Politics: Some Notes on Party Membership in Congress, II." *American Political Science Review* 43, no. 3 (1949): 492–508.

Bingham, Hiram. *An Explorer in the Air Service*. New Haven: Yale University Press, 1920.

Bradley, Omar N. *A Soldier's Story*. New York: Henry Holt, 1951.

Boyne, Walter J. *Beyond the Wild Blue: A History of the United States Air Force, 1947–1997*. New York: St. Martin's Press, 1997.

Brown, Richard C. *Social Attitudes of American Generals, 1898–1940*. New York: Arno Press, 1979.

Burns, James MacGregor. *Roosevelt: The Lion and the Fox*. New York: Harcourt, Brace and World, 1956.

Cameron, Rebecca H., and Barbara Wittig, eds. *Golden Legacy, Boundless Future: Essays on the United States Air Force and the Rise of Aerospace Power*.Proceedings of a symposium held in Crystal City, Virginia, May 28–29, 1997. Washington DC: Air Force History and Museums Program, 2000.

Carll, George S., Jr. "Congress Struggling with the Air Problem." *U.S.Air Services* 11, no. 3 (1926): 45.

Clausewitz, Carl von. *On War*. Ed. and trans. Michael Howard and Peter Paret. 1976. Reprint, Princeton: Princeton University Press, 1984.

Clodfelter, Mark. *The Limits of Air Power: The American Bombing of North Vietnam*. New York: Free Press, 1989.

———. "Molding Airpower Convictions: Development and Legacy of William Mitchell's Strategic Thought." Air Command and Staff College. Vol. 4, *Military Studies*. Maxwell Air Force Base AL: Air Command and Staff College, 2000.

Conkin, Paul K. *The New Deal*. New York: Crowell, 1967.

Cooke, James J. *Billy Mitchell*. Boulder CO: Lynne Rienner, 2002.

Cooper, Jerry. *The Rise of the National Guard: The Evolution of the American Militia, 1865–1920*. Lincoln: University of Nebraska Press, 1997.

Copp, DeWitt S. *A Few Great Captains: The Men and Events That Shaped the Development of U.S. Air Power*. New York: Doubleday, 1980.

Corn, Joseph J. *The Winged Gospel: America's Romance with Aviation, 1900–1950*. New York: Oxford University Press, 1983.

Corum, James S. *The Luftwaffe: Creating the Operational Air War, 1918–1940*. Lawrence: University of Kansas Press, 1997.

Craven, Wesley Frank, and James Lea Cate, eds. *The Army Air Forces in World War II*. Vol. 1, *Plans and Early Operations, January 1939 to August 1940*. 1948. Reprint, Washington DC: Office of Air Force History, 1983.

———. *The Army Air Forces in World War II*. Vol. 6, *Men and Planes*. 1955. Reprint, Washington DC: Office of Air Force History, 1983.

Dallek, Robert. *Franklin D. Roosevelt and American Foreign Policy, 1932–1945*. New York: Oxford University Press, 1979.

Daso, Dik Alan. *Hap Arnold and the Evolution of American Airpower*. Washington DC: Smithsonian Institution Press, 2000.

Davis, Burke. *The Billy Mitchell Affair*. New York: Random House, 1967.

Davis, Richard G. *Carl A.Spaatz and the Air War in Europe*. Washington DC: Center for Air Force History, 1993.

Desch, Michael C. *Civilian Control of the Military: The Changing Security Environment*. Baltimore: Johns Hopkins University Press, 1999.

Douhet, Giulio. *Command of the Air*. Trans. Dino Ferrari. 1942. Reprint, USAF Warrior Studies. Ed. Richard H. Kohn and Joseph P. Harahan. Washington DC: Office of Air Force History, 1983.

Dubuque, Jean H., and Robert F. Gleckner. *Air Force Historical Studies: No.6: The Development of the Heavy Bomber, 1918 to 1944*. Maxwell Air Force Base AL: Air University USAF Historical Division, 1951.

Duke, Escal Franklin. "The Political Career of Morris Sheppard, 1875–1941." Ph.D. diss., University of Texas, 1958.

Feaver, Peter Douglas. *Armed Servants: Agency, Oversight, and Civil-Military Relations*. Cambridge: Harvard University Press, 2003.

———. *Guarding the Guardians: Civilian Control of Nuclear Weapons in the United States*. Ithaca: Cornell University Press, 1992.

Feaver, Peter Douglas, and Richard H. Kohn, eds. *Soldiers and Civilians: The Civil-Military Gap and American National Security*. Cambridge: MIT Press, 2001.

Fechet, James E. *Flying*. Baltimore: Williams and Wilkins, 1933.

Finer, Samuel E. *The Man on Horseback: The Role of the Military in Politics*. 2nd ed. Boulder CO: Westview Press, 1988.

Finney, Robert T. *Air Force Historical Studies: No. 100: History of the Air Corps Tactical School, 1920–1940*. 1955. Reprint, Washington DC: Center for Air Force History, 1992.

Fogerty, Robert P. *Air Force Historical Studies: No. 91: Biographical Data on Air Force General Officers, 1917–1952*. 2 vols. 1953. Reprint, Manhattan KS: MA/AH Publishing, 1980.

Foulois, Benjamin D., with Colonel C. V. Glines. *From the Wright Brothers to the Astronauts: The Memoirs of Major General Benjamin D. Foulois*. New York: McGraw-Hill, 1968.

Freidel, Frank Burt. *Franklin D.Roosevelt: A Rendezvous with Destiny*. Boston: Little, Brown, 1952.

Frisbee, John L. *Makers of the United States Air Force*. 1987. Reprint, Washington DC: Air Force History and Museums Program, 1996.

Futrell, Robert Frank. *Ideas, Concepts, Doctrine: Basic Thinking in the United States Air Force, 1907–1960*. Vol. 1. Maxwell Air Force Base AL: Air University Press, 1989.

Glines, Carroll V. *The Saga of the Air Mail*. New York, Arno Press, 1968.

Greenstein, Fred I. *The Presidential Difference: Leadership Style from Roosevelt to Clinton*. New York: Free Press, 2000.

Greer, Thomas H. *Air Force Historical Studies: No.89: The Development of Air Doctrine in the Army Air Arm, 1917–1941*. Maxwell Air Force Base AL: Air University USAF Historical Division, 1955.

Grumelli, Michael L. "Trial of Faith: The Dissent and Court Martial of Billy Mitchell." Ph.D. diss., Rutgers–New Brunswick, 1991.

Haight, John McVickar, Jr. *American Aid to France, 1938–1940*. New York: Atheneum, 1970.

Hendrickson, David C. *Reforming Defense: The State of American Civil-Military Relations*. Baltimore: Johns Hopkins University Press, 1988.

Holley, Irving Brinton, Jr. *Buying Aircraft: Matériel Procurement for the Army Air Forces*. United States Army in World War II, Special Studies. Ed. Stetson Conn. Washington DC: Office of the Chief of Military History, Department of the Army, 1964.

———. *Ideas and Weapons*. Washington DC: Air Force History and Museums Program, 1997.

Hoover, Herbert. "Civil Aviation's Rapid Progress," *Aero Digest,* April 1928, 509, 696–97.

———. *The Memoirs of Herbert Hoover*. Vol. 2, *The Cabinet and the Presidency, 1920–1933*. Vol. 3, *The Great Depression, 1929–1941*. New York: Macmillan, 1952.

Huntington, Samuel P. *The Common Defense: Strategic Programs in National Politics*. New York: Columbia University Press, 1961.

———. *The Soldier and the State: The Theory and Politics of Civil-Military Relations*. Cambridge: Belknap Press of Harvard University Press, 1957.

Hurley, Alfred F. *Billy Mitchell: Crusader for Air Power*. New York: Franklin Watts, 1964.

Hurley, Alfred F., and Robert C. Ehrhart, eds. *Air Power and Warfare*. Proceedings of the Eighth Military History Symposium. Washington DC: Office of Air Force History, 1979.

Huzar, Elias. *The Purse and the Sword: Control of the Army by Congress through Military Appropriations, 1933–1950*. Ithaca NY: Cornell University Press, 1950.

Ickes, Harold L. *The Secret Diary of Harold L. Ickes*. Vol. 1, *The First Thousand Days, 1933–1936*. Vol. 2, *The Inside Struggle, 1936–1939*. New York: Simon and Schuster, 1953–54.

James, D. Clayton. *The Years of MacArthur*. Vol. 1, *1880–1941*. Boston: Houghton-Mifflin, 1970.

James, W. Frank. "A Five-Year Development Program for the Air Corps at Last." *U.S. Air Services* 11, no. 6 (1926): 45–46.

Janowitz, Morris. *The Professional Soldier: A Social and Political Portrait*. Glencoe IL: Free Press of Glencoe, 1960.

Johnson, David E. *Fast Tanks and Heavy Bombers: Innovation in the U.S. Army, 1917–1945*. Ithaca: Cornell University Press, 1998.

Kennedy, David M. *Freedom from Fear: The American People in Depression and War, 1929–1945*. The Oxford History of the United States, ed. C. Vann Woodward, vol. 9. New York: Oxford University Press, 1999.

Kennedy, Paul. *The Rise and Fall of the Great Powers*. New York: Random House, 1987.

Killigrew, John W. "The Impact of the Great Depression on the Army, 1929–1936." Ph.D. diss., Indiana University, 1960.

Kohn, Richard H. *Eagle and Sword: The Federalists and the Creation of the Military Establishment in America, 1783–1802*. New York: Free Press, 1975.

———. "The Erosion of Civilian Control of the Military in the United States Today." *Naval War College Review* 55, no. 3 (2002): 8–59.

———. "Out of Control: The Crisis in Civil-Military Relations." *The National Interest* no. 35 (Spring 1994): 3–17.

———, ed. *The United States Military under the Constitution of the United States, 1789–1989*. New York: New York University Press, 1991.

Kolodziej, Edward A. *The Uncommon Defense and Congress, 1945–1963*. Columbus: Ohio State University Press, 1966.

Komons, Nick A. *Bonfires to Beacons: Federal Civil Aviation Policy under the Air Commerce Act, 1926–1938*. Washington DC: U.S. Department of Transportation, 1978.

Krauskopf, Robert W. "The Army and the Strategic Bomber, 1930–1939." *Military Affairs* 22, no. 2 (1958): 83–94.

Lamont, Thomas W. *Henry P. Davison: The Record of a Useful Life*. New York: Harper, 1933.

Larrabee, Eric. *Commander in Chief: Franklin Delano Roosevelt, His Lieutenants, and Their War*. New York: Harper and Row, 1987.

Layman, Martha E. *Air Force Historical Studies: No. 39: Legislation Relating to the Air Corps Personnel and Training Programs, 1907–1939*. Washington DC: Army Air Forces Historical Office, 1945.

Leuchtenburg, William E. *Franklin D. Roosevelt and the New Deal, 1932–1940*. New York: Harper and Row, 1963.

————. *The Perils of Prosperity: 1914–1932*. 2nd ed. Chicago: University of Chicago Press, 1993.

————. *The Supreme Court Reborn: The Constitutional Revolution in the Age of Roosevelt*. New York: Oxford University Press, 1995.

Levine, Isaac Don. *Mitchell: Pioneer of Air Power*. New York: Duell, Sloan and Pearce, 1943.

MacArthur, Douglas. *Reminiscences*. New York: McGraw-Hill, 1964.

————. *A Soldier Speaks: Public Papers and Speeches of General of the Army Douglas MacArthur*. Ed. Major Vorin E. Whan Jr. New York: Praeger, 1965.

Malone, Dumas, ed. *Dictionary of American Biography*. Vol. 7. York: Scribner, 1934.

Maney, Patrick J. *The Roosevelt Presence: A Biography of Franklin Delano Roosevelt*. New York: Twayne, 1992.

Marshall, George C. *The Papers of George Catlett Marshall*. Ed. Larry I. Bland and Sharon R. Ritenour. Vol. 1, *"The Soldierly Spirit" December 1880–June 1939*. Baltimore: Johns Hopkins University Press, 1981.

Marx, Fritz Morstein. "The Bureau of the Budget: Its Evolution and Present Role, I." *American Political Science Review* 39, no. 4 (1945): 653–84.

Mauer, Mauer. *Aviation in the U.S.Army, 1919–1939*. Washington DC: Office of Air Force History, 1987.

McCarley, J. Britt. "General Nathan Farragut Twining: The Making of a Disciple of American Strategic Air Power, 1897–1953." Ph.D. diss., Temple University, 1989.

McClendon, R. Earl. *Autonomy of the Air Arm*. 1954. Reprint, Washington DC: Air Force History and Museums Program, 1996.

McFarland, Keith D. *Harry H. Woodring: A Political Biography of FDR's Controversial Secretary of War*. Lawrence: University of Kansas Press, 1975.

McFarland, Marvin W. *The General Spaatz Collection*. Washington DC: Government Printing Office, 1949.

Meilinger, Phillip S., ed. *The Paths of Heaven: The Evolution of Airpower Theory*. The School of Advanced Airpower Studies. With a foreword by General Ronald R. Fogleman, Chief of Staff, U.S. Air Force. Maxwell Air Force Base AL: Air University Press, 1997.

Mets, David R. *Master of Airpower: General Carl A.Spaatz*. Novato CA: Presidio Press, 1988.

Millett, Allan R., and Peter Maslowski. *For the Common Defense: A Military History of the United States of America*. Rev. ed. New York: Free Press, 1994.

Mooney, Chase C. *Air Force Historical Studies: No.46: Organization of Military Aeronautics, 1935–1945*. Washington DC: Army Air Force Historical Office, Headquarters, Army Air Forces, 1946.

Mooney, Chase C., and Martha E. Layman. *Air Force Historical Studies: No.25:*

Organization of Military Aeronautics, 1907–1935. Washington DC: Historical Division, Assistant Chief of Air Staff, Intelligence, 1944.

———. "Some Phases of the Compulsory Military Training Movement, 1914–1920." *Mississippi Valley Historical Review* 38, no. 4 (1952): 633–56.

Mueller, Andrew M., chief editor. *Contrails: Air Force Academy Cadet Handbook.* Vol. 31. Colorado Springs: United States Air Force Academy, 1985–86.

Murray, Robert K. *The Harding Era: Warren G.Harding and His Administration.* Minneapolis: University of Minnesota Press, 1969.

———. *The Politics of Normalcy: Governmental Theory and Practice in the Harding-Coolidge Era.* New York: Norton, 1973.

Murray, Williamson, and Allan R. Millett, eds. *Military Innovation in the Interwar Period.* Cambridge, England: Cambridge University Press, 1996.

Nalty, Bernard C., ed. *Winged Shield, Winged Sword: A History of the United States Air Force.* Vol. 1, *1907–1950.* Washington DC: Air Force History and Museums Program, 1997.

Nicolson, Harold. *Dwight Morrow.* New York: Harcourt, Brace, 1935.

Odom, William O. *After the Trenches: The Transformation of U.S. Army Doctrine, 1918–1939.* College Station: Texas A&M University Press, 1999.

Overy, R. J. *Air War, 1939–1945.* New York: Stein and Day, 1981.

Parton, James. *"Air Force Spoken Here": General Ira Eaker and the Command of the Air.* 1986. Reprint, Maxwell Air Force Base AL: Air University Press, 2000.

Patrick, Mason. *The United States in the Air.* Garden City NY: Doubleday, Doran, 1928.

Pogue, Forrest C. *George C.Marshall.* Vol. 1, *Education of a General, 1880–1939.* New York: Viking Press, 1963.

Posen, Barry R. *The Sources of Military Doctrine: France, Britain, and Germany between the World Wars.* Ithaca: Cornell University Press, 1984.

Ransom, Harry H. "The Air Corps Act of 1926: A Study of the Legislative Process." Ph.D. diss., Princeton University, 1953.

Reynolds, Clark G. "John H. Towers, the Morrow Board, and the Reform of the Navy's Aviation." *Military Affairs* 52, no. 2 (1988): 78–84.

Roosevelt, Franklin D. *The Public Papers and Addresses of Franklin D. Roosevelt.* Ed. Samuel Irving Rosenman. Vol. 7, *1938: The Continuing Struggle for Liberalism.* Vol. 8, *1939: War—And Neutrality.* Vol. 9, *1940: War—And Aid to Democracies.* New York: Macmillan, 1941.

Rutkowski, Edwin C. *The Politics of Military Aviation Procurement, 1926–1934: A Study in the Political Assertion of Consensual Values.* Columbus: Ohio State University Press, 1966.

Schlesinger, Arthur M., Jr. *The Age of Roosevelt.* Vol. 2, *The Coming of the New Deal.* 1958. Reprint, Boston: Houghton Mifflin, 1988.

Scroggs, Stephen K. *Army Relations with Congress: Thick Armor, Dull Sword, Slow Horse.* Westport CT: Praeger, 2000.

Sherry, Michael S. *The Rise of American Air Power: The Creation of Armageddon*. New Haven: Yale University Press, 1987.

Shiner, John F. *Foulois and the U.S. Army Air Corps, 1931–1935*. Washington DC: Office of Air Force History, 1983.

———. "The General and the Subcommittee: Congress and U.S. Army Air Corps Chief Benjamin D. Foulois, 1934–1935." *Journal of Military History* 55, no. 4 (1991): 487–506.

Simonson, G. R. "The Demand for Aircraft and the Aircraft Industry, 1907–1958." *Journal of Economic History* 20, no. 3 (1960): 361–82.

Smith, Louis. *American Democracy and Military Power: A Study of Civil Control of the Military Power in the United States*. Chicago: University of Chicago Press, 1951.

Spence, Benjamin A. "The National Career of John Wingate Weeks, 1904–1925." Ph.D. diss., University of Wisconsin–Madison, 1971.

Tate, James P. *The Army and Its Air Corps: Army Policy toward Aviation, 1919–1941*. Maxwell Air Force Base AL: Air University Press, 1998.

Tillett, Paul D. *The Army Flies the Mails*. Inter-University Case Program no. 24. N.p.: Published for the Inter-University Case Program by The Bobbs-Merrill Company, Inc., n.d.

Underwood, Jeffrey S. *The Wings of Democracy: The Influence of Air Power on the Roosevelt Administration, 1933–1941*. College Station: Texas A&M University Press, 1991.

Ward, Robert D. "The Origin and Activities of the National Security League, 1914–1919." *Mississippi Valley Historical Review* 47, no. 1 (1960): 51–65.

Weigley, Russell F. "The American Military and the Principle of Civilian Control from McClellan to Powell." *Journal of Military History* 57, no. 5 (1993): 27–58.

———. *History of the United States Army*. 1967. Reprint, Bloomington: Indiana University Press, 1984.

West, Michael A. "Laying the Legislative Foundation: The House Naval Affairs Committee and the Construction of the Treaty Navy, 1926–1934." Ph.D. diss., Ohio State University, 1980.

White, Robert Paul. "Air Power Engineer: Major General Mason Patrick and the United States Air Service, 1917–1927." Ph.D. diss., Ohio State University, 1999.

———. *Mason Patrick and the Fight for Air Service Independence*. Washington DC: Smithsonian Institution Press, 2001.

Wiebe, Robert H. *The Search for Order, 1877–1920*. 1967. Reprint, Westport CN: Greenwood Press, 1980.

Wilson, John R. M. "The Quaker and the Sword: Herbert Hoover's Relations with the Military." *Military Affairs* 38, no. 2 (1974): 41–47.

Winton, Harold R., and David R. Mets, eds. *The Challenge of Change: Military*

Institutions and New Realities, 1918–1941. Lincoln: University of Nebraska Press, 2000.

Wooddy, Carroll H. "Is the Senate Unrepresentative?" *Political Science Quarterly* 41, no. 2 (1926): 219–39.

Wooster, Robert. *Nelson A.Miles and the Twilight of the Frontier Army*. Lincoln: University of Nebraska Press, 1993.

Index

Army and Navy Journal, 7, 53, 98, 108, 121

Army Reorganization Act (1920), 17

Arnold, "Hap": Air Corps expansion goals (1938) of, 131; on Air Corps' "Magna Carta," 152, 154; "air-mindedness" actions by, 91–92; Andrew's appeal over B-17 bombers to, 144; on Baker Board benefits to Air Corps, 128; becomes Chief of the Air Corps, 150–51; "exiled" for improper conduct, 65–66; "exiled" over Douglas bomber crash controversy (1939), 163–67; expansion program pursued by, 136–37, 180; frustration over B-17 battle by, 143; H.R. 8533 campaign led by, 64–65; Information Division headed by, 55; LaGuardia hearing testimony by, 17; as Mitchell supporter, 6, 9, 11, 28, 59; as Mitchell trial witness, 5; Patrick's plan supported by, 52–53; political tactics and approach used by, 155, 157–59, 162–63, 171–72, 173, 179; regarding War Department's closed attitude, 63; response to Johnson's aircraft plan by, 157–58; response to letter from McSwain by, 135; support of foreign aircraft sales by, 163–67, 168–70; on Westover's comments regarding independent Air Corps, 134; Westover's new approach approved by, 134–35; wife's defense of, 214nn44–45

Assistant Secretary of War for Air, 74, 75–76, 218n2, 248n78; attempts to eliminate, 100–4; creation of, 55, 71; duties of, 218–19n7. *See also* Davison, F. Trubee; Lovett, Robert A.

B-17 Flying Fortress: ACTS high-altitude, daylight bombing raids

made possible by, 243n2; Air Corps on military benefits of, 130–31; early plans and design (1933) for, 109; increased confidence in Air Corps following testing of, 250n120; political fight (1936–1938) over, 139–50; test crashing of, 142. *See also* aircraft

B-18 bombers, 142, 148–49

Baker, Newton D.: Baker Board headed by, 126; on congressional testimonies by military, 18–19, 21, 22; disagreement with Crowell report by, 14; embarrassed by Rogers, 21; examination of air power's role in defense ordered by, 13; opposition to air autonomy by, 15, 16

Baker Board, 126–28, 131, 132, 133, 149–50

Bane, Thurman, 17, 18

Barbour, Henry E., 106–7

Begg, James T., 60

Bingham, Hiram, 47–48, 51, 52, 70, 77, 88

Bissell, Clayton, 19

Black, Hugh L., 114, 118

Blanton, Thomas, 7

Bolton, Chester C., 166

bombing tests (1921), 24–25

Borah, William E., 16, 88, 156, 160

Bowley, Albert J., 4

Bradley, Follett, 112

Branch, Harllee, 115

British aircraft sales, 163–67, 168–70, 252n25

British Royal Air Force, 14

Brown, Walter F., 114

Bruce, Bill, 91

Bryan, William Jennings, 11

Budget and Accounting Act (1921), 27

Bullitt, William C., 157

Bulow, William J., 160

"Bureau of Air Navigation," 48